Second Nature

ZOO AND AQUARIUM BIOLOGY AND CONSERVATION SERIES

Published in cooperation with the American Zoo and Aquarium Association

This series publishes innovative works in the field of zoo and aquarium biology and conservation, with priority given to books that focus on the interface between captive and field conservation and seek to merge theory with practice. Topics appropriate for this series include but are not limited to zoo- and aquarium-based captive breeding and reintroduction programs, animal management science, philosophy and ethics, public education, professional training and technology transfer, and field conservation efforts.

SECOND NATURE

Environmental Enrichment for Captive Animals

Edited by David J. Shepherdson, Jill D. Mellen, and Michael Hutchins

SMITHSONIAN INSTITUTION PRESS
Washington and London

Copy editor: Susan A. Kreml
Production editor: Deborah L. Sanders
Designer: Janice Wheeler

Library of Congress Cataloging-in-Publication Data
Second nature : environmental enrichment for captive animals / edited by David J. Shepherdson, Jill D. Mellen, and Michael Hutchins.
p. cm. — (Zoo and aquarium biology and conservation series)
Based on papers from a conference held in Portland, Oregon.
Includes bibliographical references and index.
ISBN 1-56098-745-6 (cloth : alk. paper)
1. Captive wild animals—Congresses. 2. Environmental enrichment (Animal culture)—Congresses. I. Shepherdson, David J. II. Mellen, Jill D. III. Hutchins, Michael. IV. Series.
SF408.S435 1998
636.088'9—dc21 97-37840

British Library Cataloguing-in-Publication Data available

A paperback reissue (ISBN 1-56098-397-3) of the original cloth edition

Manufactured in the United States of America
05 04 03 02 01 00 99 5 4 3 2 1

The paper used in this publication meets the minimum requirements of the American National Standard for Information Sciences—Permanence of Paper for Printed Library Materials ANSI Z39.48-1984.

CONTENTS

PART THREE: ENVIRONMENTAL ENRICHMENT IN CAPTIVE MANAGEMENT, HUSBANDRY, AND TRAINING

CONTRIBUTORS

Cheryl Aday is a student in the doctoral program in biology at Boston University. She has done behavioral enrichment research on dolphins, seals, otters, and leopards.

Stan Anderson is leader of the Wyoming Cooperative Fish and Wildlife Research Unit in Laramie.

Janet F. Baer is the director of laboratory animal medicine at the California Institute of Technology and a consultant to the Association for Assessment and Accreditation of Laboratory Animal Care International and other biomedical research organizations.

Benjamin B. Beck is associate director of biological programs at the National Zoological Park, and he maintains a database on reintroduction programs for captive-born animals.

Dean Biggins is a doctoral candidate at Colorado State University. He is also in charge of black-footed ferret research for the Biological Resources Division of the Mid-Continent Ecological Science Center (U.S. Geological Survey), Fort Collins, Colorado.

Kathy Carlstead, an ethologist, is a research associate at the National Zoological Park. She is also coordinator and a principal investigator of the Methods of Behavioral Assessment project, a collaborative effort by researchers at ten U.S. zoos to develop a methodology for collecting and comparing animal behavior data across institutions.

M. Inês Castro, an animal behaviorist, is U.S. program officer for the Golden Lion Tamarin Conservation Program at the National Zoological Park, Washington, D.C.

Carolyn M. Crockett is a research scientist at the Regional Primate Research Center of the University of Washington.

Tim Desmond, as co-owner and president of the animal behavior consulting firm Active Environments, is directing animal operations staff for the new Beijing Landa Aquarium and for a conservation-focused marine animal park to be built in Osaka, Japan.

Nina Fascione is a senior program associate at Defenders of Wildlife, where she manages recovery programs for the red wolf and eastern timber wolf.

Debra L. Forthman, a certified applied animal behaviorist, is coordinator of scientific programs at Zoo Atlanta.

Jerry Godbey is a biologist for the Biological Resources Division of the Mid-Continent Ecological Science Center (U.S. Geological Survey), Fort Collins, Colorado, where he has evaluated reintroduction techniques for black-footed ferrets.

Louis Hanebury is a biologist at Bowdoin National Wildlife Refuge and a participant in the Montana Black-Footed Ferret Recovery Program.

Marc P. Hayes, a herpetologist, teaches at Portland Community College and is adjunct assistant professor of biology at Portland State University and a research associate at the Metro Washington Park Zoo in Portland, Oregon.

Michael Hutchins is director of conservation and science for the American Zoo and Aquarium Association (AZA), Bethesda, Maryland. He and his staff administer the AZA Species Survival Plans for many endangered and threatened animal species.

Mark R. Jennings is a fish and wildlife research biologist with the U.S. Geological Survey in Davis, California, a research associate in herpetology at the California Academy of Sciences, and an adjunct assistant professor of biology at the University of California–Santa Barbara.

Devra G. Kleiman is senior research scientist at the National Zoological Park and adjunct professor at the Department of Zoology, University of Maryland. She coordinates the international Golden Lion Tamarin Conservation Program in Brazil.

Michael D. Kreger, previously an animal caretaker, is a technical information specialist for the Animal Welfare Information Center of the U.S. Department of Agriculture, Washington, D.C. He is also vice president of World Nature Association, a small organization that funds start-up international conservation projects involving local peoples.

Stan A. Kuczaj II is a professor in and the chair of the Department of Psychology at the University of Southern Mississippi. In the past decade, his studies in comparative psychology have focused on captive marine mammals.

C. Thad Lacinak is vice president and corporate curator of animal training for Sea World, Inc. He oversees the animal training departments of Sea World parks throughout the United States.

Gail Laule is director of animal behavior and a co-owner of Active Environments, an animal behavior consulting firm. She is also an instructor at the Principles of Elephant Management School of the American Zoo and Aquarium Association.

Donald G. Lindburg is research behaviorist at the Zoological Society of San Diego, California.

Scott W. Line is staff veterinarian and animal behaviorist at the Animal Humane Society in Minneapolis and a member of the faculty at the University of Minnesota College of Veterinary Medicine.

Terry L. Maple is the director of Zoo Atlanta and professor of psychology at Georgia Institute of Technology.

Hal Markowitz is a professor of biology at San Francisco State University and a fellow of the California Academy of Sciences.

Jill D. Mellen is a research biologist at Disney's Animal Kingdom in Orlando, Florida, and formerly was conservation research coordinator of the Metro Washington Park Zoo in Portland, Oregon.

Joy A. Mench, an ethologist, is a professor in the Department of Animal Science and the Department of Avian Sciences at the University of California-Davis.

Brian Miller is coordinator of conservation and research at the Denver Zoo. He recently coauthored a book on black-footed ferret conservation (*Prairie Night: Black-Footed Ferrets and the Recovery of Endangered Species,* 1996, Smithsonian Institution Press).

Kathleen N. Morgan is assistant professor of psychology at Wheaton College in Norton, Massachusetts, where she teaches courses in animal behavior and psychobiology.

John Oldemeier is assistant director of the Mid-Continent Ecological Science Center of the U.S. Geological Survey, Fort Collins, Colorado. He has studied herbivore-habitat relationships in the western United States for the past twenty years.

Trevor B. Poole recently retired from the positions of deputy director of the Universities Federation for Animal Welfare (UFAW) and scientific director of the International Academy of Animal Welfare Sciences (IAAWS).

Alfred L. Rosenberger is a physical anthropologist and a primatologist at the National Zoological Park and the National Museum of Natural History, Smithsonian Institution.

Carlos R. Ruiz-Miranda is a research collaborator in the Department of Zoological Research of the National Zoological Park, where he has been studying the behavioral development of captive, reintroduced, and wild golden lion tamarins.

John Seidensticker is a wildlife ecologist and curator of mammals at the National Zoological Park. He served as founding principal investigator of the Smithsonian-Nepal Tiger Ecology Project and as an ecologist and park planner for the Indonesian World Wildlife Program.

David J. Shepherdson is program scientist at the Metro Washington Park Zoo in Portland, Oregon.

Ted N. Turner is vice president and curator for the Animal Training Department of Sea World of Ohio. He oversees the behavioral and enrichment programs for various marine mammal, primate, and avian species.

Astrid Vargas is director of the National Black-Footed Ferret Conservation Center of the U.S. Fish and Wildlife Service, Laramie, Wyoming.

Chris Wemmer, the director of the Conservation and Research Center of the National Zoological Park, has investigated the behavioral ecology of various endangered species for nearly thirty years.

FOREWORD

The contributions to this volume demonstrate the credibility and maturity of the behavioral specialty known as *environmental enrichment.* The book is based upon an exciting conference held in Portland, Oregon, the participants of which blended science and practice in equal parts. No ivory tower pontification was permitted, as the presentations dealt with real-world problems in management and husbandry and the creative tools and techniques designed to resolve them. The work represented here has been propelled by the energy of new ideas, sustained by empirical testing, and codified by dialogue, debate, and decisive action. After two decades of focused research, we have learned a great deal about environmental enrichment. An administrator who carefully reads these chapters will be prepared to overcome institutional inertia, and working programs of environmental enrichment are certain to proliferate in laboratories, zoos, aquatic parks, and other wildlife environments.

About twenty years ago I participated in a similar symposium concerned with environmental influences on the behavior of nonhuman primates. This gathering was more modest in scope, addressing a small but enthusiastic audience in Seattle, Washington, at the 1977 meeting of the Western Psychological Association. The book that resulted from this session reported data from experimental studies of eleven different primate taxa, although thirty-seven distinct species were cited in the book's index. When you compare the two volumes, it is obvious that environmental enrichment has enjoyed a renaissance of interest and productivity affecting a much larger universe of species. In 1977, environmental enrichment could be described by a small cadre of western psychologists who were seeking to understand the behavioral effects of captivity. Today, building on a foundation of knowledge about social and environmental deprivation, we are much more focused on the beneficial effects of enriched environments. Our evaluations of enrichment strategies require input and feedback from keepers, curators, and veterinarians. Our biomedical technology provides physiological as well as behavioral indicators of performance. No longer must we rely on case studies of single subjects. We have come a long way in two decades.

The "science of environmental enrichment" is relatively new, but its roots can be found in the scholarship of psychobiologist Robert M. Yerkes and the esteemed zoo director Heini Hediger. In the 1920s Yerkes first outlined the many variables that affected the psychological well-being of apes and monkeys. Somewhat later, Hediger's observations recorded the effects of captivity on a wide variety of zoo animals. The importance of the social environment was carefully delineated in the 1960s and 1970s in a series of experiments conducted by Harry F. Harlow and his students. The contemporary birthplace of zoo environmental enrichment was Portland, Oregon, where Hal Markowitz's pioneering laboratory influenced an entire generation of zoo biologists. The stimulating conference from which these papers were derived was a fitting tribute to Markowitz and his Washington Park Zoo collaborators and successors. There must be something about the Pacific Northwest that is uniquely conducive to the luminaries of environmental enrichment.

Anyone who works with animals today—keepers and curators, scientists and technicians, veterinarians and colony managers—recognizes that husbandry and management standards continue to advance. Indeed, wherever animals are confined for use by humankind, facilities and practices are carefully scrutinized by local, state, and federal government agencies and by nonprofit animal welfare organizations. No one should be more concerned about the welfare of animals than those who work so closely with them. I am proud to say that the science of environmental enrichment techniques can be applied to a diversity of organisms in a multitude of settings. We are limited only by our imaginations and our budgets, and the former restraint has been overcome by leaps and bounds. I can only hope that committed administrators will identify the financial resources to keep pace with our opportunities. It is encouraging to note that environmental enrichment is typically simple and cost-effective to implement.

The reader will discover that environmental enrichment has become much more than an intuitive collection of anecdotes and informal deductions. The field has been enhanced by controlled experimentation, systematic field studies, collaborations with pooled data from comparable settings, an infusion of computer-based technology, and the discipline and context provided by theories of learning and behavioral ecology. The extension of environmental enrichment to a greater diversity of animals suggests that a truly comparative psychology of environmental enrichment is about to become reality. It is likely that responses to environmental enrichment will vary according to individual, ecological, and taxonomic differences. In this realm, there is room for both applied and basic science.

I am gratified by the scholarship revealed in this collection of outstanding papers. The authors are all qualified by their experience, knowledge, and training, representing an appropriate diversity of approaches and backgrounds. The editors selected well, organizing and fine-tuning each contribution into a unified whole. From frogs to whales, from tiny labs to vast ecosystems, and encompassing many disciplines, this book is inclusive and nearly exhaustive. It is, in a phrase, a virtual encyclopedia of enrichment strategies.

Looking ahead, this volume contemplates a workplace where enrichment is an integral part of daily husbandry. Caretakers are routinely taught to feed, water, observe, handle, and enrich their animals. Before long, we will be able to quantify the totality of our contributions to their psychological well-being with a kind of "environmental calculus" that will guide our work plans and influence the design of our facilities. Whereas the first book in this Smithsonian series (*Ethics on the Ark,* edited by Norton et al., 1995) defined the common ground that unifies zoo and aquarium biology and animal welfare, *Second Nature,* second in the series, is more a guidebook. Herein the reader will learn how to structure enrichment programs that get results and to apply our growing database to renew and extend a humane management culture.

As a zoo professional, I find it comforting that zoo biologists are leading the way as agents of management change. It is a grass-roots revolution, driven by the creativity of smaller zoos and advocated by enthusiastic keepers, curators, volunteers, and students. Fueled by good science and fresh ideas, the field of environmental enrichment should continue to stimulate some of the most creative thinking in all of zoo biology.

Terry L. Maple

PREFACE

This book represents the second in the Zoo and Aquarium Biology and Conservation Series published by Smithsonian Institution Press in affiliation with the American Zoo and Aquarium Association. The first, *Ethics on the Ark: Zoos, Animal Welfare, and Wildlife Conservation* (Norton et al. 1995), dissected the conflicts between animal rights activists, managers of captive populations of animals, and conservationists. Each group discussed its own distinct views on assessing the welfare and well-being of animals and the role of modern zoos and aquaria in conservation. A key point in *Ethics on the Ark* is the importance of enrichment in the enhancement of well-being for captive animals. In this book, we begin to provide a theoretical framework for the *science* of environmental enrichment.

One way of improving the well-being of captive animals is to enhance their environment, a technique often referred to as "environmental enrichment." This environment includes the physical and social environment and the management regime associated with its care, including diet. This field is not new. Heini Hediger (1950, 1966) and Robert Yerkes (1925) discussed components of enrichment long ago. Zookeepers and other animal caretakers, concerned about animal well-being, have historically provided enrichment for animals under their care. Several publications, including *The Shape of Enrichment, Animal Keepers' Forum, Zoo Biology, IAAWS Newsletter* (International Academy of Animal Welfare Sciences), *Ratel, Animal Welfare,* and *Applied Animal Behaviour Science,* report innovative and practical ideas for enhancing the environments of captive animals. The authors and editors of two books have wrestled with the difficulties of defining, measuring, and providing well-being for captive, nonhuman primates: *Through the Looking Glass* (Novak and Petto 1991) and *Housing, Care, and Psy-*

chological Well-Being of Captive and Laboratory Primates (Segal 1989). Given this history, we perceived two areas of omission in this rapidly growing field in the science of animal management: a theoretical framework within which to consider current and future enrichment and a conceptual evaluation of enrichment for a broader range of vertebrates.

Second Nature: Environmental Enrichment for Captive Animals is an outgrowth of a conference organized by David Shepherdson and Jill Mellen. In July 1993, the First Conference on Environmental Enrichment was held at the Metro Washington Park Zoo in Portland, Oregon. The chapters in this book are based on a selected subset of the papers presented at the conference.

In Chapter 1, David Shepherdson outlines the primary role of this book—to set the stage for the development of a theoretical framework—providing a foundation for evaluating the past, present, and future attempts by zoo biologists to enhance the well-being of captive animals both in the laboratory and in zoos and aquaria. The book is divided into three parts.

Part One has five chapters, each of which provides components of a theoretical framework for the evaluation and expansion of environmental enrichment. John Seidensticker and Debra Forthman discuss how our knowledge of biology and behavior can be used to enrich the environments of captive animals. They suggest that maintaining behavioral competence in captive animals is a primary goal of zoo biology. Joy Mench discusses the idea that animals engage in exploratory behaviors whose primary function is to obtain information about the environment; she terms this "intrinsic exploration." By approaching enrichment from the perspective of information-gathering by the animal, managers and caretakers are better able to develop improved strategies for enrichment. Hal Markowitz and Cheryl Aday provide a historical review of the enrichment opportunities designed and implemented by Markowitz and colleagues. The key consideration in the range of options they describe is that enrichment techniques "empower" the animals with some control over their environments. Michael Kreger and colleagues evaluate how context and ethics can influence the planning and implementation of environmental enrichment programs in zoos and aquariums using an ethical foundation they term "moral pluralism." This concept recognizes that human values can vary with context. Trevor Poole urges zoo managers and caretakers to ascertain the psychological needs of mammals by providing opportunities for a full range of normal behaviors and by offering adequate mental stimulation with some degree of novelty.

In Part Two, we attempt to expand the reader's views on the importance of environmental enrichment in a variety of contexts. The first two chapters (Brian Miller et al. and M. Inês Castro et al.) discuss the importance of enriched environments for captive-born animals that are to be reintroduced into the wild; Miller and colleagues focus primarily on carnivores while Castro and colleagues describe the role of enrichment for a primate species to be reintroduced. Carolyn Crockett emphasizes the importance of testing our preconceived notions about enrichment. She also presents lessons that can be learned about enrichment in a laboratory situation and how those lessons can be more broadly applied to other captive populations. Kathleen Morgan and colleagues present examples illustrating that untested decisions based upon common sense sometimes may be incorrect and costly. Moving away from an emphasis on nonhuman primates, Kathy Carlstead discusses the benefits of enrichment for carnivores, while Jill Mellen and colleagues describe the effects of the captive environment on the behavior of small cats in zoos.

In Part Three, we discuss the importance of integrating enrichment in captive management, husbandry, and training. Marc Hayes and colleagues discuss enrichment for amphibians and reptiles, and Debra Forthman presents ideas about the optimal care of ungulates in captivity. Donald Lindburg presents a theoretical approach to providing food to animals in captivity. We asked a veterinarian, Janet Baer, to summarize potential health risks associated with enhanced environments; this is not meant to dissuade enrichment, simply to make the process more thoughtful and safer for captive animals. We also provide two chapters that discuss the use of positive reinforcement training as an enrichment tool (Laule and Desmond and Kuczaj et al.).

In the concluding chapter, Chapter 19, we the editors suggest potential options and directions that zoo biologists may wish to pursue to enhance both the science of environmental enrichment and the execution of enrichment in the daily management of captive animals.

In summary, this volume is meant to serve as a beginning, not an end, to discussion of environmental enrichment as a testable tool for potentially enhancing the well-being of the animals under our care. In 1969, Hediger wrote: "In every good zoo the animal does not feel itself in any way a prisoner, but—as in the wild—it feels more like a tenant or owner of a piece of land." More than a quarter of a century later, this definition is still appropriate.

REFERENCES

Hediger, H. 1950. *Wild Animals in Captivity.* London: Butterworths.

———. 1966. Diet of animals in captivity. *International Zoo Yearbook* 6:37–57.

———. 1969. *Man and Animal in the Zoo.* London: Routledge and Kegan Paul.

Norton, B. G., M. Hutchins, E. F. Stevens, and T. L. Maple, eds. 1995. *Ethics on the Ark: Zoos, Animal Welfare, and Wildlife Conservation.* Washington, D.C.: Smithsonian Institution Press.

Novak, M. A., and A. J. Petto, eds. 1991. *Through the Looking Glass: Issues of Psychological Well-Being in Captive Non-human Primates.* Washington, D.C.: American Psychological Association.

Segal, E. F., ed. 1989. *Housing, Care, and Psychological Well-Being of Captive and Laboratory Primates.* Park Ridge, N.J.: Noyes Publications.

Yerkes, R. 1925. *Almost Human.* New York: Century.

ACKNOWLEDGMENTS

Virtually all successful projects involve inspiration, teamwork, and support. We were fortunate to have an abundance of all three.

This book was written as an outcome of the First Conference on Environmental Enrichment held at the Metro Washington Park Zoo in Portland, Oregon, on July 16–20, 1993. We gratefully acknowledge the support of the zoo's director, Y. Sherry Sheng, who encouraged us to "get the ball rolling." We are especially grateful to Dennis Pate, general curator, and Michael Keele, assistant curator, for facilitating the full involvement of the Metro Washington Park Zoo's keeper staff and for their long-term and wide-encompassing support of environmental enrichment. We also thank the entire keeper staff of the Metro Washington Park Zoo for the participation in their conference and for their long-term commitment to enriching the lives of animals in their care. We especially thank the conference participants; we are indebted to all who attended and helped expand our ideas about enrichment.

The innumerable components of hosting a large conference can be overwhelming. We were fortunate to have the organizational skills of Jan Barker, the conference coordinator (also known as the Conference Queen). Jan's skills at juggling dozens of details simultaneously were impressive. The Metro Washington Park Zoo's zoo guides proved to be invaluable in making the conference participants feel welcome and in addressing everything from registration to coffee breaks. They did all this efficiently and with great humor and positive energy. We especially thank Linda Waltmire, volunteer coordinator of the enrichment conference.

With regard to the book itself, we are grateful for the diligence, patience, and tenacity of the contributors, especially in light of the hurry-up-and-wait pacing of the project. We gratefully acknowledge the following individuals for their careful and insightful reviews of papers submitted for consideration: Pat Alford, Judy Ball, Lynne Baptista, Mike Beecher, Cynthia Bennett, Joel Berger, Joe Bielitzki, Mollie Bloomsmith, Gordon Burghardt, Kathy Carlstead, Carolyn Crockett, Teresa DeLorenzo, Betsy Dresser, Sue Ellis, Joe Erwin, Nina Fascione, Debra Forthman, John Fraser, Valerius Geist, Ken Gold, Becky Houck, Nancy King, Devra Kleiman, Karl Kranz, William Langbauer, Annarie Lyles, Don Lindburg, Joy Mench, Kathleen Morgan, Jackie Ogden, Dennis Pate, Trevor Poole, Miles Roberts, Andrew Rowan, Anne Savage, Dietrich Schaaf, John Seidensticker, Marty Sevenich, Alan Shoemaker, Charles Snowdon, Elizabeth Stevens, Steven Thompson, Chris Tromborg, Kris Vehrs, and Peregrine Wolff.

In terms of the editing process, we thank Susan Long for her technical assistance. Jean McConville was tireless in her detailed reviews of the manuscripts, and we thank her for teaching us a great deal about grammar. We thank Peter Cannell of Smithsonian Institution Press for his support and patience and Susan Kreml for her editorial expertise.

And finally, we thank animal keepers and caretakers for their inspiration and tireless dedication to animal care and for being the driving force behind environmental enrichment.

Second Nature

DAVID J. SHEPHERDSON

Introduction

TRACING THE PATH OF ENVIRONMENTAL ENRICHMENT IN ZOOS

Environmental enrichment is an animal husbandry principle that seeks to enhance the quality of captive animal care by identifying and providing the environmental stimuli necessary for optimal psychological and physiological well-being. In practice, this covers a multitude of innovative, imaginative, and ingenious techniques, devices, and practices aimed at keeping captive animals occupied, increasing the range and diversity of behavioral opportunities, and providing more stimulating and responsive environments. The artificial termite mound is a good example of environmental enrichment in practice. The observation that wild chimpanzees *(Pan troglodytes)* use sticks as tools to obtain termites from termite mounds spawned the idea of constructing artificial termite mounds for chimps in captivity (e.g., Gilloux et al. 1992). Many of these devices resemble their natural counterpart only in that tool use is required for animals to obtain food from them. And yet they have proved effective, not just for chimpanzees but for a wide range of primates.

Scattering an animal's daily ration around its exhibit, freezing its food into blocks of ice, or hiding it, rather than simply placing it in a bowl, is another effective form of enrichment. On a larger scale, environmental enrichment includes the renovation of an old and sterile concrete exhibit to provide a greater variety of natural substrates and vegetation, or the design of a new exhibit that maximizes behavioral opportunities. The training of animals can also be viewed as an enrichment activity because it engages the animals on a cognitive level, allows positive interaction with caretakers, and facilitates routine husbandry activities. Indeed, with correct knowledge, resources, and imagination, caretakers can enrich almost any part of the environment that the captive animal can perceive.

Research in the field of environmental enrichment is primarily focused on identifying, characterizing, and evaluating the relative importance of different environmental stimuli and finding the most effective ways of providing them. Research is also directed at identifying the underlying principles and concepts needed to guide this process. Clearly, the study of animal behavior plays a key role in helping us to understand what animals do in different captive and wild environments and why they act as they do. Studying their behavior is often as close as we can get to asking questions directly of the animals about their preferences or well-being. Indeed, the term "behavioral enrichment" is frequently used synonymously with environmental enrichment (although the latter is preferable because a change in behavior is only one of many possible consequences of improving or enriching the animal's environment). The disciplines of endocrinology, veterinary science, animal husbandry, and zoo design and management are all relevant to this approach, as are psychology, ecology, natural history, anatomy, and physiology. Environmental enrichment is truly multidisciplinary.

Although the primary purpose of environmental enrichment is enhancement of psychological and physiological well-being, there are some other important and legitimate goals for environmental enrichment, at least as far as zoos are concerned. The captive breeding of endangered species is an important conservation role that zoos and aquaria fill. Environmental enrichment can increase reproductive success, directly by providing the social and physical environments necessary for successful reproductive behavior and parental care and indirectly by providing the developmental environment required for the growth of behaviorally normal, and therefore reproductively viable, adults (Carlstead and Shepherdson 1994; Kreger et al., Chapter 5, this volume).

Perhaps the conservation value of environmental enrichment is most evident in programs in which animals born in captivity are released into the wild (Shepherdson 1994). It is critical that, while in captivity, these animals have sufficiently rich environments to allow the performance and maintenance of the species-typical behaviors necessary for survival in the wild. Although few studies have set out to quantify the relationship between the captive environment and postrelease survival, the findings of two such studies are reported in this volume for the reintroduction of golden lion tamarins *(Leontopithecus rosalia)* (see Castro et al., Chapter 8, this volume) and of black-footed ferrets *(Mustela nigripes)* (see Miller et al., Chapter 7, this volume).

Given that the main purpose of zoos is conservation education, another important goal of enrichment is education and enlightenment of zoo visitors by means of exhibits that are "enriched" and therefore more informative and interesting (Kreger et al., Chapter 5, this volume). A strong argument has been made (albeit with few empirical data) for the importance of displaying animals in naturalistic settings that visitors find interesting and in which they can see the animals' natural patterns of behavior (Hutchins et al. 1984; Coe 1985). "Naturalistic" from the visitor's point of view, however, is not necessarily naturalistic from the animal's point of view. Exhibits designed purely in the interests of aesthetic naturalism from an anthropomorphic perspective may offer as little to the animals in terms of behavioral opportunities as the sterile exhibits of old. The role of environmental enrichment is to ensure that these exhibits are naturalistic for both the viewer and the inhabitant.

Use of the term environmental enrichment with respect to the husbandry of captive animals is relatively recent. However, its popularity has increased dramatically during the last decade or so, to the point that it is rare now to find a professional publication on zoo or laboratory animal care that does not use the term in some context or another. In the United States, environmental "enhancement" for some captive nonhuman primates is no longer just an option; it is mandated by legislation (APHIS 1992). Here I review some of the background to this increased interest and try to identify some of the concepts that underlie the field.[1]

PHILOSOPHICAL BACKGROUND: CHANGES IN THE WAY RESEARCHERS PERCEIVE ANIMALS

Concern about the appropriateness of captive environments is not new. Professionals directly responsible for the care of captive animals have always been concerned about both the physical and the psychological well-being of their charges. Similarly, the importance of the physical and social environment to the well-being of captive animals has been appreciated for many years. So, what, if anything, is new about the contemporary field of environmental enrichment? I believe that some answers to this question can be found by looking at recent philosophical and conceptual changes, particularly in the fields of animal behavior and experimental and comparative psychology.

In an attempt to gain an objective, scientific understanding of how animals behave and why, psychologists such as E. L. Thorndyke and J. B. Watson developed a methodology, or paradigm, in which only events that could be directly observed and quantified were considered. Subjective states of mind, they argued, can be neither observed nor quantified objectively and are therefore outside the realm of scientific inquiry. It was believed that animal behavior and animal learning can be explained largely in terms of conditioned responses to stimuli and innate drives or instincts (Gleitman 1981). The behaviorist approach of Watson was developed and greatly expanded by Skinner with his concept of "operant conditioning" (Skinner 1974), but it was still essentially a model of animal learning that was blind and automatic. Although these approaches were both necessary and successful in their time, they did tend to encourage a somewhat mechanistic view of animals in which psychological needs were, if not exactly denied, then ignored, at least from a scientific perspective.

Research during the past few decades has revealed some of the limitations of this approach. For example, observations by Breland and Breland (1961) illustrated some of the problems with the simple behaviorist model. They cited a number of examples in which animals had been trained using operant conditioning techniques to perform certain actions for the purposes of entertainment. For example, in one study pigs *(Sus scrofa)* were taught to carry tokens to a food dispenser to obtain food. In this case, as in others, all went well for a while, but then something strange occurred. The pigs stopped taking the tokens to the dispenser and instead rooted them on the ground with their snouts. The hungrier the animals became, the more intense their rooting. They had, it seemed, reverted to their natural behavior patterns. Reversion of this sort, observed by many other studies and termed "instructive drift," indicates that conditioned response alone cannot explain all of the complex animal behavior that we observe.

Further research in animal behavior has also produced results that are difficult to explain without resorting to models that acknowledge the existence of mental states (reviewed by Griffin 1984). The subtle complexity of the lives of wild animals has been revealed by long-term field studies—by Goodall (1986) on chimpanzees *(Pan troglodytes),* by Schaller (1972) on lions *(Panthera leo),* and by Seyfarth et al. (1980) on vocal communication among vervet monkeys *(Cercopithecus aethiops),* for example. These studies and others (e.g., Byrne and Whiten 1988) demonstrated that wild animals manufacture and use tools, show tremendous diversity and adaptability in behavioral response to physical and social stimuli, and engage in complex or subtle social interactions such as deliberately deceiving others, forming alliances, and cooperating on tasks of mutual benefit.

Perhaps the most influential research, in terms of revealing the cognitive abilities of animals, has been the ape language studies (e.g., Fouts 1974; Gardner and Gardner 1978; Savage-Rumbaugh et al. 1980). These studies leave little doubt that apes are capable of mastering simple linguistic communication and concepts previously thought to be the preserve of humans. Furthermore, the findings of these studies may not be limited to apes or even to mammals, as research by Pepperberg (1991) on parrots indicates.

It has also been argued (e.g., Griffin 1984) that if the complex mental states possessed by humans are adaptive in an evolutionary sense, then it is unlikely that our ancestors, and therefore their modern-day descendants, did not share them to at least a small degree. If complex mental states confer an adaptive advantage on humans, it seems likely that the same would be true for a wide range of other species.

As a consequence of these studies and ideas, something of a paradigm shift has taken place within the field of animal behavior. It is now accepted by many behavioral scientists that models of animal behavior acknowledging and discussing the existence of mental states are not only acceptable but are necessary to explain many of their observations (but see Kennedy 1992 for some of the pitfalls of this approach). These findings have clearly influenced the way that we think about animal well-being and the care of animals in captivity. If animals do have complex mental states, analogous in principle to our own, then they also may have similarly complex psychological needs. Of course, the extent and nature of these needs are likely to be highly species specific and very different from those of humans.

INFLUENTIAL RESEARCH: FINDINGS AND CONCEPTS

Against this overall background of philosophical change, there have been some specific areas of study that have had, and continue to have, a strong influence on the ideas, approaches, and fundamental concepts considered important by those involved in the study and application of environmental enrichment.

Harry Harlow and his colleagues conducted a series of experiments that documented the importance of specific environmental and social stimuli for normal primate behavioral development (Harlow and Harlow 1962; Gluck et al. 1973). These researchers were among the first to quantify the effects of environmental variables on the behavior of captive animals. The work by Pfaffenberger and Scott (1976) revealed similar findings for domestic dogs raised as guide dogs for the blind. In this study, dogs *(Canis familiaris)* raised in a sterile kennel environment for longer than the first twelve months of life were incapable of adapting to a more complex environment (such as an urban environment) for the rest of their adult lives. Similar research was done in zoos. As curator at the London Zoo, Desmond Morris (1964) studied the psychological needs of captive animals and described some of the abnormal behaviors that can develop when these needs are ignored. Meyer-Holzapfel (1968) also conducted research that drew attention to the role of environmental factors in the development of abnormal behaviors, such as stereotypic behavior, in zoo animals.

An important concept to come out of research is that of behavioral needs. Dating back to Konrad Lorenz's (1950) pioneering research, a number of researchers have postulated that animals are motivated to perform some natural behaviors even in the absence of any need to do so. Indeed this concept has been an influential factor in animal welfare legislation (Thorpe 1969). The common observation that domestic cats *(Felis sylvestris catus)* that have already received their daily food requirement will still perform prey-catching behavior toward live prey (Leyhausen 1979) is the kind of evidence that supports this view, as does the oft-repeated finding by Neuringer (1969) that rats *(Rattus norvegicus)*, pigeons *(Columba livia)*, and many other species continue to "work" (peck a light or push a bar, for example) for food even when "free" food is available. This opened the possibility that preventing animals from performing these appetitive behaviors may, under some circumstances, be frustrating or stressful (reviewed by Hughes and Duncan 1988). Shepherdson et al. (1993) attempted to test the applicability of this concept to zoo animals by providing food to small cats (leopard cats, *Prionailurus*, and a fishing cat, *Felis viverrina*) in a way that required them to search for and find their food. The reductions in abnormal behavior that resulted from this experiment offer strong support for the concept. The validity of the concept of behavioral needs, and deciding what these needs are for different species, remains to be satisfactorily resolved. In the meantime, however, the concept exerts a strong influence on thinking in the field of environmental enrichment.

The concept of behavioral needs is one of the reasons that the behavior of an animal in the wild is often cited as a benchmark for evaluating captive well-being (e.g., Hediger 1969). While helpful in some respects, this concept is not without its problems (for critical discussion, see Veasy et al. 1996). A fundamental problem is that we do not yet know how many species behave in the wild. From those species that have been relatively well studied, it is clear that behavior in the wild is often highly variable and dependent upon local environmental conditions. Thus, it is difficult to determine to what extent behavior altered by captivity is simply another adaptive change to a different environment rather than an indication of adversity. Furthermore, animals perform behaviors in the wild that do not seem to be consistent with enhanced well-being in captivity, such as predator avoidance and behaviors related to poor health. It would be a mistake to assume, just because a behavior is performed in the wild, that a zoo animal of the same species is necessarily suffering if it does not have the opportunity to express that same behavior in captivity.

Studies of the physiology and psychology of stress in farm and laboratory animals have indicated how important the ability to predict and the opportunity to control aspects of their environment are to their well-being. Weiss (1972), for example, found that rats were less stressed when they could predict the occurrence of an aversive event or when they could control its occurrence with an appropriate behavior (see Dantzer and Mormede 1983 for a review). Similarly, Carlstead et al. (1993) found that providing hiding places for leopard cats *(Prionailurus bengalensis)* that were stressed by the nearby presence of large cat species (*Panthera* spp.) resulted in less pacing and reduced cortisol concentration.

The effects of environmental enrichment have been examined in different contexts since the early 1960s by psychologists studying the influence of environmental stimuli on learning. Most of these investigations compared the behaviors, learning abilities, and physiological variables between animals (usually rodents) reared in "enriched" environments and those reared in "impoverished" environments. In general, enriched captive environments result not only in improved learning abilities but also in increased cortical thickness and weight, increased size, number, and complexity of nerve synapses, and an increased ratio of RNA to DNA (reviewed by Widman et al. 1992). Behavioral changes include qualitative and quantitative increases in exploratory behavior (Renner 1987). We have yet to appreciate the full implications of these findings for environmental enrichment in the context of captive animals. In particular, these findings and others, such as those by Carlstead et al. (1993), which showed an inverse correlation between cortisol levels and exploration in leopard cats, point to the importance of exploratory behavior as an indicator of environmental quality (see Mench, Chapter 3, this volume). Furthermore, Inglis and Fergusson (1986) have suggested on the basis of their puzzle feeding research with songbirds that animals are strongly motivated to seek new information from their environment. They have proposed that information-gathering is the primary activity of all animals, subsumed only when specific motivations (to satisfy hunger, for example) reach high levels. This would help to explain why some of the most effective enrichment techniques are often centered around the hiding of food items (e.g., Shepherdson et al. 1993).

A SYNTHESIS

The broad concept of environmental enrichment has been around at least since the early part of this century, when Robert Yerkes (1925) wrote, "[T]he greatest possibility for improvement in our provision for captive primates lies with the invention and installation of apparatus which can be used for play or work." In 1950 Hediger echoed this conviction: "Clearly one of the most urgent problems in the biology of zoological gardens arises from the lack of occupation of the captive animal." As a solution he proposed that training and play should be used as a form of occupational therapy for captive animals (Hediger 1955), and that zoo animal enclosures should contain everything that is important for an animal to be able to behave the same way as its free-living counterpart (Hediger 1969). In 1960 Morris, then curator of mammals at the London Zoo, described an enrichment device that carried fish around the seal pool before releasing them to the chasing seals (*Halichoerus* spp.). During the same period, Reynolds and Reynolds (1965) and Kortland (1960) were among the first field biologists to follow Hediger's lead and apply their knowledge of animal behavior in the wild (chimpanzees in this case) to the design of zoo environments. In 1973 Freeman and Alcock published one of the first quantitative evaluations of a zoo environment (for gorillas, *Gorilla gorilla gorilla,* and orangutans, *Pongo pygmaeus*).

For many people however, Hal Markowitz (1982; see also Chapter 4, this volume) stands out as one of the first and most influential people to adopt the systematic, empirical approach to improving zoo animal environments based on scientific concepts that we now call environmental enrichment. Since the late 1970s when Markowitz began his research at the Portland Zoo (now called the Metro Washington Park Zoo), the field has grown rapidly and has evolved into a number of different directions. The behavioral engineering approach pioneered by Markowitz was largely based on the operant conditioning techniques that psychologists had developed as a tool to investigate animal learning. For example, Markowitz devised a system of stimulus lights and levers that allowed a white-handed gibbon *(Hylobates lar)* to activate a food dispenser by brachiating high in its enclosure (Markowitz 1982; see also Chapter 4, this volume). Such techniques, although successful within their own terms of reference, were criticized for being impractical (owing to the time and money required to service the apparatus) and artificial (in terms of both the stimuli provided and the behavioral responses they initiated) (Hutchins et al. 1979). A more naturalistic approach was advocated whereby stimuli similar to those found in the wild were used in attempts to evoke natural behavior patterns in captive animals (Hancocks 1980; Hutchins et al. 1984). This latter approach owes more to the field of ethology, the study of animal behavior in the wild, than it does to psychology. Researchers have since come to understand that the two approaches are not necessarily incompatible (Forthman-Quick 1984), and most are in general agreement about the validity of the major concepts if not about the relative importance, mode, and context of their application (Carlstead et al. 1991; Kreger at al., Chapter 5, this volume).

Returning now to the question posed at the start of this chapter: What is new about the field of environmental enrichment? I believe that the contemporary study of environmental enrichment differs from its past incarnations in that it is a systematic, scientific approach to understanding and providing for the psychological and behavioral needs of captive animals. By drawing on recently derived knowledge, particularly from the fields of ethology, psychology, and animal science, the new discipline of environmental enrichment offers an alternative and exciting way of looking at the environments that we provide for animals.

CONCLUSION

Despite the extensive research background described here, the study of subjectively defined mental states and well-being in animals is in its infancy, and many key concepts have yet to be refined or discovered. Therefore, judging what constitutes successful enrichment can still be difficult. Desirable results of enrichment efforts, based on the concepts just discussed, include increases in the amount of stimulation and complexity of the environment, reductions in stressful stimuli, opportunities for species-appropriate behavior and for exercising control, and appropriate contingencies between behaviors and their consequences or effects. Certainly further research in the field is needed to refine our ideas and increase the effectiveness of our attempts to enrich zoo captive animal environments. Simple descriptive case studies of the kind that predominate in the literature can, when done well, provide useful data. However, these studies must be complemented by experiments that attempt to reveal the underlying causes for the change in behaviors seen as a consequence of enrichment and which attempt to develop and validate concepts to explain them.

Furthermore, much of the research into cognitive abilities, mental states, and environmental effects on development has been conducted on primates. This, combined perhaps with our taxonomic affinity and anthropocentrism, has lent a primate bias to enrichment research and protocols. It is time now to diversify and consider the needs of other taxa far more carefully. Just because the sensory and communicatory modes of other taxa are more difficult to understand is not a reason for ignoring them.

We must not however forget that ultimately the effectiveness of environmental enrichment relies on its implementation. Such implementation frequently involves changing attitudes, priorities, and working practices at all levels of an organization. This is a topic that concerns people more than it does the animals in our care and, as a consequence, remains one of the more challenging aspects of successful enrichment.

ACKNOWLEDGMENTS

My thanks to the institutions that have supported my research in zoos, which include the Zoological Society of London, the Universities Federation for Animal Welfare, Metro Washington Park Zoo (Portland, Oregon), and the Friends of Metro Washington Park Zoo. Also thanks to the many friends and colleagues who have worked with and shared their ideas with me. A number of people have made comments and given me ideas for this chapter, notably Kathleen Morgan, Leslee Parr, Mike Hutchins, and Susan Long. I particularly thank Jill Mellen and Kathy Carlstead for discussion, information, and ideas.

REFERENCES

APHIS (Animal and Plant Health Inspection Service). 1992. *Animal Welfare Regulations.* Document 311-364/50538. Washington, D.C.: U.S. Government Printing Office.

Breland, K., and M. Breland. 1961. The misbehavior of organisms. *American Psychology* 16:681-684.

Byrne, R., and A. Whiten. 1988. *Machiavellian Intelligence: Social Expertise and the Evolution of Intellect in Monkeys, Apes, and Humans.* Oxford: Clarendon Press.

Carlstead, K., and D. J. Shepherdson. 1994. Effects of environmental enrichment on reproduction. *Zoo Biology* 13:447-458.

Carlstead, K., J. L. Brown, and J. Seidensticker. 1993. Behavioral and adrenocortical responses to environmental changes in leopard cats *(Felis bengalensis). Zoo Biology* 12:321-331.

Carlstead, K., J. Seidensticker, and R. Baldwin. 1991. Environmental enrichment for zoo bears. *Zoo Biology* 10:3-16.

Coe, J. C. 1985. Design and perception: Making the zoo experience real. *Zoo Biology* 4:197-208.

Dantzer, R., and P. Mormede. 1983. Stress in farm animals: A need for a re-evaluation. *Journal of Animal Science* 75 (1): 6-18.

Forthman-Quick, D. L. 1984. An integrative approach to environmental engineering in zoos. *Zoo Biology* 3:65-78.

Fouts, R. S. 1974. Language: Origins, definition, and chimpanzees. *Journal of Human Evolution* 3:475-482.

Freeman, H. E., and J. Alcock. 1973. Play behavior of a mixed group of juvenile gorillas and orangutans. *International Zoo Yearbook* 13:189-194.

Gardner, R. A., and B. T. Gardner. 1978. Comparative psychology and language acquisition. *Annals of the New York Academy of Sciences* 309:37-76.

Gilloux, I., J. Gurnell, and D. J. Shepherdson. 1992. An enrichment device for great apes. *Animal Welfare* 1:279-289.

Gleitman, H. 1981. *Psychology.* New York: W. W. Norton.

Gluck, J. P., H. F. Harlow, and K. A. Schiltz. 1973. Differential effect of enrichment and deprivation on learning in the rhesus monkey *(Macaca mulatta). Journal of Comparative Physiology and Psychology* 84:598–604.

Goodall, J. 1986. *The Chimpanzees of Gombe: Patterns of Behavior.* Cambridge: Harvard University Press.

Griffin, D. R. 1984. *Animal Thinking.* Cambridge: Harvard University Press.

Hancocks, D. 1980. Naturalistic solutions to zoo design problems. In *Third International Symposium on Zoo Design and Construction,* ed. P. Stevens, 166–173. Paignton, U.K.: Whitley Wildlife Trust.

Harlow, H. F., and M. K. Harlow. 1962. Social deprivation in monkeys. *Scientific American* 207:137–146.

Hediger, H. 1950. *Wild Animals in Captivity.* London: Butterworths.

———. 1955. *The Psychology and Behaviour of Animals in Zoos and Circuses.* London: Butterworths.

———. 1969. *Man and Animal in the Zoo.* London: Routledge and Kegon Paul.

Hughes, B. O., and I. J. H. Duncan. 1988. The notion of ethological "need," models of motivation, and animal welfare. *Animal Behaviour* 36:1696–1707.

Hutchins, M., D. Hancocks, and T. Calip. 1979. Behavioral engineering in the zoo: A critique. *International Zoo News* Part I, 25 (7): 18–23; Part II, 25 (8): 18–23; Part III, 26 (1): 20–27.

Hutchins, M., D. Hancocks, and C. Crockett. 1984. Natural solutions to the behavioral problems of captive animals. *Zoologische Garten* 54:28–42.

Inglis, I. R., and N. J. K. Fergusson. 1986. Starlings search for food rather than eat freely-available, identical food. *Animal Behaviour* 34:614–617.

Kennedy, J. S. 1992. *The New Anthropomorphism.* Cambridge: Cambridge University Press.

Kortland, A. 1960. Can lessons from the wild improve the lot of captive chimpanzees. *International Zoo Yearbook* 2:76–81.

Leyhausen, P. 1979. *Cat Behavior: The Predatory and Social Behavior of Domestic and Wild Cats.* New York: Garland Press.

Lorenz, K. 1950. The comparative method in studying innate behaviour patterns. *Symposia of the Society for Experimental Biology* 4:221–268.

Markowitz, H. 1982. *Behavioral Enrichment at the Zoo.* New York: Van Nostrand Reinhold.

Meyer-Holzapfel, M. 1968. Abnormal behavior in zoo animals. In *Abnormal Behavior in Animals,* ed. M. W. Fox, 476–504. Philadelphia: W. B. Saunders.

Morris, D. 1960. Automatic seal feeding apparatus at London Zoo. *International Zoo Yearbook* 2:70.

———. 1964. The response of animals to a restricted environment. *Symposia of the Zoological Society of London* 13:99–118.

Neuringer, A. J. 1969. Animals respond to food in the presence of free food. *Science* 166:399–401.

Pepperberg, I. M. 1991. A communicative approach to animal cognition: A study of conceptual abilities of an African grey parrot *(Psittacus erithacus)*. In *Cognitive Ethology: The Minds of Other Animals,* ed. C. Ristau, 153–186. Hillsdale, N.J.: Lawrence Erlbaum.

Pfaffenberger, C. J., and J. P. Scott. 1976. Early rearing and testing. In *Guide Dogs for the Blind: Their Selection and Training. Developments in Animal and Veterinary Sciences,* Vol. 1. Amsterdam: Elsevier.

Renner, M. J. 1987. Experience dependent changes in exploratory behavior in the adult rat *(Rattus norvegicus):* Overall activity level and interactions with objects. *Journal of Comparative Psychology* 101 (1): 94–100.

Reynolds, V., and F. Reynolds. 1965. The natural environment and behavior of chimpanzees *(Pan troglodytes schweinforth)* and suggestions for their care in zoos. *International Zoo Yearbook* 5:141–144.

Savage-Rumbaugh, E. S., D. M. Rumbaugh, and S. Boysen. 1980. Do apes use language? *American Scientist* 68:49–61.

Schaller, G. B. 1972. *The Serengeti Lion: A Study of Predator–Prey Relations.* Chicago: University of Chicago Press.

Seyfarth, R. M., D. L. Cheney, and P. Marler. 1980. Vervet alarm calls: Semantic communication in a free-ranging primate. *Animal Behaviour* 28:1070–1094.

Shepherdson, D. 1994. The role of environmental enrichment in captive breeding and reintroduction of endangered species. In *Creative Conservation: Interactive Management of Wild and Captive Animals,* ed. G. Mace, P. Olney, and A. Feistner, 167–177. London: Chapman & Hall.

Shepherdson, D. J., K. Carlstead, J. D. Mellen, and J. Seidensticker. 1993. The influence of food presentation on the behavior of small cats in confined environments. *Zoo Biology* 12:203–216.

Skinner, B. F. 1974. *About Behaviorism.* New York: Random House.

Thorpe, W. H. 1969. Welfare of domestic animals. *Nature* (London) 224:18–20.

Veasy, J. S., N. K. Waran, and R. J. Young. 1996. On comparing the behaviour of zoo housed animals with wild conspecifics as a welfare indicator. *Animal Welfare* 5:13–24.

Weiss, J. M. 1972. Psychological factors in stress and disease. *Scientific American* 226:104.

Widman, D. R., G. C. Abrahamson, and R. A. Rosellini. 1992. Environmental enrichment: The influence of restricted daily exposure and subsequent exposure to uncontrollable stress. *Physiology and Behavior* 51:309–318.

Yerkes, R. M. 1925. *Almost Human.* New York: Century.

Part One

THEORETICAL BASES OF ENVIRONMENTAL ENRICHMENT

JOHN SEIDENSTICKER AND DEBRA L. FORTHMAN

2

EVOLUTION, ECOLOGY, AND ENRICHMENT

Basic Considerations for Wild Animals in Zoos

Animal considerations should be, but are not always, the highest priority in the planning and operation of modern zoo exhibits. There is no industry-wide standard, and no general procedure or process has emerged to identify the matrix of variables that influence the lives of wild animals living in zoos. As an initial step in zoo animal enrichment programs, we explore in this chapter some ways and means of initiating a process to identify ecological and behavioral characteristics of animals that will be or presently are maintained in zoos and show how knowledge of an animal's natural history can be used to enrich that animal's zoo environment. We conclude with a plea to include the maintenance of behavioral competence in zoo animals as a primary goal of zoo biology.

Today we think about wild animals living in zoos in different ways than we did a century ago, and, in a decade, our thinking will differ from what it is today. Identifying and meeting a wild animal's needs is like trying to capture an elusive butterfly. The subject is largely formed by the perceptions of both zoo professionals and zoo visitors. Our perception of the impact that zoo environments have on their wild animal inhabitants and our presentation of wild animals to visitors have become more comprehensive in the century since many of our zoos were established. This change is the result of at least three factors: (1) a transformation in American cultural and economic circumstances that has strongly influenced animal welfare issues; (2) an expanded understanding of the natural history of wild animals, a refined understanding of the ways natural systems and wild animals are affected by human activity, and an enhanced ability to assess the impacts that zoo environments can have on wild animals; and (3) advances in zoo exhibition technologies, especially those employing "natural habitats." Aligning these factors is an immediate challenge and responsibility we, as zoo professionals, face as the stewards of the wild animals that live in our zoos.

The care and exhibition of zoo animals is not a cloistered process. Zoos and their animal care systems are strongly influenced by the socioeconomic and political environment in which they develop and function. Because this is a dynamic process, perceptions of what wild animals maintained in zoos need also change. A primary theme in American cultural and economic circumstance has been a shift from a largely rural life to urban and suburban society. Zoo visitors today are isolated from agriculture and its utilitarian view of domestic animals and wildlife. These visitors also have fewer opportunities to experience nature (Conway 1969). Since the Second World War, improving environmental quality has become an integral part of our search for greater health and a higher standard of living (Hayes 1987). It should be no surprise to zoo animal caretakers that zoo visitors—supporters and critics alike—are calling for better environments, greater health, and higher standards of living for animals maintained in zoos and in confinement generally (Holden 1988). Animal welfare issues raised by this increasingly demanding public are defining research needs for zoo exhibition programs (Bostock 1993).

The Victorians were collectors of things and strove for the artistic (Rybczynski 1992). The Victorian zoological garden of a century ago sought to display the animal as a beautiful, interesting, or unusual object, or, as May and Lyles (1987, 642) phrased it, as "living Latin binomials." In this living Latin binomial paradigm, zoo husbandry can be likened to the maintenance of a greenhouse plant: feed and water it, clean the waste, propagate it when you need more, and replace it when it dies. This paradigm makes basic assumptions about how wild animals respond to zoo environments and confinement in general, the most fundamental of which is that an animal provided with sufficient food and shelter will have no motivation to seek them as it would in the wild. This assumption, as Dawkins (1988) pointed out, results in environmental deprivation, one of the central problems in animal welfare. The careers of numerous psychologists, such as Harry Harlow and his intellectual offspring—Erwin, Maple, and Mitchell (Erwin et al. 1979)—have been devoted to characterizing the effects of environmental deprivation on the development and adult expression of behavior in a small number of species sufficiently flexible to survive in severely restricted environments. Thus, few psychobiologists or field biologists were surprised by recent findings that environmental deprivation, as practiced in traditional zoos, causes profound problems for many species, such as bears, that are not normally housed in laboratories (Carlstead et al. 1991; Forthman et al. 1992).

The relevant stimuli and precise nature of the response of a species to zoo environments will, of course, vary with its natural history. Because our background is as field researchers studying mammalian behavioral ecology, we restrict our discussion here to what we know best. From our initial perspective as outsiders to zoo biology, we were acutely aware that zoo animal management paradigms were not actively incorporating the exciting results from contemporary field studies. This was a surprise and disappointment because as graduate students we had both read the books by the founding father of zoo biology, H. Hediger (Hediger 1950, 1955), and therefore thought there was an advanced and evolving system of zoo animal care that included not only advanced medicine and nutrition (Fowler 1986) but also up-to-date natural history information.

Hediger (1950) recognized that wild animals in confinement are subjected to different selection pressures than their free-ranging conspecifics, but the substance of his books is directed at how this could be at least partially ameliorated by matching facilities with requirements drawn from an understanding of their natural history. He characterized zoo animal facilities as either "kennels" or "territories." Hediger (1969, 1970) pointed out that the old-style zoo cages or kennels, even if they had glass or wire for public viewing and painted backdrops, were a kind of anteroom where the animal was stuffed with food before it died and stuffed with preservative afterward. Territory, on the other hand, was a natural division of space with species-specific habitat and social organization for an animal or a social unit of animals (Hediger 1970). Recognition of this dichotomy at mid-century was a major paradigm shift for the maintenance of zoo animals and was, in a large measure, the result of an emerging understanding of the natural history of animals, of which Hediger was an apt student. For the next two decades, however, Hediger's (1950) vision of the future of zoo animal management remained largely unrealized, as Hediger himself later noted (1970).

When the concept of endangered species preservation was incorporated into American thought as a national and international resource management and conservation objective at the end of the 1960s and in the early 1970s, zoos were quick to point out that they had long been players in these kinds of conservation activities (Conway 1969). Reintroduction of animals from zoos to the wild became a potential conservation tool and an objective widely embraced by the zoo community. At the same time, thoughtful zoo biologists carefully examined paradigms of zoo animal care and found them deficient. For example, Dolan (1977) addressed shortcomings in captive management that affected social behavior in endangered ungulates, while Eisenberg and Kleiman (1977) discussed spatial, structural, social, dietary, and demographic requirements in primates. If there were significant shortcomings in the management of wild animals maintained in zoos, where did we stand using zoo animals as a source of animals for reintroduction back into the wild?

During the past two decades, there has been a greening and softening of zoo animal exhibition (e.g., Jones 1982; Coe 1985, 1989; Maple and Finlay 1989). There is less emphasis on exhibiting animals as objects and more emphasis on showing them engaged in natural behaviors and living in naturalistic environments (Forthman-Quick 1984; Hutchins et al. 1984; Greene 1987). Thus, zoo animal care seemed comfortably on track until Robert May—a giant in late twentieth-century ecology—and A. M. Lyles in 1987 challenged zoos (in the pages of the international journal *Nature*) to recognize that much of the husbandry and exhibition technology zoos employ is still suitable only for the maintenance of the "living Latin binomials." May and Lyles (1987, 643) wrote: "Serious attempts to preserve biological species must deal with the vexing question of preserving not just the species but its pristine range of behaviors." To us, the criticism expressed by May and Lyles was particularly pointed. Even more than an impetus for our own work, we detected in their words an underlying tone of disappointment in zoos, which are or should be one of the last bastions of organismal biology rather than mere repositories of genetic material and future museum specimens.

We propose that any progress zoo professionals make toward the goal of maintaining the pristine range of behaviors of a species will be achieved by virtue of a dynamic process of assessment for all species we maintain in zoos (Forthman and Ogden 1992). This process would be similar to the environmental impact statements required to assess the consequences of proposed land development (Erickson 1979) but would be conducted at the level of the individual animal and the species rather than at the ecosystem level (Lubchenco et al. 1991). Such a first-order lens would allow us to (1) identify and match an animal's characteristics with those of existing or planned zoo environments, (2) focus studies in the zoo setting on improving conditions of confinement, (3) design environmental enrichment strategies targeted at specific outcomes, and (4) develop strategies for the maintenance of behavioral competence in zoo animals slated for reintroduction.

MATCHING ZOO SPACES AND ANIMAL NEEDS

Traditional conservation science has been based on autecological studies of individual ecological systems and the interdependencies of habitat requirements of component species (Simberloff 1988). Resource managers have invested heavily in understanding how wild species respond to changing environments. Because this is essentially what we do when we bring a species into the zoo, an interdisciplinary approach seems to us more efficient. Accordingly, we sought insight from such work as studies in cross-scaling in ecosystems (Holling 1992), because one of the things we do by maintaining a large mammal, for example, in confinement is to impose a greatly reduced scale on its life. We believe that the foundation to understanding and planning for the future of species in zoos lies in modern field studies of each species in question. As one axis in this analysis, we propose the model used by Eisenberg (1981) in his studies of mammalian life history strategies. He applied the variables of behavioral systems and macroniches to identify the range of adaptive syndromes common to the class. With this model (or its variations as proposed by ourselves and others) it is possible to assess how we influence and affect those adaptive syndromes in zoo environments and in altered or confined conditions generally. We also suggest that there is much that zoos can learn from environmental restoration science.

We argue that the lot of mammals living in zoos cannot be improved by simply investigating the range of their phenotypic behavioral responses to either confined environments or native environments. In general, behavior in the wild demonstrates the "best-fit" or modal repertoire; it does not usually illustrate an animal's adaptive limitations or potential. Meaningful insights for improving zoo environments and understanding how specific environments affect behavior and health require a synthesis of information obtained from field and laboratory (Forthman-Quick 1984). For example, when you are planning the exhibition and care of a species, first review detailed field studies of that species (Maple and Finlay 1989). It is encouraging to note that some field biologists have recently made specific recommendations to improve conditions for confined wild mammals such as cheetahs *(Acinonyx jubatus)* (Laurenson 1993) and gorillas *(Gorilla gorilla)* (Harcourt 1987; Watts 1990). Second, review experimental behavioral research conducted with a species or its close relatives using either classic laboratory techniques or manipulations performed on the species. There are some ingenious examples. For example, who would have guessed, without testing the notion, that a wild lion *(Panthera leo)* would perform a highly stereotyped attack on a stuffed specimen set in front of a loudspeaker? (Grinnell et al. 1995).

While this convergent approach seems obvious to us, we realize that we are simply reiterating what has been said by others who have traveled the same path, including Eisenberg and Kleiman (1977), Hutchins et al. (1984), Maple (1981), and Maple and Finlay (1989). Perhaps this is because the process is much easier said than done. If that is so, how do we distill and synthesize the wealth of information about a species obtained from field and laboratory into practical solutions that can be used in the zoo context?

Table 2.1
Categories Used in Eisenberg's (1981) Feeding and Substrate Matrix for Mammals

Dietary type	Substrate type
Piscivore and squid-eater	Fossorial
Carnivore	Semifossorial
Nectarivore	Aquatic
Gumivore	Semiaquatic
Crustacivore and clam-eater	Volant
Myrmecophage	Terrestrial
Aerial insectivore	Scansorial
Foliage-gleaning insectivore	Arboreal
Insectivore/omnivore	
Frugivore/omnivore	
Frugivore/herbivore	
Herbivore/browser	
Herbivore/grazer	
Planktonivore	
Sanguivore	

EISENBERG'S ANALYTICAL MODEL

Detailed field studies, combined with those facts of natural history that can only be learned from examinations of specimens in zoos and museums, are at the core of Eisenberg's (1981) analysis of mammalian life history strategies. He included the following variables in this analysis: adult body size, relative brain weight, basic metabolic rate, geographic range, diet type, prey size, dietary diversity, food-finding strategy, activity patterns, substrate use, vegetation and habitat type(s) used, mating system, rearing system, foraging system, dispersal system, refuging, and antipredator system. Eisenberg folded these variables into the two broad concepts of macroniche and behavioral systems. First, he sought to define modal adaptive strategies among mammals by classifying them in a matrix of macroniches according to their use of environmental substrates and dietary preferences (Table 2.1). In all, he identified sixty-four such macroniche combinations in the Mammalia and showed that for each of these categories there are consequences and constraints imposed in the process of adaptation. For any species, an understanding of the major macroniche dimensions of substrate use and feeding specialization, and the activity cycle, is essential in moving beyond what Hediger (1970) termed the kennel mode of wild animal care.

Eisenberg's second area of emphasis is what he called the primary behavioral systems: mating, rearing, dispersal, foraging, refuging, and antipredator. Eisenberg (1981) teases these out for many species of mammals. These behavioral systems and their various phases must be identified and recognized for appropriate maintenance and exhibition of wild mammals in zoo environments. Elsewhere, we have provided examples (Forthman et al. 1995; Seidensticker and Doherty 1996) of how Eisenberg's model or similar analytical models can be used in matching zoo environments and animal needs.

We can begin to use Eisenberg's model by crafting these ideas into elements for exhibit design and improvement. A useful starting point is to fold the seventeen factors just listed into six primary design considerations to ensure optimal animal activity and visitor viewing opportunity.

1. Select species that may be seen in the available facility. An asocial, nocturnal, burrowing mammal, for example, is not suitable in an outdoor enclosure open for daytime viewing.
2. Ensure that the macroniche of the species is considered in the exhibit design and that the exhibit is environmentally appropriate for the species.
3. Establish which behavioral system or systems are to be featured while the animal is in the exhibit space. Optimize opportunities for the animals to engage in nonaggressive social interactions during public viewing hours.
4. Provide appropriate species-specific, as well as age- and sex-specific, resting or refuging sites for the animals in the exhibit space. Recent studies have shown that careful consideration of this factor is important in reducing stress in animals living in zoo environments (Carlstead et al. 1993).
5. Manipulate food type, amount, distribution, and timing of delivery to optimize vigilance, foraging, and, for some species, feeding (Shepherdson et al. 1993).
6. Finally, think beyond the zoo environment to the question of the behavioral competence of a species that one day may live in the wild.

SCALE: MORPHOLOGICAL, SPATIAL, AND TEMPORAL

In zoos, we usually keep hyrax in smaller spaces than elephants without much thought about what this means in terms of what each species needs as an outcome of its evolutionary history. This is the basis for another approach in fine-tuning zoo environments for wild mammals that comes from studying the linkages between animal morphology and the processes which structure ecosys-

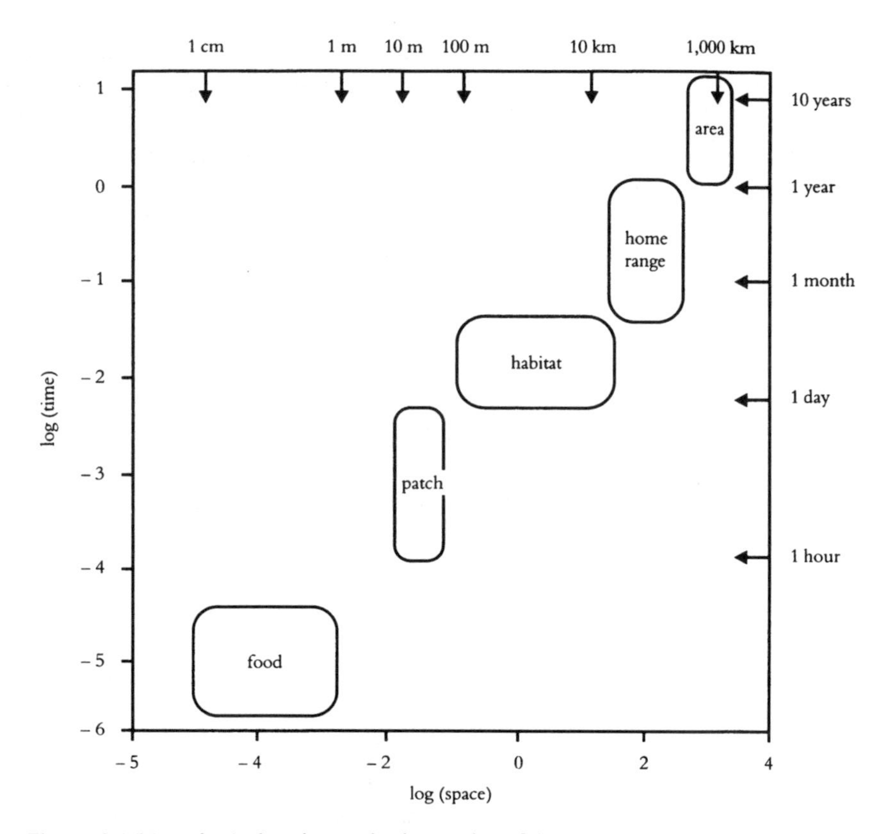

Figure 2.1 Hypothetical scales in the hierarchy of decision making by large mammals such as bears. (Adapted from Holling 1992.)

tems. Allometry or scale—in time, space, and body size—is the operative word here and is an increasingly central concept in ecology, especially as researchers attempt to develop programs to evaluate, monitor, and predict ecosystem change.

In Holling's (1992) analysis of scales in ecosystems, he used a hierarchy of decisions made by animals that increase in scale from food to foragable patch to habitat to home range and to region (Figure 2.1). The size and form of an animal defines the spatial grain at which a species samples a landscape, and Holling used allometric relationships to convert body mass into absolute space and time equivalents. Because landscapes are discontinuous in their spatial and temporal properties (or grain) as these apply to species or individual requirements, animals have attributes of size and behavior that are scaled by the discontinuous architecture of the landscapes in which they live. In Holling's approach (1992, 474), objects an animal encounters in its environment are either "edible" (and we would add: "or poisonous"), "frightful, lovable, ignorable, or novel." The first three of Holling's terms define the resources required to provide food, protection, and conditions for reproduction, and the latter two describe resources essential for long-term survival in strong seasonal environments. We would also point out that what amounts to the "emotional valence" of objects may change by virtue of the animal's experience over time; in confinement, this experience is frequently, although sometimes unwittingly, dictated by us, the animal managers.

As our example, we will focus on one resource in a bear's life: the edible. After a period in the young bear's life, the dispersal phase, in which it travels widely in the region (the area in Figure 2.1), it settles into a portion of that area that will provide both its survival and reproductive needs—its home range. The home range includes areas of habitat that provide its seasonal needs and which contain areas or patches that are far more abundant in food than areas between patches. In the old system of keeping bears in zoos, only food was provided, usually once a day and in one spot. Attempts to enrich bear exhibits were a matter of turning the entire exhibit into a food patch by hiding or scattering the food throughout the exhibit (Carlstead et al. 1991; Forthman et al. 1992). We can envision an exhibit design that takes the hierarchy of decision making in bear foraging farther than a single patch by including food patches within a habitat. In attempting to enrich a bear's zoo space, it is important to use the hierarchy of decisions that bears use in the real world. Bears apparently are "hard-wired" for this cognitive process, resulting in what have been referred to as an animal's psychological needs (Poole 1992).

In planning for mammals living in confined environments, we believe that another important axis of the assessment process is a consideration of scale in the hierarchy of decision making (e.g., for those objects that are edible as well as for those that are poisonous, frightful, lovable, ignorable, or novel). Indeed, the lack of a lovable object (i.e., a mate) was found to be the cause of stereotypic behavior in a confined American black bear *(Ursus americanus)* (Carlstead and Seidensticker 1991). We recognize that Holling's concept of scale in animal decision making (1992) is new and will undergo refinement. In the meantime, we think it shows great promise as an approach for thinking about the changes we make in an animal's life when moving it from the wild to a zoo environment.

LESSONS FOR ZOOS FROM ENVIRONMENTAL RESTORATION SCIENCE

In our fast-changing world, much wildlife habitat is destroyed daily by extractive industries. If we hope to retain many wildlife species in some semblance of their historical distribution, environmental restoration is essential. This requires a clear understanding of wildlife habitat requirements. The most advanced thinking in this area can be found in environmental restoration projects for ungulates in western North America. The case of bighorn sheep *(Ovis canadensis)* has been particularly well documented (MacCullum and Geist 1992).

With three decades of continuous comparative study of bighorns and other wild sheep in both disturbed and relatively undisturbed environments, Geist and his associates have narrowed the critical elements of an environmental restoration program to (1) escape terrain, (2) mineral licks, (3) lambing shelter, (4) vegetation cover types, and (5) food quality and distribution. These workers have been able to provide increasingly precise criteria for each of these factors. When we look at these elements, they also seem to encompass the design criteria for this species in zoo environments, leading us to suggest that zoos can learn much from this field that can be applied to exhibit design. Although we do not address it here, we note that behavioral management plays an important and unappreciated role in maintaining mammals in zoo environments, as it can in the conservation of wild mammals (Geist 1971).

DISCUSSION: ROUNDING OUT ZOO BIOLOGY

Zoo environments can and usually do influence animals maintained in them at the individual, group, population, and species level, and our understanding of these potential impacts has advanced unevenly. For example, our ability to conceptualize and evaluate what animals belong to which populations or even species has evolved rapidly. Using the increased power of molecular genetics to resolve phylogenetic partitioning, O'Brien and Mayr (1991) have offered guidelines for subspecies considerations in endangered species conservation. Such an understanding is a primary axis in the conservation of biodiversity (Wilson 1985) and in the linkage of ex situ with in situ conservation efforts. The discovery by Ralls and her associates (Ralls et al. 1979; Ralls and Ballou 1983) that the deleterious effects of inbreeding are widespread in zoos has motivated a concentrated effort to address this problem in captive breeding programs (Seal 1985) and to develop the technologies, particularly those involving assisted reproduction, to counter shortcomings (Wildt et al. 1992). For the zoo biologist working at the level of the individual and group, however, these considerations offer little guidance for improving the life of wild animals living in zoo environments or in maintaining their behavioral competence. We must not forget that this is the level at which our visitors see the animals maintained in zoos—nor that the individual is also the level at which natural or artificial selection operates.

Behavioral competence is not included in widely used planning models linking ex situ and in situ conservation of endangered species (e.g., Seal 1985). For example, the establishment of a self-sustaining captive population of golden lion tamarins *(Leontopithecus rosalia)* rested largely on determining the appropriate mating and rearing systems for the species and employing these in the captive setting (Eisenberg and Kleiman 1977). Successful captive breeding was one thing; survival in the wild requires other skills. Evaluation of initial reintroduction attempts demonstrated that deaths resulted from deficits in food-finding, locomotion, and orientation (Beck et al. 1991; see Castro et al., Chapter 8, and Miller et al., Chapter 7, this volume). Important environmental features such as substrate and behaviors such as foraging and refuging had largely been ignored in captive rearing. The stewards of wild mammals living in zoos must pay close attention to all behavioral systems to ensure behavioral competence. Although it is not the only goal of living collections, the acid test for zoo biology is reintroduction.

In zoo biology, as in so many other areas of environmental science, we are attempting to cope with major and accelerated environmental changes for animals when we define a species needs and match those to zoo environments. We can expect a certain tension in discussions about placement and maintenance of wild animals in zoos and the behavioral and even physical competence of captive-bred animals returned to natural environments (Lyles and May 1987; May and Lyles 1987; Schaller 1993; see Castro et al., Chapter 8, and Miller et al., Chapter 7, this volume). In the pages of *Zoo Biology* during the past decade, we see a record of an evolving system for zoo animal care intended both to improve the life of wild animals maintained in zoos and to improve the visitor experience.

But we need more. To empower zoos to maximize their participation in reintroduction and environmental restoration efforts, we must, as stewards of the wild animals we maintain in zoos, set our sights beyond Hediger's (1969, 54) definition of zoo biology as providing "the scientific basis for the maintenance of wild animals in the zoo under optimal and appropriate conditions[.]" Zoos, like other research, education, and environmental decision-making institutions, face the challenge of advancing conservation in the face of wide-scale and massive ecological change (Western and Pearl 1989; Lubchenco et al. 1991). Strengthening the bond between ex situ and in situ conservation demands that we understand how the zoo environment affects the behavior of the wild animals in captivity. It is not enough to produce and maintain genetically diverse offspring; we must also produce and maintain behaviorally competent animals that can thrive in the wild.

CONCLUSION

Our approach to zoo animal care, and our perception of the needs of wild animals living in zoos, are changing because of changes in the cultural and economic circumstances of American life, expanded understanding of natural history, refined understanding of the ways ecosystems and animals are affected by human activities, greater ability to assess the impacts of confined environments on wild animals, and improved zoo exhibition technologies. We suggest using natural history, psychobiology, Eisenberg's analytical model, Holling's concept of scale and allometry, and restoration ecology to assist in defining species needs and matching those to zoo environments. These approaches can be used as a first-order lens to identify and match an animal's characteristics with those of existing or planned zoo environments. The fine-tuning between an animal's needs and exhibit design is an ongoing process based on detailed studies of natural history and experimentation in the zoo setting. We propose that zoo biology explicitly recognizes the need to maintain the behavioral competence of wild animals maintained in zoos at a level such that their reintroduction back to the wild is feasible.

ACKNOWLEDGMENTS

Many of the ideas expressed here were developed for a chapter in *Wild Mammals in Captivity* (Kleiman et al. 1996), and J.S. thanks J. D. Doherty (Seidensticker and Doherty 1996) for his discussions and contributions. J.S. presented a paper on this topic at the First Environmental Enrichment Conference. Seidensticker and Forthman (1994) revised and expanded that discussion in a session at the American Zoo and Aquarium Association Annual Meeting, and this chapter is a synthesis of both presentations. J.S. thanks S. Lumpkin for more than a decade of discussions on these topics, and K. Carlstead and R. Baldwin, co-members in the Active Animal Project at the National Zoological Park, where we first developed and experimented with some of these ideas. This work was supported by the Friends of the National Zoo, Smithsonian Scholarly Studies Program, Geraldine R. Dodge Foundation, and the Smithsonian Women's Committee. Our thinking about wild animals living in zoos has also benefited from ongoing discussions with B. Beck, J. Eisenberg, J. Garcia, D. Kleiman, K. Kranz, F. Koontz, T. Maple, H. Markowitz, J. Mellen, D. Shepherdson, and S. Wells.

We dedicate this chapter to John F. Eisenberg in thankful recognition of the enormous insights he has provided in the understanding of wild animals, both in and out of the zoo.

REFERENCES

Beck, B. B., D. G. Kleiman, J. M. Dietz, I. Castro, C. Carvalho, A. Martins, and B. Rettenberg-Beck. 1991. Losses and reproduction in reintroduced golden lion tamarins *Leontopithecus rosalia. Dodo* 27:50–61.

Bostock, S. S. C. 1993. *Zoos and Animal Rights.* New York: Routledge.

Carlstead, K., J. Brown, and J. Seidensticker. 1993. Behavioral and adrenocortical responses to environmental changes in leopard cats *(Felis bengalensis). Zoo Biology* 12:1–11.

Carlstead, K., and J. Seidensticker. 1991. Seasonal variation in stereotypic pacing in an American black bear *Ursus americanus. Behavioural Processes* 25:155–161.

Carlstead, K., J. Seidensticker, and R. Baldwin. 1991. Environmental enrichment for zoo bears. *Zoo Biology* 10:3–16.

Coe, J. C. 1985. Design and perception: Making the zoo experience real. *Zoo Biology* 4:197–208.

———. 1989. Naturalizing habitats for captive primates. *Zoo Biology Supplement* 1:117–125.

Conway, W. G. 1969. Zoos: Their changing roles. *Science* 163:48–52.

Dawkins, M. S. 1988. Behavioural deprivation: A central problem in animal welfare. *Applied Animal Behaviour Science* 20:209–225.

Dolan, J. M. 1977. The saiga *(Saiga tatarica):* A review for the management of endangered species. *International Zoo Yearbook* 17: 25–32.

Eisenberg, J. F. 1981. *The Mammalian Radiations.* Chicago: University of Chicago Press.

Eisenberg, J. F., and D. G. Kleiman. 1977. The usefulness of behavioral studies in developing captive breeding programs for mammals. *International Zoo Yearbook* 17:81–89.

Erickson, P. A. 1979. *Environmental Impact Assessment.* New York: Academic Press.

Erwin, J., T. L. Maple, and G. Mitchell, eds. 1979. *Captivity and Behavior: Primates in Breeding Colonies, Laboratories, and Zoos.* New York: Van Nostrand Reinhold.

Forthman, D. L., S. D. Elder, R. Bakeman, T. W. Kurkowski, C. C. Noble, and S. W. Winslow. 1992. Effects of feeding enrichment on behavior of three species of captive bears. *Zoo Biology* 11:187–195.

Forthman, D. L., R. McManamon, U. A. Levi, and G. Y. Bruner. 1995. Interdisciplinary issues in the design of mammal habitats (excluding marine mammals and primates). In *Captive Conservation of Endangered Species,* ed. E. Gibbons, Jr., J. Demarest, and B. Durrant, 377-399. Albany: State University of New York Press.

Forthman, D. L., and J. J. Ogden. 1992. The role of applied behavior analysis in zoo management: Today and tomorrow. *Journal of Applied Behavioral Analysis* 25:647–652.

Forthman-Quick, D. L. 1984. An integrative approach to environmental engineering in zoos. *Zoo Biology* 3:65–77.

Fowler, M., ed. 1986. *Zoo and Wild Animal Medicine.* Philadelphia: W. B. Saunders.

Geist, V. 1971. A behavioral approach to the management of wild ungulates. In *The Scientific Management of Animal and Plant Communities for Conservation,* ed. E. Duffy and A. S. Watt, 413–424. Oxford: Blackwell.

Greene, M. 1987. No rms, jungle vu. *The Atlantic Monthly* 260 (6): 62–78.

Grinnell, J., C. Packer, and A. E. Pusey. 1995. Cooperation in male lions: Kinship, reciprocity, or mutualism? *Animal Behavior* 49:95–105.

Harcourt, A. H. 1987. Behaviour of wild gorillas *(Gorilla gorilla)* and their management in captivity. *International Zoo Yearbook* 26:248–255.

Hayes, S. P. 1987. *Beauty, Health, and Permanence: Environmental Politics in the United States, 1955–1985.* Cambridge: Cambridge University Press.

Hediger, H. 1950. *Wild Animals in Captivity.* London: Butterworths.

———. 1955. *Studies of the Psychology and Behaviour of Captive Animals in Zoos and Circuses.* London: Butterworths.

———. 1969. *Man and Animal in the Zoo.* London: Routledge and Kegan Paul.

———. 1970. The development of the presentation and the viewing of animals in zoological gardens. In *Development and Evolution of Behavior,* ed. L. P. Aronson, E. Tobach, D. S. Lehrman, and J. S. Rosenblatt, 519–528. San Francisco: W. H. Freeman.

Holden, C. 1988. Uncle Sam wants happy chimps. *The Washington Post* (16 Oct.): C3.

Holling, C. S. 1992. Cross-scaling morphology, geometry, and dynamics of ecosystems. *Ecological Monographs* 62:446–502.

Hutchins, M., D. Hancocks, and C. Crockett. 1984. Naturalistic solutions to behavioral problems of captive animals. *Zoologische Garten* 54:28–42.

Jones, G. R. 1982. Design principles for presentation of animals and nature. In *Proceedings of the American Association of Zoological Parks and Aquariums Annual Conference,* 184–192. Wheeling, W.Va.: AAZPA.

Kleiman, D. G., M. A. Allen, K. V. Thompson, S. Lumpkin, and H. Harris, eds. 1996. *Wild Mammals in Captivity: Principles and Techniques.* Chicago: University of Chicago Press.

Laurenson, M. K. 1993. Early maternal behavior of wild cheetahs: Implications for captive husbandry. *Zoo Biology* 12:31–43.

Lubchenco, J., A. M. Olson, L. B. Brubaker, et al. 1991. The sustainable biosphere initiative: An ecological research agenda. *Ecology* 72:371–412.

Lyles, A. M., and R. M. May. 1987. Problems in leaving the ark. *Nature* (London) 326:245–246.

MacCullum, B. N., and V. Geist. 1992. Mountain restoration: Soil and surface wildlife habitat. *GeoJournal* 27:23–46.

Maple, T. L. 1981. Evaluating captive environments. In *Proceedings of the Annual Meeting of the American Association of Zoo Veterinarians,* 4–6. Philadelphia: AZAA.

Maple, T. L., and T. W. Finlay. 1989. Applied primatology in the modern zoo. *Zoo Biology Supplement* 1:101–116.

May, R. M., and A. M. Lyles. 1987. Living Latin binomials. *Nature* (London) 326:643–643.

O'Brien, S. J., and E. Mayr. 1991. Bureaucratic mischief: Recognizing endangered species and subspecies. *Science* 251:1187–1188.

Poole, T. B. 1992. The nature and evolution of behaviour needs in mammals. *Animal Welfare* 1:203–220.

Ralls, K., and J. Ballou. 1983. Extinctions: Lessons from zoos. In *Genetics and Conservation,* ed. C. M. Schonewald-Cox, S. M. Chambers, and B. MacBryde, 164–184. Menlo Park, Calif.: Benjamin Cummings.

Ralls, K., K. Brugger, and J. Ballou. 1979. Inbreeding and juvenile mortality in small populations of ungulates. *Science* 206:1101–1103.

Rybczynski, W. 1992. *Looking Around: A Journey through Architecture.* New York: Penguin Books.

Schaller, G. B. 1993. *The Last Panda.* Chicago: University of Chicago Press.

Seal, U. S. 1985. The realities of preserving species in captivity. In *Animal Extinctions,* ed. R. J. Hoage, 71–95. Washington, D.C.: Smithsonian Institution Press.

Seidensticker, J., and J. G. Doherty. 1996. Integrating animal behavior and exhibit design. In *Wild Mammals in Captivity: Principles and Techniques,* ed. D. G. Kleiman, M. E. Allen, K. V. Thompson, S. Lumpkin, and H. Harris, 180–190. Chicago: University of Chicago Press.

Seidensticker, J., and D. L. Forthman. 1994. Planning for the species: Incorporating behavioral and ecological data. In *Proceedings of the American Zoo and Aquarium Association Annual Conference,* 39–45. Wheeling, W.Va.: AZAA.

Shepherdson, D. J., K. Carlstead, J. Mellen, and J. Seidensticker. 1993. The influence of food presentation on the behavior of small cats in confined environments. *Zoo Biology* 12:203–216.

Simberloff, D. 1988. The contribution of population and community biology to conservation science. *Annual Review of Ecology and Systematics* 19:473–511.

Watts, D. P. 1990. Mountain gorilla life histories, reproductive competition, and sociosexual behavior and some implications for captive husbandry. *Zoo Biology* 9:185–200.

Western, D., and M. Pearl, eds. 1989. *Conservation for the Twenty-First Century.* Oxford: Oxford University Press.

Wildt, D. E., S. L. Monfort, A. M. Donoghue, L. A. Johnson, and J. Howard. 1992. Embryogenesis in conservation biology—or, how to make an endangered species embryo. *Theriogenology* 37:161–184.

Wilson, E. O. 1985. The biological diversity crisis: A challenge to science. *Issues in Science and Technology* 2:20–29.

JOY A. MENCH

ENVIRONMENTAL ENRICHMENT AND THE IMPORTANCE OF EXPLORATORY BEHAVIOR

In 1985, the Animal Welfare Act in the United States was amended to require research laboratories to carry out provisions to promote the "psychological well-being" of nonhuman primates. Examples of methods to improve well-being suggested in the regulations included the provision of perches, swings, mirrors, manipulable objects, and task-oriented feeding methods (USDA 1991). This represented the first legislative recognition in this country of the importance of environmental enrichment for confined animals. It also stimulated research into the types of objects and environmental modifications that are most appropriate for species of primates commonly used in laboratories. Zoo researchers and caretakers, of course, have been pioneers in the area of environmental enrichment by demonstrating the importance of naturalistic or complex habitats to the well-being of captive animals (Markowitz 1982). Interest in enrichment strategies for farm animals, particularly swine and poultry, has also increased.

WHAT IS ENVIRONMENTAL ENRICHMENT?

Experimental psychologists have studied environmental enrichment extensively, primarily to evaluate the effects of enrichment during development on learning, social behavior, and neuroanatomy and neurophysiology (Renner and Rosenzweig 1987). Enriched environments have usually been defined in these studies as cages containing social companions in which stimulus complexity is increased by the addition of a variety of objects such as tubes, blocks, or plastic toys. These objects are exchanged frequently to ensure continuing novelty.

A person scrutinizing modern laboratory, zoo, and agricultural animal housing, however, might find it difficult to discern what constitutes an "enriched" environment using a definition based primarily on complexity. Many agricultural animals are kept in completely barren environments in which their ability to make even normal postural adjustments is curtailed (Mench and van Tienhoven 1986). For example, to prevent nursing sows from lying on and crushing piglets, the sows are often housed in narrow enclosures (farrowing crates) that do not allow them to turn around. Veal calves and laying hens are also closely confined, for reasons of both animal health and production efficiency. Social contact may also be extremely limited in some production systems to prevent aggression and transmission of diseases. Simply providing bedding materials such as straw to crated sows or wood shavings to caged laying hens can be considered an important form of environmental enrichment under these conditions (Fraser 1975). A standard laboratory cage for rodents that contained similar bedding materials, however, would be viewed as an impoverished environment (Renner and Rosenzweig 1987). In turn, even enriched laboratory and farm animal housing is typically less complex and variable (particularly environmentally, visually, and socially) than the average unenriched zoo enclosure.

For these reasons, practical environmental enrichment for captive animals is often defined in terms of its purposes rather than simply as a process or a phenomenon. Chamove and Anderson (1989), for example, suggested that enrichment should decrease abnormal behaviors, increase the behavioral repertoire, facilitate a more normal temporal patterning of behavior, and enable the animal to cope with challenges in a more normal way. Poole (1992) recommended that captive environments be designed to provide the animal with "stability and security" and "opportunities to achieve goals," in addition to complexity and unpredictability.

Rather than clarifying what enrichment is, these definitions invoke several problematical concepts that have stimulated much discussion among farm animal welfare researchers (Moberg 1985; Hughes and Duncan 1988; Dawkins 1990; Mench and Mason, n.d.). What, for example, are "normal" behaviors or typical temporal patternings of behavior for a species of animal that shows a high degree of behavioral plasticity in the wild? Which behaviors does the animal really need to perform so as to ensure psychological well-being? Which kinds of coping responses show that the animal is suffering and which show that the animal is simply responding appropriately to environmental challenges? How can "goals" in animals be understood and assessed?

The weight given to answering these questions will probably in part depend on whether enrichment strategies are being formulated for animals on the farm, in the laboratory, or in the zoo. Although there is an emerging consensus in

Western society that people have important ethical obligations to animals (Mench and Kreger 1996), these obligations are still viewed differently depending on the purpose for which the animal is being kept in captivity (Kreger et al., Chapter 5, this volume). Thus enrichment strategies for farm animals are constrained primarily by their potential economic impact on food or fiber production, while strategies for zoo animals are constrained by factors such as space, available resources, and visitor acceptability. Other constraints, such as the effect of enrichment on animal health, are common to all three environments. The primary constraints on enrichment programs in farm, zoo, and laboratory environments are shown in Table 3.1. When the costs of enrichment (whether those costs are increased egg prices on a laying hen farm or decreased control of the spread of infectious diseases in the laboratory) must be kept within narrow limits, determining which behaviors are most crucial for the animal to perform for its well-being will have a high priority.

Although implementation constraints may differ, the study of environmental enrichment for all animals could benefit from increased discussion among farm, laboratory, and zoo animal researchers about common theoretical foundations. I wish to discuss one such theoretical foundation, the idea that animals engage in exploratory behaviors whose primary function is to obtain information about the environment (Barnett and Cowan 1976; Inglis 1983), a type of exploration that Berlyne (1960) called intrinsic exploration. I also suggest some methods and considerations for studying enrichment that follow from this perspective.

EXPLORATORY BEHAVIOR AND ENVIRONMENTAL ENRICHMENT

There has recently been a renewal of interest in exploratory behavior and its relationship to information-gathering in animals, particularly in terms of its importance to improvements in the welfare of captive animals (Archer and Birke 1983; Poole 1992; Shepherdson et al. 1993; Wemelsfelder and Birke 1997). An increased emphasis on animals as information gatherers parallels the development of the field of cognitive science. There is an emerging view that "mind" is a property of the functioning of the nervous system, although a complex and as yet poorly understood property. Advances in the study of artificial intelligence and information processing have lent support to this view (Churchland 1986). The philosopher Daniel Dennett (1991) has even likened human consciousness to a "virtual machine, a sort of evolved (and evolving) computer program that shapes the activities of the brain."

Table 3.1
Primary Constraints on Environmental Enrichment Strategies in Different Settings

Zoo	Laboratory	Farm
Resource availability	Experimental protocols	Economics
Social companions	Hygiene and disease control	Product cost
Materials from natural habitat	Regulatory requirements	Product quality
Animal health	Space	Animal health
Aesthetics and acceptability to visitors	Cost	Worker health and safety
Space		Environmental impact
Conservation mission		

Given the similarities in the organization and function of the central nervous system among vertebrates, it seems likely that there is also an evolutionary continuity of mental experience. This assumption has formed the basis for the development of the field of cognitive ethology (Bekoff and Jamieson 1990). The ethologist Donald Griffin, perhaps the best-known exponent of the view that animals have subjective experiences, has offered the following definition of animal consciousness:

> An animal may be considered to experience a simple level of consciousness if it subjectively thinks about objects and events. Thinking about something in this sense means attending to the animal's internal mental representations. . . . These may represent current situations confronting the animal, memories, or anticipations of future situations. Such thinking often leads to comparisons between two or more representations and to choices and decisions about behavior that the animal believes is likely to attain desired results or avoid unpleasant ones (Griffin 1991).

Dennett (1989), while disagreeing with Griffin that animals have consciousness, did argue that a productive approach to explaining animal behavior is to view animals as rational beings with beliefs and desires that show intentionality in their actions. Toates (1983) also emphasized the goal-directed nature of animal behavior in his discussion of the motivational basis for intrinsic exploration. Toates suggested that exploration serves to establish and continuously refine the animal's "cognitive map" with respect to the location of food sources, aversive stimuli, and other relevant dimensions of the environment. This cognitive map then enables the animal to carry out goal-directed behaviors by using the various sites encoded in the map as the basis for a succession of decisions. Preferred goals are those that currently have the highest incentive value to the animal because of interactions between environmental stimuli and the animal's internal state.

Explanations of behavior based on an assumption that animals can form and be aware of mental representations are not without controversy. Kennedy (1992), for example, has criticized studies that have been interpreted to show that animals have "search images" or engage in intentional or goal-directed activities, arguing that more mechanistic interpretations are adequate to explain the observed behaviors. Toates (1983) was also careful to point out that his incentive theory of exploratory motivation does not rely on "mentalistic" constructs such as boredom.

While the precise role that mental representations play in the motivation of intrinsic exploratory behavior is still unclear, there are several lines of evidence that provide support for the "information primacy" theory of exploration elaborated by Inglis (1983). The first is that animals will preferentially search for food even in the presence of readily available food. Inglis and Ferguson (1986) found that starlings presented with mealworms that were either hidden randomly under flaps in a board or placed in an open dish obtained a significant proportion (up to 81 percent) of their mealworms from under the flaps. Searching behavior of this type seems to serve to provide information about the location and quality of future potential foraging sites in patchy environments. Chipmunks *(Tasmias sibiricus)*, which accumulate the seeds of deciduous trees and store them in their burrows, increase exploratory behavior when seed density begins to decrease in the particular food patch that they are currently exploiting (Kramer and Weary 1991).

The second line of evidence is drawn from studies showing that animals will explore familiar or novel environments even when those environments contain no resources used by the animal during the period of exploration. (Most studies of locomotor exploration have involved an analysis of the behavior of animals placed in an open field. Because the open field test has been widely criticized as a method for examining exploration [Daly 1973; Russell 1983; Renner 1990], those studies are not discussed in this chapter.) Rodents living in a cross-maze regularly visit all arms of the maze even when one is empty (Barnett and Cowan 1976). Similarly, hens placed in an enclosure with abundant resources still spend a proportion of their time exploring an empty tunnel attached to the enclosure (Nicol and Guilford 1991), and pigs will leave their home pen to explore an adjacent pen (Wood-Gush et al. 1990). This type of exploration, called patrolling, may serve to provide information about the general spatial patterning of resources in the environment (Albert and Mah 1972). The investigation of novel aspects of environments appears to have particular value for animals. Pigs more readily explore a pen next to their home pen if it contains a novel object instead of a familiar one (Wood-Gush and Vestergaard 1991). Rats also voluntarily leave their home cage to enter an arena where they can actively investigate a variety of novel objects (Renner and Seltzer 1991).

Again, one function of investigatory behavior of this type appears to be to provide predictive cues that can help the animal gauge future resource availability. An example can be seen in the behavior of Japanese macaques *(Macaca fuscata)* toward a highly preferred food source, akebi fruit. After macaques are presented with ripe akebi fruit for the first time, they subsequently inspect akebi vines, explore locations that might contain vines, and manipulate and taste fruit at different times of the year when it is at various stages of ripeness (Menzel 1991). Investigative exploration has also been shown to provide information important for predator avoidance strategies. Rats that spend more time investigating a box that leads to an escape hole in a testing arena escape from a simulated predator more quickly than rats that show less active investigation of the box (Renner 1988).

If the information primacy theory of exploration is correct, there are important implications for the practical implementation of environmental enrichment. An effort is needed to provide not only complexity but continuing novelty and variability to highly exploratory animals. But which animals are most likely to have information-gathering needs?

Largely on the basis of data showing taxonomic differences in brain size, Poole (1992) has suggested that only mammals have what he has termed "psychological needs." I contend that the complexity and variability of the animal's natural environment are more important predictors of cognitive needs than brain size, at least among homeotherms. Similar conclusions were drawn by Glickman and Sroges (1966), who compared the reactions of many different species of mammals to novel objects introduced into their zoo enclosures. While all animals need stimulation and enrichment to maintain arousal, I suggest that providing opportunities for exploration be high priority for the following types of captive animals:

- *species that are generalists or are adapted to environments that are highly variable in terms of resource availability*

  The ability to make predictions about the location and relative quality of resources, and to determine effective strategies for gaining access to those resources, is of the greatest importance to animals from habitats with patchy or seasonally variable resources.
- *species that exhibit complex antipredator behaviors*

  For example, some animals learn and use various escape routes within their ranges.
- *species that have a complex social structure*

> It has been suggested that social behavior is a primary factor driving increases in cognitive ability because a social animal needs to communicate information accurately, predict the behavior of other individuals by assessing their mental states, and detect deception during social interactions (Cheney and Seyfarth 1990; Trivers 1991).

Patterns of exploratory behavior in such species can be used as the basis for the design of enrichment devices to satisfy information-gathering needs.

TRANSLATING THEORY INTO METHODOLOGY

Environmental enrichment devices sometimes appear be chosen more for their durability, safety, availability, cost, or appeal to the investigator (or to the company marketing them) than for any properties they might have that are salient to the animal. Consequently, animals may not use the enrichment devices provided or may not show the beneficial effects expected from enrichment. Examples of "failed" enrichment devices are commercial pet toys for patas monkeys *(Erythrocebus patas)* (Weld 1992) and "nylaballs" for rhesus monkeys *(Macaca mulatta)* and cynomologous monkeys *(M. fascicularis)* (Line 1987). I suggest that viewing enrichment from the perspective of the information-gathering needs of the animal can enable us to develop enrichment strategies more systematically and effectively. In the following sections, I discuss several factors that might affect exploratory behavior and the resulting efficacy of enrichment.

Properties of the Enrichment Device

To avoid a continuing trial-and-error approach in enrichment programs, we need more information about the attributes of environmental features (e.g., visual characteristics, odors, sounds, and tactile characteristics) and enclosures (e.g., type of three-dimensional space including height, angles, crevices, floor area, and number of levels) that elicit exploratory behaviors in different species of animals. To some extent, this information can be inferred from observations of animals in wild or naturalistic habitats. A variety of enrichment devices have been developed for capuchins (*Cebus* spp.), for example, based on information about patterns of foraging and object manipulation in nature (Fragaszy and Adams Curtis 1991). Similarly, Stolba and Wood-Gush (1984) used their observations of the exploratory behavior of pigs in a naturalistic setting as the basis for designing an enriched intensive swine production system. It is often difficult, however, to observe and characterize exploratory behavior in the wild in sufficient detail to provide a basis for designing enrichment devices. Furthermore, it is not clear what constitutes a truly naturalistic environment for many domesticated or highly selected strains of animals, or for animals maintained in captivity for several generations. An alternative means of determining the types of stimulation that will elicit exploratory behaviors is to present the animal with objects (or environments) that vary systematically, allowing the animal to choose those which it prefers.

This latter approach, which would allow us to discover general principles about species-relevant properties of enrichment devices, is used too infrequently in enrichment research. Studies of this type can yield results that have important consequences for ideas about enrichment priorities and strategies. For example, in contrast to previous studies showing that prosimians show little interest in novel nonfood objects (Jolly 1964), Renner et al. (1992) found that bushbabies (*Galago* spp.) do indeed investigate objects. However, their propensity to do so varies markedly depending on the characteristics of the objects presented, with larger, more manipulable objects preferred.

Influence of Age, Sex, Individual Variation, Social Context, and Genetics

Several factors have been shown to influence exploration and responses to novelty in animals, including age, sex, genetics, and individual variation (Jones 1987; Renner and Rosenzweig 1987; Jones et al. 1991; Lawrence et al. 1991; Renner et al. 1992). However, there have been few attempts to examine these factors systematically in different species of animals and to determine their significance to the design of enrichment devices or housing environments. We need more studies such as the one by O'Neill et al. (1990) in which the exploratory behaviors of rhesus monkeys of different ages and sexes were characterized in both large and confined environments to provide information pertinent to developing individualized and developmentally specific enrichment programs. An important consideration for group-housed animals is that social interactions can result in increased interest in enrichment devices because of social facilitation (Renner et al. 1992). However, more assertive animals can also restrict the access of other individuals to preferred enrichment devices (Fragaszy and Adams-Curtis 1991). Thus, the relationship between exploration and social interaction also requires further study.

Developmental History and Prior Experience

The previous experience of an animal with enriched or impoverished environments, particularly during early development, exerts profound effects on neural organization and later patterns of emotional behavior and responses to stressors (Moberg 1985; Renner and Rosenzweig 1987). Whether enrichment always results in beneficial behavioral changes (Newberry 1995), however, is a matter of some controversy. It has been argued that many enrichment studies fail to take into account the animal's natural history when the outcome of the enrichment is interpreted. Daly (1973), for example, criticized what he called the anthropocentric view that decreased emotionality in a rodent tested in an open field test is adaptive, and stated: "It should be obvious that any small rodent who unhesitatingly enters a brightly lit novel environment is pathologically fearless." The effects of early enrichment on coping strategies must therefore be evaluated cautiously from the perspective of the natural history of neophilia and neophobia in a species, particularly for animals that might be reintroduced into their natural habitat (Shepherdson 1994).

Another consideration is that early enrichment can have effects on different facets of exploratory behavior that are subtle and age dependent. Renner and Rosenzweig(1986) reported that juvenile rats from enriched environments are no more likely than unenriched rats to contact novel objects in an arena. Enriched juveniles, however, show a greater diversity of behavior than unenriched juveniles when they encounter objects that are movable. Adults from enriched environments, on the other hand, demonstrate more complex behaviors toward both manipulable and nonmanipulable objects, and also show more general exploration of the arena (Renner 1987). In contrast, Wood-Gush et al. (1990) found that juvenile pigs from enriched environments spent less time examining an unfamiliar arena and the novel objects that it contained than pigs from unenriched environments. Whether these differences represent species differences or are related to the testing measures recorded or enrichment procedures used is unclear.

The likelihood that there are developmental constraints on cognitive abilities and on the animal's expectations about environmental uncertainty raises difficult questions about when, how, and if enrichment programs should be implemented. Inglis (1983) and Wemelsfelder (1993) have suggested that the "belief structure" of animals exposed to monotonous environments degenerates, eventually causing the animal to stop seeking external stimulation. The extent to which an animal's expectations and goals (or lack of expectations and goals) might become fixed during development is unknown. Juvenile or adult animals reared in monotonous or barren environments may require less input and thus benefit less from enrichment strategies based on information primacy models than will animals exposed to environmental variability from the time they are young (Glanzer 1958). However, such animals may also have difficulty evaluating environmental stimuli and thus respond to unusual situations with excessive fear or arousal (Wemelsfelder and Birke 1997).

ENRICHMENT FOR PRACTICAL ENDS

Environmental enrichment has been investigated as a means for achieving practical goals and resolving short-term welfare problems. In farm animals, for example, handling and object enrichment have been studied in an attempt to improve disease resistance and growth rates and to decrease fear responses to acute stressors such as capture and transport before slaughter (Gvaryahu et al. 1989; Pearce et al. 1989; Nicol 1992; Reed et al. 1993).

In these studies, enrichment sometimes produces apparently paradoxical (and certainly undesirable) results. For example, enrichment has been reported to increase aggression and cannibalism in chickens (Reed et al. 1993). McGregor and Ayling (1990), who found that male mice introduced into an enclosure containing novel objects fought more than mice introduced into a more barren enclosure, suggested that the objects were perceived as defensible resources, stimulating aggression (see Jones 1992 for a critique of the methods used in this study). Another explanation, however, is that the degree or type of enrichment used in these studies was inappropriate to achieve the desired goals.

Many focused enrichment studies have followed the standard enrichment paradigm derived from experimental psychology, described previously, in which randomly chosen novel objects are introduced into the animal's enclosure. Besides objects, enrichment treatments may also include handling, music, or visual or olfactory stimulation (Newberry 1995). While novelty stimulates exploratory behavior and can be associated with temporary increases in arousal that have beneficial effects (Chamove and Moodie 1990), it can also be a potent fear elicitor and therefore act as a stressor. Careful studies are needed to discover the type and degree of novelty that produce the most beneficial effects on behavior and coping skills. To do so will require detailed analysis of the behaviors shown by animals in response to individual enrichment stimuli and groups of stimuli as well as determination of the rates of response habituation to those stimuli.

One critical factor influencing the effectiveness of enrichment may be the degree of control that the animal has with respect to seeking and interacting with, or conversely avoiding, novel stimulation in the environment. Mice that are either placed in a novel environment, or are prevented by a barrier from returning to their familiar environment after they have voluntarily entered the novel environment, have increased levels of corticosterone. If the mice are allowed to move freely between the novel and familiar environments, however, they do not show increased corticosterone levels (Misslin and Cigrang 1986). Giving the animal control thus appears to decrease the stress associated with novelty. Wemelsfelder and Birke (1997) emphasized the importance of the voluntary and interactive aspects of exploration and suggested that animals need to "do" as well as "learn." In a widely cited paper, Maier and Seligman (1976) argued that a state of learned helplessness occurs when an animal lacks control over environmental stimulation and thus has difficulty in perceiving the relationship between its own behavior and the results of its behavior. Allowing the animal to choose the intensity and duration of novel stimulation to which it is exposed might help decrease paradoxical effects of enrichment.

An additional problem in using enrichment for practical ends lies in determining whether different types of environmental enrichment have similar effects on coping abilities. For example, object enrichment leads to a decreased tonic immobility reaction in chickens, but handling does not have similarly predictable effects (Nicol 1992). Although most researchers have reported that early enrichment leads to fear reduction in farm animals (Jones 1982; Nicol 1992; Pearce and Paterson 1993; Reed et al. 1993), there is still too little information about the generalizability of enrichment effects to permit accurate predictions as to whether or not (or in which situations) enrichment will be the most appropriate strategy for stress reduction.

CONCLUSIONS

In discussing the relationship between exploration and environmental enrichment, I do not intend to suggest that meeting the information-gathering needs of animals should be the primary goal of environmental redesign programs. Although there is evidence that animals will pay a price (such as crossing a shock grid or learning an operant task) to engage in exploration or to elicit stimulus change (Nissen 1930; Myers and Miller 1954; Barnes and Baron 1961), intrinsic exploration still appears to be a comparative luxury. Searching behaviors decrease significantly under conditions of food deprivation (Inglis and Ferguson 1986), as does another form of exploration—play behavior (Müller-Schwarze et al. 1982; Mench 1988). This suggests that preference should be given in enrichment programs to the satisfaction of higher priority behaviors that are strongly internally motivated, such as the complex of behaviors associated with feeding and acquiring food (Dawkins 1990).

The satisfaction of both the appetitive and consummatory phases of these behaviors may be important to the animal. Providing ready-made nests to sows or hens, for example, changes but does not inhibit the performance of nest-building behaviors before to parturition or egg laying (Duncan and Kite 1989; Hughes et al. 1989; Arey et al. 1991). Sows will, however, redirect their nest-building behaviors to cloth tassels hung in their enclosure if no nesting materials are available (Widowski and Curtis 1990). Enrichment that allows animals to carry out the appetitive components of behavior is also being used successfully for both laboratory and zoo animals (Carlstead 1996). Providing fleece-covered foraging boards to primates and live fish to fishing cats *(Prionailurus viverrinus)* are examples of enrichments that allow animals to carry out the appetitive components of feeding behavior (Bayne et al. 1991; Shepherdson et al. 1993; Carlstead 1996).

As we reconsider our housing and care systems for animals in captivity, satisfying the high-priority behavioral needs of animals should be the central consideration. Wherever possible, however, housing systems should also incorporate enrichment that allows animals to engage in their individual and species-typical patterns of information-gathering behavior.

ACKNOWLEDGMENTS

I thank Clare Knightly for her assistance in the preparation of this chapter, and Bryan Jones and Devra Kleiman for their helpful comments on an earlier draft of this manuscript.

REFERENCES

Albert, D. J., and C. J. Mah. 1972. An examination of conditioned reinforcement using a one-trial learning procedure. *Learning and Motivation* 3:369–388.

Archer, J., and L. I. A. Birke, eds. 1983. *Exploration in Animals and Humans.* Berkshire, U.K.: Van Nostrand Reinhold.

Arey, D. S., A. M. Petchey, and V. R. Fowler. 1991. The preparturient behaviour of sows in enriched pens and the effect of pre-formed nests. *Applied Animal Behaviour Science* 31:61–68.

Barnes, G. W., and A. Baron. 1961. Stimulus complexity and sensory reinforcement. *Journal of Comparative and Physiological Psychology* 54:466–469.

Barnett, S. A., and P. E. Cowan. 1976. Activity, exploration, curiosity, and fear: An ethological study. *Interdisciplinary Science Reviews* 1:43–62.

Bayne, K., H. Mainzer, S. Dexter, G. Campbell, F. Yamada, and S. Suomi. 1991. The reduction of abnormal behaviors in individually housed rhesus monkeys *(Macaca mulatta)* with a foraging/grooming board. *American Journal of Primatology* 23:23–35.

Bekoff, M., and D. Jamieson. 1990. Cognitive ethology and applied philosophy: The significance of an evolutionary biology of mind. *Trends in Ecology and Evolution* 5:156–159.

Berlyne, D. E. 1960. *Conflict, Arousal, and Curiosity.* New York: McGraw-Hill.

Carlstead, K. 1996. Effects of captivity on the behavior of wild mammals. In *Wild Mammals in Captivity: Principles and Techniques,* ed. D. G. Kleiman, M. E. Allen, K. V. Thompson, and S. Lumpkin, 317–333. Chicago: Chicago University Press.

Chamove, A. S., and J. R. Anderson. 1989. Examining environmental enrichment. In *Psychological Well-Being of Primates,* ed. E. Segal, 183–202. Philadelphia: Noyes Publications.

Chamove, A. S., and E. M. Moodie. 1990. Are alarming events good for captive monkeys? *Applied Animal Behaviour Science* 27:169–176.

Cheney, D. L., and R. M. Seyfarth. 1990. *How Monkeys See the World.* Chicago: University of Chicago Press.

Churchland, P. S. 1986. *Neurophilosophy: Toward a Unified Science of the Mind-Brain.* Cambridge: MIT Press.

Daly, M. 1973. Early stimulation of rodents: A critical review of present interpretations. *British Journal of Psychology* 64:435–460.

Dawkins, M. S. 1990. From an animal's point of view: Motivation, fitness, and animal welfare. *Behavioral and Brain Sciences* 13:1–61.

Dennett, D. C. 1989. Cognitive ethology: Hunting for bargains or a wild goose chase? In *Goals, No-Goals, and Own Goals,* ed. A. Montefiore and D. Noble, 101–116. London: Unwin Hyman.

———. 1991. *Consciousness Explained.* Boston: Little, Brown.

Duncan, I. J. H., and V. G. Kite. 1989. Nest site selection and nest-building behaviour in domestic fowl. *Animal Behaviour* 7:215–231.

Fragaszy, D. M., and L. E. Adams-Curtis. 1991. Environmental challenges in groups of capuchins. In *Primate Responses to Environmental Change,* ed. H. O. Box, 239-264. London: Chapman & Hall.

Fraser, D. 1975. The effect of straw on the behavior of sows in tether stalls. *Animal Production* 21:59–68.

Glanzer, M. 1958. Curiosity, exploratory drive, and stimulus satiation. *Psychological Bulletin* 55:307–315.

Glickman, S. E., and R. W. Sroges. 1966. Curiosity in zoo animals. *Behaviour* 6:151–188.

Griffin, D. R. 1991. Progress toward a cognitive ethology. In *Cognitive Ethology,* ed. C. A. Ristau, 3–17. Hillsdale, N.J.: Lawrence Erlbaum.

Gvaryahu, G., D. L. Cunningham, and A. van Tienhoven. 1989. Filial imprinting, environmental enrichment, and music application. *Poultry Science* 68:21–217.

Hughes, B. O., and I. J. H. Duncan. 1988. The notion of ethological "need," models of motivation, and animal welfare. *Animal Behaviour* 36:1696–1707.

Hughes, B. O., I. J. H. Duncan, and M. F. Brown. 1989. The performance of nest building by domestic hens: Is it more important than the construction of a nest? *Animal Behaviour* 37:210–214.

Inglis, I. R. 1983. Towards a cognitive theory of exploratory behaviour. In *Exploration in Animals and Humans,* ed. J. Archer and L. I. A. Birke, 72–116. Berkshire, U.K.: Van Nostrand Reinhold.

Inglis, I. R., and N. J. K. Ferguson. 1986. Starlings search for food rather than eat freely-available, identical food. *Animal Behaviour* 34:614–617.

Jolly, A. 1964. Prosimians' manipulation of simple object problems. *Animal Behaviour* 12:560–570.

Jones, R. B. 1982. Effects of early environmental enrichment upon open-field behavior and timidity in the domestic chick. *Developmental Psychobiology* 15:105–111.

———. 1987. Social and environmental aspects of fear in the domestic fowl. In *Cognitive Aspects of Social Behaviour in the Domestic Fowl,* ed. R. Zayan and I. J. H. Duncan, 82–149. Amsterdam: Elsevier.

———. 1992. Varied cages and aggression. *Applied Animal Behaviour Science* 33:295–296.

Jones, R. B., A. D. Mills, and J.-M. Faure. 1991. Genetic and experiential manipulation of fear-related behavior in Japanese quail chicks *(Coturnix coturnix japonica). Journal of Comparative Psychology* 105:15–20.

Kennedy, J. S. 1992. *The New Anthropomorphism.* Cambridge: Cambridge University Press.

Kramer, D. L., and D. M. Weary. 1991. Exploration versus exploitation: A field study of time allocation to environmental tracking by foraging chipmunks. *Animal Behaviour* 41:443–449.

Lawrence, A. B., E. M. C. Terlouw, and A. W. Illius. 1991. Individual differences in behavioural responses of pigs exposed to non-social and social challenges. *Applied Animal Behaviour Science* 30:73–86.

Line, S. W. 1987. Environmental enrichment for laboratory primates. *Journal of the American Veterinary Medical Association* 190:854–859.

Maier, S. F., and M. E. Seligman. 1976. Learned helplessness: Theory and evidence. *Journal of Experimental Psychology: General* 105:3–46, 105.

Markowitz, H. 1982. *Behavioral Enrichment in the Zoo.* New York: Van Nostrand Reinhold.

McGregor, P. K., and S. J. Ayling. 1990. Varied cages result in more aggression in male CFLP mice. *Applied Animal Behaviour Science* 26:277–281.

Mench, J. A. 1988. The development of aggressive behaviour in male broiler chicks: A comparison with laying-type males and the effects of feed restriction. *Applied Animal Behaviour Science* 21:233–242.

Mench, J. A., and M. D. Kreger. 1996. Ethical and welfare issues associated with keeping wild mammals in captivity. In *Wild Mammals in Captivity,* ed. D. G. Kleiman, M. E. Allen, K. V. Thompson, and S. Lumpkin, 5–15. Chicago: Chicago University Press.

Mench, J. A., and G. Mason. 1997. Behaviour. In *Animal Welfare,* ed. M. C. Appleby and B. O. Hughes. Wallingford, U.K.: CAB International.

Mench, J. A., and A. van Tienhoven. 1986. Farm animal welfare. *American Scientist* 74:598-603.

Menzel, C. R. 1991. Cognitive aspects of foraging in Japanese monkeys. *Animal Behaviour* 41:397-402.

Misslin, R., and M. Cigrang. 1986. Does neophobia necessarily imply fear or anxiety? *Behavioural Processes* 12:45-50.

Moberg, G. P. 1985. Biological response to stress: Key to assessment of animal well-being? In *Animal Stress,* ed. G. P. Moberg, 27-49. Bethesda, Md.: American Physiological Society.

Müller-Schwarze, D., B. Stagge, and C. Müller-Schwarze. 1982. Play behavior: Persistence, decrease, and energetic compensation during food shortage in deer fawns. *Science* 215:85-87.

Myers, A. K., and N. E. Miller. 1954. Failure to find a learned drive based on hunger: Evidence for learning motivated by "exploration." *Journal of Comparative and Physiological Psychology* 47:428-436.

Newberry, R. 1995. Environmental enrichment: Increasing the biological relevance of captive environments. *Applied Animal Behaviour Science* 44:229-243.

Nicol, C. J. 1992. Effects of environmental enrichment and gentle handling on behaviour and fear responses of transported broilers. *Applied Animal Behaviour Science* 33:367-380.

Nicol, C .J., and T. Guilford. 1991. Exploratory activity as a measure of motivation in deprived hens. *Animal Behaviour* 41:333-341.

Nissen, H. W. 1930. A study of exploratory behavior in the white rat by means of the obstruction method. *Journal of Genetic Psychology* 37:361-376.

O'Neill, P. L., C. Price, and S. J. Suomi. 1990. Designing captive primate environments sensitive to age- and gender-related activity profiles for rhesus monkeys *(Macaca mulatta).* In *Proceedings of American Association of Zoological Parks and Aquariums Regional Conference,* 546-554. Wheeling, W.Va.: AAZPA.

Pearce, G. P., and A. M. Paterson. 1993. The effect of space restrictions and provision of toys during rearing on the behaviour, productivity, and physiology of male pigs. *Applied Animal Behaviour Science* 36:11-28.

Pearce, G. P., A. M. Paterson, and A. N. Pearce. 1989. The influence of pleasant and unpleasant handling and the provision of toys on the growth and behaviour of male pigs. *Applied Animal Behaviour Science* 23:27-37.

Poole, T. B. 1992. The nature of evolution of behavioural needs in mammals. *Animal Welfare* 1:203-220.

Reed, H. J., L. J. Wilkins, S. D. Austin, and N. G. Gregory. 1993. The effect of environmental enrichment during rearing on fear reactions and depopulation trauma in adult caged hens. *Applied Animal Behaviour Science* 36:39-46.

Renner, M. J. 1987. Experience-dependent changes in exploratory behavior in the adult rat *(Rattus norvegicus):* Overall activity level and interactions with objects. *Journal of Comparative Psychology* 101:94–100.

———. 1988. Learning during exploration: The role of behavioral topography during exploration in determining subsequent adaptive behavior. *International Journal of Comparative Psychology* 2:43–56.

———. 1990. Neglected aspects of exploratory and investigatory behavior. *Psychobiology* 8:16–22.

Renner, M. J., A. J. Bennett, M. L. Ford, and P. J. Pierre. 1992. Investigation of inanimate objects by the greater bushbaby *(Otolemur garnettii). Primates* 33:315–328.

Renner, M. J., and M. R. Rosenzweig. 1986. Object interactions in juvenile rats *(Rattus norvegicus):* Effects of different experimental histories. *Journal of Comparative Psychology* 100:229–236.

———. 1987. *Enriched and Impoverished Environments.* New York: Springer-Verlag.

Renner, M. J., and C. P. Seltzer. 1991. Molar characteristics of exploratory and investigatory behavior in the rat *(Rattus norvegicus). Journal of Comparative Psychology* 105:326–339.

Russell, P. A. 1983. Psychological studies of exploration in animals: A reappraisal. In *Exploration in Animals and Humans,* ed. J. Archer and L. I. A. Birke, 2–54. Berkshire, U.K.: Van Nostrand Reinhold.

Shepherdson, D. 1994. The role of environmental enrichment in the captive breeding and re-introduction of endangered species. In *Creative Conservation: Interactive Management of Wild and Captive Animals,* ed. G. Mace, P. Olney, and A. Feistner, 167–175. London: Chapman & Hall.

Shepherdson, D. J., K. Carlstead, J. D. Mellen, and J. Seidensticker. 1993. The influence of food presentation on the behavior of small cats in confined environments. *Zoo Biology* 12:203–216.

Stolba, A., and D. G. M. Wood-Gush. 1984. The identification of behavioural key features and their incorporation into a housing system for pigs. *Annales de Recherches Veterinaires* 15:287–289.

Toates, F. M. 1983. Exploration as a motivational and learning system: A cognitive incentive view. In *Exploration in Animals and Humans,* ed. J. Archer and L. I. A. Birke, 55–71. Berkshire, U.K.: Van Nostrand Reinhold.

Trivers, R. 1991. Deceit and self-deception. In *Man and Beast Revisited,* ed. M. H. Robinson and L. Tiger, 175–191. Washington, D.C.: Smithsonian Institution Press.

USDA (United States Department of Agriculture). 1991. Animal welfare standards, final rule (Part 3, Subpart D): Specifications for the humane handling, care, treatment, and transportation of nonhuman primates. *Federal Register* 56 (32): 6495–6505.

Weld, K. P. 1992. Environmental enrichment of laboratory-housed nonhuman primates. Master's thesis, University of Maryland, College Park.

Wemelsfelder, F. 1993. The concept of animal boredom and its relationship to stereotyped behaviour. In *Stereotypic Animal Behaviour: Fundamentals and Applications to Welfare,* ed. A. B. Lawrence and J. Rushen, 65–96. Wallingford, U.K.: CAB International.

Wemelsfelder, F., and L. Birke. 1997. Environmental challenge. In *Animal Welfare,* ed. M. C. Appleby and B. O. Hughes. Wallingford, U.K.: CAB International.

Widowski, T. M., and S. E. Curtis. 1990. The influence of straw, cloth tassel, or both on the prepartum behaviour of sows. *Applied Animal Behaviour Science* 27:53–71.

Wood-Gush, D. G. M., and K. Vestergaard. 1991. The seeking of novelty and its relation to play. *Animal Behaviour* 42:599–606.

Wood-Gush, D. G. M., K. Vestergaard, and H. V. Petersen. 1990. The significance of motivation and environment in the development of exploration in pigs. *Biology of Behavior* 15:39–52.

HAL MARKOWITZ AND CHERYL ADAY

4

POWER FOR CAPTIVE ANIMALS

Contingencies and Nature

In 1972, an extensive program in behavioral enrichment for captive animals was begun in the Portland Zoo, now the Metro Washington Park Zoo in Portland, Oregon. Initially this work was referred to as "behavioral engineering" because environmental components were engineered to provide increased behavioral opportunities for animals. The first devices installed encouraged brachiation and leaping in white-handed gibbons *(Hylobates lar)* by providing opportunities for them to deliver food to themselves by moving between stations high in their enclosure (Markowitz 1982). Even though they were given any leftover food toward the end of the day on the same schedule as the other primates, these apes chose to feed themselves actively for seven years. When it became clear that some people, who otherwise liked the concept of enriching the environments of animals, misunderstood and thought that these new designs were intended to engineer the behavior of animals, the nomenclature "behavioral enrichment" was adopted.

As described in detail elsewhere (Markowitz 1982), there was never an opportunity in the early years of this work to design entirely new environments. Instead, efforts in the early years were largely dedicated to finding ways to give animals the power to more actively control parts of their own feeding schedules in largely outmoded exhibits. A few examples should suffice to provide the flavor of these efforts.

After the success of the work with gibbons in increasing species-typical behaviors, a second primate environment, which housed diana monkeys *(Cercopithecus diana)*, was modified. As with the gibbon apparatus, the new design included stations to move between, but this time the primates earned plastic chips that they could exchange at an "automat" for varied food whenever they wished. Thus, these animals were given more control of their own schedules and could choose to exchange the tokens for food immediately, hoard them, steal them from each other, give them away, etc. The remarkable variety of responses to these opportunities has been described in detail elsewhere (Markowitz 1982). One example serves to illustrate that an actively responsive environment provided chances for the diana monkeys to exhibit their ability to derive clever and unique solutions ensuring that they maintained their share of the food production.

Butch, the adolescent male diana monkey, was by far the most proficient at earning tokens and frequently shared the outcome of his efforts by leaving tokens around for other monkeys to redeem. However, during one period his mother was being particularly demanding by pushing him away and taking the fruit or chow directly from the automat every time that he ordered food. Butch's solution was to fake his mother out by clanging the token into the automat slot without actually depositing it and then palming the token. When his mother showed apparent chagrin at the failure of food to be delivered and eventually left Butch alone at the feeding station, he would then actually deposit the token and consume his reward.

This behavior resembles in many ways some of the instances of "tactical deception" recorded in wild populations (Whiten and Byrne 1990; Byrne and Whiten 1992) and in that sense it is naturalistic. However, as has been emphasized in the past (Markowitz 1982), this behavior is obviously not species typical, there being no monkey automat or token-dispensing apparatus in the wilds of Africa. It is not what would be designed given budgets for producing more naturalistic contingencies, especially recognizing that the major contemporary defense for zoos is conservation education. What this temporary solution to providing power for animals in antiquated enclosures did illustrate to zoo visitors as well as researchers, however, was the remarkable flexibility and creative learning skills of these nonhuman primates. Animals with well-developed cerebral cortices are clearly a joint product of their genetic makeup and their history of learning. Keeping such animals in an environment that offers no substantial opportunities to exercise their learning skills and be rewarded for their efforts is not humane.

Nature is full of contingencies to which animals must learn to respond in effective ways. Where captive environments cannot include the replication of natural contingencies, unnatural ones may serve to provide animals with power. After all, the behavior of most human primates is controlled by few *natural* contingencies. When we are deprived of the power to control most aspects of our own daily lives, we suffer. In institutional settings, the helplessness accompanying behavioral deprivation is typically debilitating, with long-lasting consequences. Hobfoll (1989) has written a compelling conceptualization of stress, emphasizing the importance to human well-being of maintaining control of resources. Some of the examples provided here serve to illustrate that being able to maintain some control, even over limited aspects of their environment, is also of critical importance to nonhuman primates.

ANIMAL HEALTH

The importance of developing wellness models when designing behavioral enrichment components to enhance both planned and existing animal environments has received increasing emphasis (Markowitz and Line 1989, 1990; Markowitz 1990; Mizuhara 1993). These efforts focus on establishing environments and husbandry practices that increase species-typical behaviors which are healthful aspects of natural behaviors. It was established earlier (Schmidt and Markowitz 1977; Markowitz et al. 1978) that significant benefits result from providing active behavioral contingencies for zoo animals, such as detecting any illness and helping the captives overcome some of the effects of living in impoverished environments.

In one example, giving servals *(Leptailurus serval)* the opportunity to pursue artificial prey not only provided healthful exercise but allowed detection of a chronic diaphragmatic hernia in one of the cats. Providing an active feeding paradigm for polar bears *(Ursus maritimus)* reduced aggression, and the male polar bear was also able to develop a full physiological reserve of fat during the appropriate season. In contrast, the work with mandrills *(Mandrillus sphinx),* described next, exemplifies using equipment as a temporary ameliorative solution to encourage active behaviors, thus eliminating severe aggression in a group in which this had been a significant problem.

ACTIVE TYPES OF ENRICHMENT COMPARED WITH PASSIVE TYPES

With the new federal regulations requiring provisions for the psychological well-being of primates came challenges in finding meaningful ways to measure whether animals were being housed in conditions meeting this requirement. Although there has been some controversy about the meaning of the term "psychological well-being" (Novak and Suomi 1988; Markowitz and Line 1989,

1990), most investigators report behavioral or physiological measures and compare these to species-typical ranges under conditions that are not highly stressful.

Using both types of measures (Line et al. 1987a,b, 1989, 1991a,b; Markowitz and Line 1989, 1990), we have provided evidence that older animals that have been individually housed for substantial periods of their lives do not show continued use of equipment or signs of improved well-being when they are provided with passive toys on a long-term basis. In contrast, when even limited environmental control is offered to primates, such as providing touch controls with which they can access food treats and music or silence, they show sustained use of the equipment. Those primates having responsive devices that produce differential outcomes as a function of their behavior also show much quicker physiological recovery from routine sources of stress in their everyday lives (Line et al. 1991a).

While it is true that such measures as routinely changing available passive toys or regularly changing the location of food hidden in trees may reduce stress for captive animals, many facilities, including zoos and aquariums, are not adequately staffed to allow labor-intensive solutions such as daily rotation of enrichment items. Furthermore, the enrichment needs of animals are often the last to be served when there are personnel shortages because of illnesses or other unscheduled events. In our experience, although the behavioral engineering approach is initially more costly, the availability of responsive elements in the environment on a continuous basis both increases the animal's power and reduces the number of occasions on which animals will be unstimulated.

"BAND-AID" SOLUTIONS

Historically, the most inappropriate exhibits in zoos and aquariums (from the standpoint of being species-appropriate environments) frequently have been the ones selected for improvements based on behavioral enrichment techniques. Although there was no hope of transforming these exhibits into appropriately naturalistic contemporary facilities, the fact that budgets dictated that mammals would live in them for an indeterminate number of years led us to search for temporary methods to provide the animals with some power and to learn what we could about their special abilities along the way. An early example was a speed game invented for mandrills to play against zoo visitors, or against a computer when there were no visitors with which to compete. This device allowed us to learn that the mandrill had a terrifically fast reaction time, sufficient to win contests with most human beings (Markowitz et al. 1982). It also greatly reduced levels of aggression in the enclosure (Yanofsky and Markowitz 1978) and generally entertained visitors and resident animals alike (Markowitz 1982).

A more recent set of studies in a less than ideal environment involved marine mammals in the Steinhart Aquarium at the California Academy of Sciences (Aday 1993). In this enrichment project, one of the few involving a mixed-species group, a number of different reinforcers were made available to two Pacific white-sided dolphins *(Lagenorhynchus obliquidens)* and three harbor seals *(Phoca vitulina richardii)*. The reinforcers included fish and simple toys, tactile stimulation by humans, activation of a water jet, and playback from any of three different sound channels. Each of these items was associated with a key on a xylophone-like apparatus, which had eight keys made of PVC pipes of graduated lengths.

Discrimination trials were used to determine whether the animals could associate visual availability of reinforcers with the use of particular keys. The animals as a group were able to reach the 90 percent correct level. All the mammals were seen to activate the keys of this apparatus; however, two animals, one dolphin and one seal, largely monopolized use of the enrichment device and accounted for 80 percent of all key presses. There was no significant decrease in use of the apparatus over a one-year period, during which it was available to the marine mammals three days a week for two 25-minute sessions.

As indicated by behavioral measures, the dolphins in this study showed a significant increase in well-being during those sessions when the apparatus was present. Even though one of the dolphins used the apparatus approximately five times as much as the other, both dolphins showed increased time spent in active behavior, decreased frequency of agonistic behavior, and decreased frequency of wall-touching, a stereotypic behavior. The behavior of the seals, on the other hand, did not indicate any change in the well-being of these animals, with no significant change in duration of active behavior or in frequency of agonistic behavior. These data suggest that this temporary approach to enrichment was an effective one for dolphins but that it was not appropriately designed to enhance the well-being of seals. Future equipment intended to provide more stimulating opportunities for these harbor seals should probably be designed to emphasize their special haul-out and feeding styles and allow for activities independent of the dolphins.

NATURALISTIC SCHEDULES AND ENVIRONMENTS

When sufficient money has been available, it has sometimes been possible to incorporate elements that do a better job of zoo education by combining naturalistic appearances with stimulation of as much naturalistic behavior as the public will tolerate. It deserves repeated mention that in general the public and regulatory agencies will not tolerate totally natural zoos, with mammals consuming each other on a routine basis and dying from various diseases as they would in the wild. This does not mean that zoo animals must necessarily be deprived of all elements of the chase or active food gathering in exhibits that look natural but are stripped of all contingencies which motivate species-appropriate behaviors (Forthman-Quick 1984). We note with much sincere pleasure that today almost everyone recognizes that there is a need for environments which are both naturalistic in appearance and rich in stimulating opportunities for the resident animals.

In the mid-1970s, a set of artificial prey and foraging devices controlled by a computer program was designed and installed in the Panaewa Rain Forest Zoo near Hilo, Hawaii (Markowitz 1982). Budgetary shortfalls for the county and state of Hawaii eventually led to such understaffing that there was no one available to load food into the feeding devices. During the time of their use, however, these mechanical devices provided active exercise for tigers *(Panthera tigris)* and primates, and their development clearly illustrated that artificial prey made to look naturalistic could be stimulating for zoo visitors and was regularly used by the animals on exhibit. The computer program incorporated elements of randomness in controlling the artificial prey in both location and time, and it also provided scrolling graphics so that the public could learn about the rain forest and buttons that allowed visitors to determine the hunting sequence.

On a more modest scale, live-prey-hunting opportunities were provided for small-clawed Asian river otters *(Aonyx cinerea)* in Marine World-Africa USA. This equipment provided artificial hunting opportunities for the otters while allowing the collection of data about their behavior and some aspects of public response to the exhibit. In addition to generally convincing some management personnel who were previously skeptical that the otters would actively respond to the hunting opportunity in this noisy milieu, the enrichment work brought much positive response from visitors and the media. A comparative study using items ranging from live crickets to gelatin capsules as rewards for hunting showed that the ability to control the delivery of even nonnutritive items was apparently entertaining for the otters, but that their greatest motivation resulted when live prey could be captured (Foster-Turley and Markowitz 1982). Even though this theme park uses very large numbers of dead fish as food for animals and rewards for show behaviors, there was reluctance to use live fish on the grounds that it might offend visitors to see "fishies" captured and eaten (Markowitz 1982) and we never progressed to the stage of adding additional kinds of live prey.

At the San Francisco Zoo in California, we have just completed a pilot study involving North American river otters *(Lontra canadensis)* in preparation for installing a live-fish dispenser and a programmed timer that will provide the two resident otters the chance to capture fish on a random naturalistic schedule. The pilot study included two weeks of daily sessions in which live fish were manually dispensed into the otters' pool; their behaviors and those of the visitors were monitored. As might well be predicted, the otters readily responded to the opportunity to pursue live fish and entertained themselves by chasing them under logs in the pool and finding ways to compete with each other to capture the most fish. Visitors gathered around the otter exhibit at times when they otherwise typically ignored this rather outmoded pool with a wooden otter shack on the island in the center. During several sessions we used fish that looked very much like the goldfish that people often keep in home aquariums. Children would root for the fish to escape and prolong the chase, but then, when the fish were captured, they would cheer for the otters. We were surprised that there were no negative comments from the spectators, even when their opinions were actively solicited.

SOME NEW IDEAS FOR ENRICHMENT: REDUCING COSTS AND SERVICE NEEDS

There has been considerable concern about the costs of producing effective, naturalistic opportunities for captive animals (Markowitz and Woodworth 1978; Markowitz 1982; Forthman-Quick 1984; Markowitz and Spinelli 1986). When substantially naturalistic surroundings are designed to include mechanical devices that emulate prey, the design and construction costs may exceed the available budgets of many institutions, and the devices typically require servicing that may necessitate adding technical staff to often-limited zoo personnel budgets (Markowitz 1982). In the San Francisco Zoo, the next series of major changes will involve the elimination of outmoded feline enclosures, the development of an off-exhibit feline conservation center, and the construction of more naturalistic modern environments for cats that are to be maintained on exhibit. In preparation for these changes, we have begun studies of ways to provide healthful activities for the cats and to promote naturalistic behaviors. The first such work was developed in an existing enclosure and designed to be transportable and easily reinstalled in new facilities with minimum difficulty. The use of moving sounds rather than moving objects allowed less-expensive construction and easier portability.

Although acoustic devices ranging from key- or coin-operated public information systems through naturalistic sounds to enhance the visitor experience have been used in zoos for some time, there are only limited examples of sound used as part of an active environmental enrichment protocol (Warner et al. 1979). Because naturalistic sounds can be produced without substantial moving parts and with standard hardware, the service requirements for a properly designed acoustic enrichment apparatus are minimal compared with the costs of maintaining mechanical prey. It is also considerably easier to arrange speakers in safe locations in animal environments than it is to install the large mechanical devices associated with mechanical prey and to provide protection for these devices from destruction by animals or environmental factors such as moisture.

The rapidly growing resources for computer-generated and -controlled sounds make this a propitious time to develop acoustic enrichment because it is no longer necessary to use high-maintenance equipment such as tape playback units. Instead, it is possible to use a single, moderately priced computer to control more than one enrichment system and collect on-line data about the use of devices by the animals. This first effort in the San Francisco Zoo has been developed with an eye to making the acoustic enrichment system as flexible as possible, so that as additional exhibits are added to the enrichment scheme they can be widely varied without having to redesign the control and data collection systems.

The animal provided with the first new acoustic enrichment apparatus was a sixteen-year-old African leopard *(Panthera pardus)* named Sabrina. Because leopards consume such a wide variety of prey in the wild, including antelope, wild pig, water buffalo, deer, dassies (rock hyrax), small rodents, birds, and insects (Bertram 1982; Norton et al. 1986; Amerasinghe et al. 1990), opportunities are plentiful to change the acoustic prey should the cats become less interested in the same sounds. For several months before the new equipment was installed, behavioral data were collected for comparison with behavior after the start of the acoustic prey regimen. Many attempts at environmental enrichment have not been accompanied by sufficient evaluation of outcomes (Markowitz and Line 1990). This research was designed to allow some assessment of general changes in behavior as well as those changes specific to use of the apparatus.

Sabrina was quick to learn the contingencies involved in obtaining special food treats, and her daily activity and apparent general demeanor were significantly enhanced as a function of this enrichment paradigm. This leopard shows no indication that the opportunity to engage in this artificial hunt has become less appealing to her over time; she continued to actively hunt after more than eight months of exposure to the equipment.

Among the anecdotal findings that are rewarding and interesting for those of

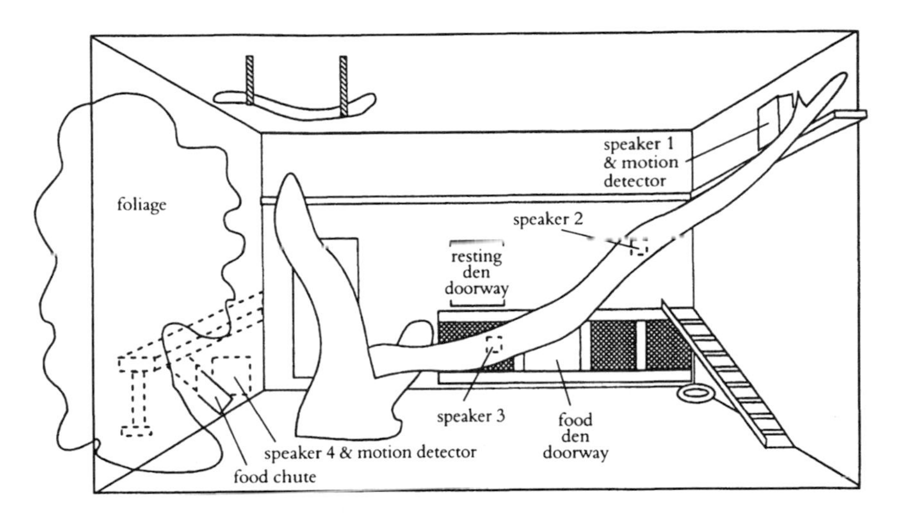

Figure 4.1. Layout of acoustic prey device for leopards. When the bird sound is playing at the upper speaker *(speaker 1)*, a forage at that location causes the bird to "move" sequentially from speaker to speaker to the opposite side of the enclosure. If the leopard forages at the final location *(speaker 4)* when the bird sound is heard there, a piece of food is delivered to the leopard.

us who observe her behavior are the varied ways in which she satisfies the prerequisites to "capturing" the bird. Sabrina sometimes races up the tree segment, bounding on to it, thereby shaking the shelf to which the tree is anchored at the top with sufficient force to trigger the motion detector before she reaches the foraging station (Figure 4.1). In her active phase she then uses one of several short routes to bound down and trigger the motion detector in the bushes that delivers her special food. At other times when she is more lethargic, Sabrina chooses to use her new-found power in more slow-moving ways, moving out of her den to the top level, softly triggering the motion detector with the weight of her body on the top platform, and then taking the easiest route down to the bushes at her leisure, often barely arriving within the prescribed time to capture her prey.

Even with this prototypic acoustic prey enrichment paradigm, it is clear that, as with many more complex and expensive mechanical alternatives used in the past (Foster-Turley and Markowitz 1982; Markowitz 1982; Markowitz and LaForse 1987), providing animals with broad contingencies and allowing them to "invent" ways to use them lead to interesting and varied behavior rather than excessively redundant stereotypic outcomes. Engineering ways in which animals can control some aspects of their environment should, we believe, always be done with the goal of allowing the animal to decide when and, as broadly as possible, how to use these opportunities.

In designing behavioral enrichment equipment and methods for introducing the equipment to animals, another principle of some importance is making the contingencies as clear and yet as interesting as possible for the individual animals. Here we have tried to keep the "shaping" requirements to a minimum. Our only coaxing of the leopard to use the equipment was to present initial opportunities where just hunting in the bushes (the last response in the chase sequence) was required to deliver food, and a limited number of trials in which we held chicken parts up near the foraging device at the top platform (see Figure 4.1) to show Sabrina that movement to that position would trigger the "flight" of the acoustic bird through a series of steps to the bushes at the bottom of the enclosure where she already knew how to earn treats. If we had waited for an indeterminate amount of time, it is very possible that Sabrina would have eventually learned these contingencies without any help from us. However, in the wild, animals learn some of their hunting skills as a function of watching what works for others, and in this sense it seemed appropriate to us to show her what worked and then let her use the apparatus as she wished.

The fact that this leopard often ignores the sound of the birds when they occur at random times and clearly chooses when she wishes to hunt is exactly the outcome for which we had hoped. It frustrates some zoo personnel and visitors that we cannot tell them exactly when to come to watch Sabrina hunt. We explain that this is not something that she is compelled to do, but something that empowers her to control part of her environment whenever she wishes to seize the opportunity. This satisfies all those who really care about the leopard's well-being—including us.

CONCLUSION

Designs for truly naturalistic-appearing environments combined with responsive environmental designs should ultimately gain more respect and support for zoos. Zoos thus will be enabled to accomplish more completely their goal of conservation education by producing environments that encourage animals to engage in species-typical behaviors which resemble those in the wild. Many of the current criticisms of zoos characterizing them as places of unkind incarceration can only be properly addressed by showing that caretakers can provide richer lives with more power for the animals. Empowered animals serve as emissaries for their species by conveying to people the beauty of animal diversity. Such efforts allow zoos to more effectively increase the number of people that recognize the urgency of reducing our own numbers and our exploitation of the planet if our children are to share the world with other capable, wonderful, powerful creatures.

REFERENCES

Aday, C. R. 1993. Environmental enrichment for dolphins and seals. Master's thesis, San Francisco State University, San Francisco.

Amerasinghe, F. P., U. B. Ekanayake, and R. D. A. Burge. 1990. Food habits of the leopard *(Panthera pardus fusca)* in Sri Lanka. *Ceylon Journal of Science: Biological Sciences* 21 (1): 17-24.

Bertram, B. C. R. 1982. Leopard ecology as studied by radio tracking. *Symposia of the Zoological Society of London* 49:341-352.

Byrne, R. W., and A. Whiten. 1992. Cognitive evolution in primates: Evidence from tactical deception. *Man* (London) 27:609-627.

Forthman-Quick, D. L. 1984. An integrative approach to environmental engineering in zoos. *Zoo Biology* 3:65-77.

Foster-Turley, P., and H. Markowitz. 1982. A captive behavioral enrichment study with Asian small-clawed river otters *(Aonyx cinerea)*. *Zoo Biology* 1:29-43.

Hobfoll, S. E. 1989. Conservation of resources: A new attempt at conceptualizing stress. *American Psychologist* 44:513-524.

Line, S. W., A. S. Clarke, G. Ellman, and H. Markowitz. 1987a. Behavioral and physiologic responses of rhesus macaques to an environmental enrichment device. *Laboratory Animal Science* 37:509.

Line, S. W., A. S. Clarke, and H. Markowitz. 1987b. Plasma cortisol of female rhesus monkeys in response to acute restraint. *Laboratory Primate Newsletter* 26 (4): 1-4.

Line, S. W., H. Markowitz, K. Morgan, and S. Strong. 1991a. Cage size and environmental enrichment: Effects upon behavioral and physiological responses to the stress of daily events. In *Through the Looking Glass: Issues of Psychological Well-Being in Captive Non-human Primates,* ed. M. A. Novak and A. Petto, 160-180. Washington, D.C.: American Psychological Association.

Line, S. W., K. N. Morgan, and H. Markowitz. 1991b. Simple toys do not alter the behavior of aged rhesus monkeys. *Zoo Biology* 10:473-484.

Line, S. W., K. Morgan, H. Markowitz, and S. Strong. 1989. Influence of cage size on heart rate and behavior in rhesus monkeys. *American Journal of Veterinary Research* 50:1523-1526.

———. 1990. Increased cage size does not alter heart rate or behavior in female rhesus monkeys. *American Journal of Primatology* 20:107-113.

Markowitz, H. 1982. *Behavioral Enrichment in the Zoo.* New York: Van Nostrand Reinhold.

———. 1990. Environmental opportunities and health care. In *CRC Handbook of Marine Mammal Medicine: Health, Disease, and Rehabilitation,* ed. L. Dierauf, 483–488. Boca Raton, Fla.: CRC Press.

Markowitz, H., and S. LaForse. 1987. Artificial prey as behavioural enrichment for felines. *Applied Animal Behaviour Science* 18:31–43.

Markowitz, H., and S. W. Line. 1989. Primate research models and environmental enrichment. In *Housing, Care, and Psychological Well-Being for Laboratory Primates,* ed. E. Segal, 203–212. Park Ridge, N.J.: Noyes Publications.

———. 1990. The need for responsive environments. In *The Experimental Animal in Biomedical Research,* Vol. 1, ed. B. E. Rollin and M. L. Kesel, 153–170. Boca Raton, Fla.: CRC Press.

Markowitz, H., M. Schmidt, and A. Moody. 1978. Behavioral engineering and animal health in the zoo. *International Zoo Yearbook* 18:190–194.

Markowitz, H., and J. Spinelli. 1986. Environmental engineering for primates. In *Primates: The Road to Self-Sustaining Populations,* ed. K. Benirschke, 489–498. New York: Springer-Verlag.

Markowitz, H., V. J. Stevens, J. D. Mellen, and B. C. Barrow. 1982. Performance of a mandrill *(Mandrillus sphinx)* in competition with zoo visitors and computer on a reaction-time game. *Acta Zoologica et Pathologica Antverpiensia* 76:169–180.

Markowitz, H., and G. Woodworth. 1978. Experimental analysis and control of group behavior. In *Behavior of Captive Wild Animals,* ed. H. Markowitz and V. J. Stevens, 107–131. Chicago: Nelson Hall.

Mizuhara, C. 1993. Evaluation of three environmental enrichment devices for research primates. Master's thesis, San Francisco State University, San Francisco.

Norton, P. M., A. B. Lawson, S. R. Henley, and G. Avery. 1986. Prey of leopards in four mountainous areas of the south-western Cape Province, South Africa. *South African Journal of Wildlife Research* 16 (2): 47–52.

Novak, M. A., and S. J. Suomi. 1988. Psychological well-being of primates in captivity. *American Psychologist* 43:765–773.

Schmidt, M., and H. Markowitz. 1977. Behavioral engineering as an aid in the maintenance of healthy zoo animals. *Journal of the American Veterinary Medical Association* 171:966–969.

Warner, A., H. Markowitz, and M. McBride. 1979. Environmental enrichment for polar bears in the zoo. Paper presented to the Western Psychological Association, San Diego, Calif., April 1979.

Whiten, A., and R. W. Byrne. 1990. Tactical deception in primates. *Behavioral and Brain Sciences* 13:412–414.

Yanofsky, R., and H. Markowitz. 1978. Changes in general behaviors of two mandrills *(Papio sphinx)* concomitant with behavioral testing in the zoo. *Psychological Record* 28:369–373.

5

MICHAEL D. KREGER, MICHAEL HUTCHINS, AND NINA FASCIONE

CONTEXT, ETHICS, AND ENVIRONMENTAL ENRICHMENT IN ZOOS AND AQUARIUMS

Early zoos and aquariums sought to exhibit a wide variety of species, and the space available to each was comparatively small. In an effort to prevent disease, captive wild animals were commonly kept in sterile tile and concrete enclosures to allow for easy cleaning (Hancocks 1971). In such cramped and inappropriate environments, animals often developed species-atypical behavior, such as stereotypic pacing, lethargy, and regurgitation and reingestion of food (Morris 1964; Myer-Holzapfel 1968; Hediger 1969; Erwin and Deni 1979).

With the advent of modern ethological and ecological studies, zoo biologists and architects began to understand the behavioral, dietary, veterinary, and environmental needs of captive wild animals, so that not only their physical but also their psychological health could be taken into consideration (Hediger 1969). As a result, there has been a trend toward constructing larger, more naturalistic exhibits intended to replicate many aspects of an animal's natural habitat (Hancocks 1971; Hutchins et al. 1984; Coe 1989; Tarpy 1993; Maple et al. 1995). We have also seen the evolution of environmental enrichment programs using various techniques to create even more interesting and interactive captive environments for zoo and laboratory animals (Markowitz 1982; Shepherdson 1988). According to Shepherdson (1988, 47): "The aim of environmental enrichment in zoos and aquariums is to . . . provide an environment in which [captive] animals behave as closely as possible to their wild counterparts."

Many complex ethical and practical issues can arise in the planning and implementation of environmental enrichment programs in zoos and aquariums. For example, if financial and human resources are limited which individual animals, groups, or species should receive the highest priority for enrichment?

Should an expensive technique be applied to a single species as opposed to applying a less expensive technique to several species? Should the welfare of individual animals always have the highest priority, or are there other factors (e.g., the intended goal or purpose for holding the species in captivity) that can influence the type of enrichment employed? Should the forms of enrichment always be the same for a given species, or can they vary based on context? Such decisions can be extremely complex from both an ethical and a practical point of view. Our intent, therefore, is to provide a conceptual tool to aid zoo managers in this difficult and sometimes controversial decision-making process.

The ethical foundation for our approach relies heavily on "moral pluralism," a concept that recognizes that human values can vary with context (Stone 1987). For example, it is generally considered unethical to take another person's life, but there are certain contexts in which society deems it permissible (e.g., in self-defense). Norton (1991, 198–199) stated: "Practical problem-solving and ethical ideas and principles are enlisted, not so much as *a priori* principles that enforce consensus, but as useful means to recognize similar features of varied cases—as useful tools, in other words, to aid in the development of a solution to moral quandaries."

A CONTEXTUAL APPROACH TO ENVIRONMENTAL ENRICHMENT

Because the goals of modern zoos and aquariums are diverse and their economic and human resources are limited, we contend that the optimal use and form of environmental enrichment depend largely on context. To develop an institutional environmental enrichment strategy that both maximizes animal welfare and recognizes other important concerns, zoological institutions must consider at least the following seven factors:

- the intended goal or use (i.e., purpose) of the animal or group of animals
- the physical environment in which the animals are maintained and the alternatives, if any, that are currently available
- the social environment in which the animals are maintained and the alternatives, if any, that are currently available
- the species-specific needs of the animals, such as those related to diet, locomotion, territoriality, and social contact with their own and other species
- individual variation in behavioral repertoires
- economic concerns, including available human and financial resources

- guidelines concerning animal welfare, including minimum housing and care standards, when available or legally mandated

None of these categories is mutually exclusive, and flexibility is necessary to allow for changes in the relative importance of each factor in any given context. The following section describes how context and ethics can affect strategies of environmental enrichment in zoos and aquariums. Our intent is not to provide an exhaustive list of various contexts, ethical considerations and solutions, an impossible task. Rather, our goal is to show how moral pluralism and pragmatism can be used to derive acceptable solutions to animal management problems, especially in those cases in which significant ethical questions exist.

INTENDED USE OR GOAL

Every animal in the collection of a zoo or aquarium should have a specific purpose or goal that is compatible with the institution's mission statement (Wiese and Hutchins 1994; Wiese et al. 1994; Hutchins et al. 1995). The missions of modern, professionally managed zoological institutions typically include conservation, education, research, and recreation (Conway 1969). All these goals, to some extent, depend as much on human perceptions as they do on the quality of animal care. In the context of public education and display, for example, environmental enrichment techniques must be as sensitive to the perceptions of the visitors as they are to the behavioral needs of the animals (Hutchins et al. 1978–1979; Coe 1985). When the goal of maintaining an animal in captivity is conservation, research, or even recreation, however, the context changes.

Education

One of the primary goals of modern zoos and aquariums is public education (Block 1991; Delapa 1994). Indeed, if conservation is to be successful, then the general public must develop a deeper appreciation for wildlife and their habitats. The primary threat confronting wildlife today is habitat alteration caused by the activities of humans (Ehrlich and Ehrlich 1981). Thus, conservation education programs often stress the interdependencies of wild animals and their habitats. Zoos and aquariums can promote public awareness of this concept through the design and construction of naturalistic exhibits, the intent of which is to simulate (to the extent possible in captivity) an animal's natural environment.

If an animal or group of animals is intended to serve an educational role, then the perception of the visitor is a high priority. A premium is placed on the naturalistic appearance of both the exhibit and the animals it contains (i.e., the "maximum option" after van Hooff 1986). In such a context, atypical behaviors, such as stereotypic pacing or regurgitation and reingestion of food, can give the visitor a negative impression of the exhibit, and any educational message could easily be lost (Akers and Schildkraut 1985). Environmental enrichment thus could clearly benefit both the animals and the zoo visitors, especially if it were effective in eliminating or reducing the frequency of undesirable behaviors. Appropriate enrichment however will depend largely on the message the institution intends to convey. For example, in naturalistic exhibits where the intent is to portray the animal in its natural habitat, enrichment options should be not only functional but should also appear to be a natural and integral part of the simulated natural habitat (Hutchins et al. 1978–1979, 1984; Shepherdson 1992).

In our opinion, an exhibit's appearance is not a trivial issue because it can affect not only the visitors's perceptions but also the quality of their educational experience in the zoo or aquarium setting (Sommer 1972; Hutchins et al. 1978–1979; Coe 1985; Maple and Finlay 1987). Studies conducted by Bitgood et al. (1988), Finlay et al. (1988), and Shettel-Neuber (1988) showed that zoo visitors prefer naturalistic exhibits and spend significantly more time viewing naturalistic displays than traditional displays. Visitors should also develop an appreciation and understanding of the animal for its own sake and not because of some vague similarity to humans. In more natural displays, obviously artificial objects or those that promote anthropomorphic perceptions are clearly out of place. Examples include computerized games (e.g., Markowitz 1982), circus tricks that mimic human behavior or reinforce negative or incorrect images about animals (e.g., bears begging; Hediger 1964), children's toys (Watson et al. 1989), and television (Maple and Hoff 1982).

This concept does not imply however that environmental enrichment opportunities can be ignored, but rather that they should be context appropriate (Shepherdson 1992). In naturalistic zoo or aquarium enclosures, alternative enrichment methods, such as deliberately hiding or scattering food to induce natural foraging, would be appropriate. These methods do not detract from the naturalistic appearance or from the educational experience of zoo visitors (Hutchins et al. 1978–1979, 1984; Hancocks 1980; Carlstead and Seidensticker 1991a,b). It is important however to realize that no amount of ingenuity can completely recreate the natural habitat of a species in the zoo or aquarium setting (Forthman-Quick 1984). Animal welfare should ultimately take precedence over aesthetics, but whenever possible artificial devices used in naturalistic exhibits should be hidden from public view or disguised so as not to destroy the illusion of nature.

The context changes when animals are held in off-exhibit holding facilities, night quarters, or quarantine facilities, because they are no longer in public view. The same argument could be made if animals are on exhibit but maintained in more traditional zoo enclosures. In such situations, the emphasis shifts to purely functional, rather than aesthetic, considerations.

Conservation

The conservation activities of modern zoological parks are numerous and varied (Wiese and Hutchins 1994; Hutchins and Conway 1995). However, we focus primarily on captive breeding for reintroduction because this is a unique way in which zoos and aquariums have contributed to wildlife conservation (Tudge 1991). Some examples of recent successes include the black-footed ferret *(Mustela nigripes),* Arabian oryx *(Oryx leucoryx),* and red wolf *(Canis rufus).*

If a particular species is being held in the zoo or aquarium for purposes of species conservation, then a whole new set of ethical considerations come into play. The elimination of atypical behaviors, especially as they affect courtship, mating, parental care, and reintroduction, is an important goal, and environmental enrichment can therefore contribute to conservation. Reintroduction of captive-bred animals presents some difficult ethical and practical challenges for zoo and aquarium biologists and animal welfare advocates (Beck 1995). The many threats faced by released animals include predators, competitors, starvation, inclement weather, parasites, and disease. Unlike their wild counterparts, captive-bred animals are often poorly prepared for such challenges, in that they may lack an ability to find, recognize, or acquire appropriate food items, interact socially with other conspecifics, or identify and avoid potential predators (Kleiman 1989; see also Castro et al., Chapter 8, and Miller et al., Chapter 7, this volume). It is therefore important to note that management strategies for captive-born animals intended for release should not seek to eliminate all stress, as stress is an adaptive response often critical to individual survival (Moodie and Chamove 1990; Sapolsky 1990; Snowdon 1994). Alternatively, the goal should be to maintain skills in the captive population that the animals would need to survive in the wild (Snowdon 1994).

The mortality of reintroduced captive-born animals is often high, and the responsible preparation of captive-born animals for reintroduction might require enrichment options that are comparatively stressful or even risky (Beck 1995). In this context, environmental "enrichment" might involve maintaining animals in large, naturalistic enclosures where species-typical modes of locomotion are possible (Beck et al. 1988), exposing them to real or artificial predators (Miller et al., Chapter 7, this volume), requiring them to forage for food much as they would in nature (Kleiman 1989), or even exposing them to pathogens so as to improve their immune systems by challenge (Coe and Scheffler 1989). Similarly, to survive in the wild, some predators need to gain experience in hunting and killing live prey (Eaton 1972; see Miller et al., Chapter 7, this volume).

Other issues are relevant to this discussion. For example, when deciding on enrichment priorities, should endangered animals receive a higher priority than common ones? In an ideal world, all captive animals would be provided with optimal conditions conducive to their welfare (Mason 1979). However, economic and other practical considerations may limit the number of species that can be maintained under optimal conditions. If a choice must be made, then we believe that endangered animals should receive the highest priority for environmental enrichment. Modern zoos have a tremendous responsibility in that some species have become extinct in the wild and now exist only in captive populations. If such species are to survive and have any chance of being reestablished in nature, it is critical that behaviorally intact populations be maintained (Wiese and Hutchins 1994).

Research

Zoos and aquariums have been increasingly cast into their new role as conservators of wildlife. With this responsibility has come a realization that little is known about the basic biology of many wild animals and their behavioral and environmental requirements in captivity (Hutchins 1988; Hutchins et al. 1996). Ongoing research in animal behavior, nutrition, reproduction, genetics, and clinical veterinary medicine are conducted by zoo and aquarium scientific staff and by collaborating scientists from local colleges and universities. Some zoo research may involve temporary social isolation of animals or transfer to small holding cages so that experimental manipulations can be performed (Mason 1979; Visalberghi and Anderson 1993). When animals must be placed in these less than optimal conditions, their need for environmental enrichment is likely to be high.

Veterinary considerations or experimental protocols may limit the range of enrichment options that can be utilized (i.e., in terms of social group composition, housing conditions, and space allocation) (Moor-Jankowski and Mahoney 1989). However, the options available for off-exhibit research animals can actually be more diverse and cost-effective because the emphasis can be purely on functional rather than on aesthetic considerations. In this context, puzzleboards, toys, grooming boards, and other obviously artificial devices are context appropriate. Such devices have been used effectively in biomedical research laboratories (Fajzi et al. 1989; Watson et al. 1989). Public perceptions are not as critical when animals are designated for research purposes and the research facility is not on public view. Environmental enrichment is still important, however, because atypical behavior and associated physiological stress can add unwanted variation to the experimental design, thereby confounding the results and jeopardizing the validity of the study (Snowdon 1994).

Recreation

Most professionally managed zoos and aquariums are nonprofit organizations supported by local governments and visitor admissions. Many visitors are attracted to zoos for recreational purposes, and although it may not be the primary purpose of the institution, recreation is often the avenue of support for other zoo interests such as education, research, and conservation (Hutchins and Fascione 1991). Maintaining animals for recreational purposes may also have educational benefits, and the two are seldom mutually exclusive (Hutchins and Fascione 1991). However, it can be argued that when the focus is primarily on human entertainment, rather than education, then the animals used in such activities should receive high priority for environmental enrichment. Because the goal of holding such animals in captivity is not conservation related, it can also be argued that they should receive a lower priority than endangered animals.

An example of species used primarily for recreation are ride animals, most of which are tamed or domesticated social mammals (e.g., ponies [*Equus caballus*], camels [*Camelus* spp.], and Indian elephants [*Elephas maximus*]). Because these animals may not be able to interact during working hours, the trainer or caretaker can substitute human social interaction at those times. Opportunities for social contact with other conspecifics can be provided when the animal is not working. An ability to perform other species-typical behaviors is also important. For example, the elephant's normal behavioral repertoire includes dust-bathing. While this is not something that would benefit visitors on an elephant ride, it could be permitted during rest breaks. Variation in training techniques can also be used to reduce or eliminate stress and the monotony of routine. For example, foods used as reinforcers, reinforcement schedules, and work hours and routines can be varied to introduce more unpredictability. Training animals to perform natural behaviors can be both entertaining to the public and potentially beneficial (therapeutic) for the animals (Hediger 1964, 1969; Laule 1993; Kreger and Mench 1995).

Enrichment can be used to facilitate visibility and thus contribute to an exhibit's recreational and educational potential. For example, artificial rocks containing heating coils or radiant heat lamps can be used to increase the visibility of reptiles by encouraging basking behavior. Various feeding techniques can also be employed to induce animals to use the areas of their enclosure that provide ideal viewing for the visiting public (Ogden et al. 1990; Ogden 1992).

THE PHYSICAL ENVIRONMENT

The need for enrichment and the type of enrichment employed must also be considered in the context of the physical environment. The relative need for enrichment is affected by the existing degree of complexity within an enclosure (both spatial and temporal), the type and age of an enclosure, and the enclosure's location (on- or off-exhibit, outdoors or indoors).

As explained, zoo exhibits can be a continuum of size and complexity. Either extreme could prove to be detrimental to individual animals, however, at least initially. For example, a sterile enclosure could lead to boredom and atypical behaviors in an animal raised in nature, while a highly complex, novel enclosure could initially induce fear, stress, or even aggression in a captive-born animal raised in a comparatively stark environment (Menzel 1963; Renner 1987). Gradual transitions to either extreme may therefore be necessary. For example, western lowland gorillas *(Gorilla g. gorilla)* raised indoors in traditional concrete and tile enclosures have needed several days or even weeks to adjust to outdoor, naturalistic exhibits (Maple and Hoff 1982; Ogden et al. 1990).

Behavioral stereotypies are often associated with housing conditions that deviate greatly from the natural environment (Wechsler 1991). Thus, the size and type of enclosure are important in determining enrichment priorities and options. Smaller, less complex enclosures are known to result in stereotypic, self-destructive, and other atypical behaviors in a variety of mammals and birds (Keiper 1969; Erwin and Deni 1979; Oldberg 1987; Wechsler 1991). These effects however can often be ameliorated through transfer to larger, more complex, and more naturalistic environments (Henning and Dunlap 1978; Clark et al. 1982; Ogden et al. 1990; O'Neil et al. 1991). In general, the older, and less naturalistic, complex, and species-appropriate an enclosure is, the higher its priority for environmental enrichment (Table 5.1). The same relationship holds true for the form of enrichment. The less naturalistic, complex, and species appropriate is an enclosure, the greater should be the emphasis on function rather than on aesthetics. For example, stereotypic route tracing (following a precise and invariable path within an enclosure) was reduced in captive canaries *(Serinus canarius)* by introducing a swinging perch to their flight cage (Keiper 1969). However, a similar solution would be out of place and probably unnecessary in a complex, naturalistic, outdoor aviary containing many natural plantings.

Table 5.1
A Conceptual Framework for Determining the Relative Need for Environmental Enrichment for Different Captive Animals

Context	High level of enrichment required if:	Low level of enrichment required if:
Physical environment	Traditional-style exhibit Inappropriate for species	Naturalistic exhibit Appropriate for species
Social environment	Inappropriate for species Unstable	Appropriate for species Stable
Species' characteristics	High neural complexity and cognitive ability Generalist habits Wide-ranging habits	Low neural complexity and cognitive ability Specialist habits Sedentary habits
Individual animal's behaviors	Atypical of its species	Typical of its species
Legal and professional guidelines	Not being met	Being met

Note: This table is not intended to provide solutions for the prioritization of environmental enrichment requirements; no simple formula exists, particularly when various contexts overlap or affect each other. It is intended solely as a conceptual tool to aid animal managers in making complex decisions regarding enrichment priorities.

THE SOCIAL ENVIRONMENT

For social species, the presence of other conspecifics can be a critical aspect of enrichment because appropriate group size, sex ratio, and age ratio are known to encourage species-typical behavior and reproduction (Eisenberg and Kleiman 1975; Kleiman 1980; Novak and Suomi 1988; Bayne et al. 1991). In contrast, the absence of a stable, appropriate group structure or rearing environment can lead to behavioral abnormalities (Erwin and Deni 1979; Hannah and Brotman 1990). In many cases, seasonal or temporal variations in group structure may also need to be replicated (Hutchins et al. 1984; Caro 1993).

The psychological needs or behavioral deficits resulting from the removal of an individual from its social group create new ethical considerations. In general, animals living in inappropriate social situations (i.e., in age and sex structures not typical for the species) may have a higher enrichment requirement than those in appropriate social situations (see Table 5.1). Thus, a social primate recently isolated from its group may have a higher priority for enrichment than a group-living conspecific. Because some forms of environmental enrichment have been effective in reducing aggression, this may also be true of animals in newly formed, unstable social groups (Bloomsmith et al. 1988; Boccia 1988). It should be noted, however, that removing an animal from its group, altering group membership, or introducing strangers into an unfamiliar social environment tend to be highly unpredictable and stressful events that can result in injury or even death (Bernstein 1989; Visalberghi and Anderson 1993). In the case of mixed-species exhibits, enrichment can occur as a result of interspecific interactions; however, as in single-species exhibits, attention must be paid to the potential for aggression-related injuries (Popp 1984).

There are good reasons for keeping some animals apart, including prevention of breeding, amelioration of serious aggression, or isolation of diseased animals so they do not infect others. In those cases in which animals must be isolated, social stimulation can come from visual, auditory, or olfactory stimuli rather than the actual physical presence of other conspecifics or species. Carlstead and Seidensticker (1991a,b) showed that one of the ways to reduce stereotypic pacing and increase exploratory or foraging behavior in a solitary male American black bear *(Ursus americanus)* was to place bear odors on the walls of his exhibit during the spring; in the wild, free-ranging males use olfactory cues in spring to detect potential mates.

Socially housed animals should also be provided with enrichment options that take group dynamics, such as dominance relationships, into account. For example, enrichment techniques that encourage foraging for food (e.g., puzzleboards, artificial termite mounds) can be monopolized by dominant individuals, thus providing little benefit for subordinates (Bloomstrand et al. 1986). Aggression and social dominance have additional implications for environmental enrichment requirements and strategies. For example, if a subordinate individual is seriously injured as a result of fighting, or so psychologically or physiologically stressed that its welfare or survival is jeopardized, it is often necessary to remove that animal from the group and isolate it in a holding area.

SPECIES-TYPICAL REQUIREMENTS

An examination of the species' natural history and preferences in the field are often essential in determining what environmental factors are important in captivity (Hediger 1969; Hutchins et al. 1984; see also Seidensticker and Forthman, Chapter 2, this volume). For example, capybara *(Hydrochaeris hydrochaeris),* which are large South American rodents, normally defecate in water and must be provided with a pool or tub of water for this purpose in captivity. Arboreal animals such as sloths (Family Bradypodidae) and tree kangaroos (*Dendrolagus* spp.) require trees and branches or similar structures on which to climb and rest. Similarly, burrowing animals such as meerkats *(Suricata suricatta)* and prairie dogs (*Cynomys* spp.) require natural substrates in which to dig.

Feeding behavior and ecology have been shown to be particularly important in formulating enrichment strategies for captive animals. Zoo animals are often fed the same type of foods at a given time of the day and in a given place, with emphasis being placed on nutritional requirements, economy, and ease of cleanup. This practice allows animals only the opportunity to consume their food, not to search for, pursue, or process it, which can lead to boredom and the development of stereotypies (Hutchins et al. 1984). In fact, several studies have shown that captive animals prefer to work for their food, rather than to be fed ad libitum, thus meeting their needs for both appetitive and consummatory behavior (Carder and Berkowitz 1970; Neuringer 1970; Inglis and Fergusson 1986). Many of the most successful enrichment projects have therefore been focused on "the provision of food in more interesting, challenging, and naturalistic ways" (Shepherdson 1992).

The inherent ranging patterns and dietary specializations of a species may influence the tendency to develop certain atypical or stereotypic behaviors. For example, Morris (1964) suggested a dichotomy between "extreme specialists" and "extreme generalists." He characterized dietary specialists, such as the giant panda *(Ailuropoda melanoleuca)* and koala *(Phascolarctos cinereus),* as having a comparatively low level of curiosity and activity. Once their basic nutritional needs have been met, the animals spend much of their time resting. Opportunists, on the other hand, are constantly on the move and forever exploring their environments, even when their basic needs have been satisfied. Examples include wolves *(Canis lupus),* rhesus macaques *(Macaca mulatta),* and raccoons (*Procyon* spp.). Morris further speculated that specialists could therefore adapt more readily to rigid captive environments than opportunists. Thus, abnormal behaviors were viewed as a form of compensation or adaptation to a restricted, monotonous environment. Similarly, Wechsler (1991) suggested that stereotypic behavior in polar bears *(Ursus martimus)* was caused by frustrated appetitive behavior. Wild polar bears often move great distances but are prevented from doing so in captivity. Thus, species that are wide ranging and opportunistic might be expected to have a greater tendency to develop certain atypical behaviors, such as stereotypic pacing. Following this line of reasoning, wide-ranging opportunists may have a greater need for some forms of enrichment than the relatively sedentary specialists, and thus should receive a comparatively higher priority (see Table 5.1).

The complexity of environmental enrichment required may also be taxon specific. Some investigators (e.g., Poole 1992a,b; see also Chapter 6, this volume) have argued that species with a high degree of neural complexity and advanced cognitive abilities have a greater need for environmental enrichment. For example, chimpanzees *(Pan troglodytes),* which have relatively large, complex brains, sophisticated cognitive abilities, and complex social behavior, probably require comparatively more mental stimulation to maintain their psychological well-being than a bird, snake, frog, or snail. This is not to say that animals with less neural complexity and less cognitive ability do not require appropriate enrichment (Dawkins 1992; King 1993); it does mean that if a choice must be made, mammals may be of higher priority than birds and birds of higher priority than reptiles, fish, or amphibians (Poole 1992b). This does not imply however that all mammal species should have a higher priority than all birds (e.g., shrews, Family Soricidae, versus scarlet macaws, *Ara macao;* see King 1993).

The amount of space required in captivity can also vary by species. For example, Hediger (1969) noted that species differed with regard to their "flight distance," the minimum distance at which an animal tended to flee from danger. This observation has important implications for the design of zoo and aquarium habitats in that animals maintained in small enclosures are often prevented from fleeing from people. It has been hypothesized that inability to flee a threatening situation can result in frustration and stress and lead to the development of stereotypic and other atypical behaviors (Mason 1991). Glatston et al. (1984), Hosey and Druck (1987), and Chamove et al. (1988) showed that the presence of zoo visitors negatively affected the behavior of a variety of primate species. Chamove found that the effect was strongest when visitors attempted to interact with the animals, suggesting that some captive primates do not readily habituate to humans, although all the animals studied were housed close to the public in traditional zoo enclosures. In contrast, naturalistic enclosures tend to place the animals farther from the public or at least offer opportunities for the animals to visually isolate themselves.

Some species restrict their movements to relatively small areas in nature. An animal's attachment to its territory can be very strong. For example, the Miami Metrozoo reported that many animals remained in their territories even when Hurricane Andrew had destroyed their exhibit barriers (W. Zeigler, personal communication). When planning an environmental enrichment strategy, it is therefore important to realize that enlarging an exhibit may not necessarily result in expansion of ranges or guaranteed use of new areas (Ogden 1992). Depending on the species in question, a lack of space does not always result in behavioral abnormalities. Marmie et al. (1990) showed that captive-born rattlesnakes (*Crotalus* sp.) suffered no behavioral deficiencies when reared in small, clear plastic boxes versus larger terrariums. Similarly, solitary adult marmosets (*Callithrix* spp.) raised in small laboratory cages did not develop stereotypic behaviors as readily as did other nonhuman primates under similar conditions (Berkson et al. 1966).

All other things being equal, individual animals or groups of animals whose basic species-typical needs are not being met should receive higher priority for environmental enrichment than those whose needs are being met (see Table 5.1). More specifically, if the current enclosure housing an arboreal animal does not contain appropriate climbing structures (and there are no other alternative enclosures to which the animal can be transferred), then zoo management should give this species high priority for environmental enrichment.

INDIVIDUAL VARIATION

Not only are general species considerations critical to determining priorities and forms of enrichment, but the background and characteristics or "personalities" of individual animals are also important. Although this concept has not been tested empirically, wild-caught animals may have a greater need for enrichment than captive-born individuals or vice versa because an individual's current needs and perceptions may be related to its previous experience. Because wild-born animals once lived in environments that were temporally, physically, and socially complex, they may experience relatively more psychological stress when confined in restricted, less complex, and unfamiliar captive environments. The close presence of humans is likely to be especially stressful for wild-caught animals, particularly if the species was hunted in nature and thus perceives humans as a threat.

The conditions under which animals are reared are known to have a significant impact on adult behavior. Captive-bred animals sometimes develop in inadequate social and physical environments, and enrichment may be needed to compensate for earlier deficiencies (Fritz 1986; Bayne et al. 1991). For example, hand-reared sloth bears *(Melursus ursinus)* showed significantly higher frequencies of stereotypic and self-directed behaviors such as masturbation, self-stimulation, and pacing as compared with mother-reared individuals (Forthman and Bakeman 1992). Similar patterns are well documented in primates (Erwin and Deni 1979). Zoos occasionally acquire exotic animals that were once household pets; these animals are more likely to exhibit atypical behavior as the result of an inadequate rearing environment or care.

Of primary importance is whether or not an individual is currently showing measurable signs of atypical behavior or physiological stress. In general, it would seem reasonable to assume that animals currently exhibiting signs of atypical behavior should receive the highest priority for environmental enrichment (see Table 5.1). The more frequent and atypical the behavior, the higher the priority.

ECONOMIC REALITIES

Long-range planning for an institutional environmental enrichment program must be done on a species-by-species basis, and economic constraints often determine both short- and long-term priorities. When human and financial resources are limited, the way in which they are allocated becomes an ethical question. In fact, it is conceivable that overemphasis on environmental enrichment could have a negative impact on other institutional goals, including conservation and education.

The costs of environmental enrichment should be compared to its benefits to help determine how best to allocate available resources (see Crockett, Chapter 9, this volume). Costs include time, materials, and labor. More specifically, there may be costs involved in designing, developing, and evaluating a device, in loading, cleaning, and maintaining apparatus, and in training personnel in its use. There are, of course, economic and other benefits to enrichment that might help offset these costs. Enrichment can improve general animal health and welfare, which may in turn decrease veterinary costs (Snyder 1975). Improved health and reproduction also help reduce the need to import animals from the wild or purchase them from other zoos and aquariums, dealers, or suppliers.

The cost-effectiveness of various enrichment options should be analyzed on the basis of experience at other facilities and published reports. It is not cost-effective to spend funds on devices or techniques that have been shown to be of limited effectiveness (Crockett, Chapter 9, this volume). For example, Line et al. (1991) found that toys that stimulate interest in young rhesus monkeys did not alter general activity or atypical behavior in aged animals.

New, complex, naturalistic exhibits that provide enrichment opportunities for animals are expected to attract more admission-paying visitors and generate positive media coverage (Bitgood et al. 1988; Finlay et al. 1988; Shettel-Neuber 1988). A healthy, well-enriched collection should also provide an improved image of modern zoos and aquariums. This new image could lead to increased financial support from the private sector and greater support from all levels of government. Some zoos have successfully launched fund-raising campaigns specifically to support environmental enrichment programs (Maas 1993).

LEGAL REQUIREMENTS AND PROFESSIONAL GUIDELINES

Welfare considerations affect all other considerations to the extent that zoo managers must define what levels of enrichment are considered to be minimum for each species in the collection. If an institution cannot meet such minimum requirements, then it has an ethical obligation to either immediately improve the situation or find another home for the animal(s). One factor that distinguishes a professionally managed zoological park or aquarium from a roadside animal attraction is that management continually strives to create better conditions for the animals in their care. In this regard, nearly all professionally managed zoos and aquariums in North America are accredited by the American Zoo and Aquarium Association (AZA). Representatives of the Association conduct an inspection of all member institutions on a regular basis, one goal of which is to assess the quality of animal care. While accreditation guidelines do not include detailed minimum housing and care standards for various species, many organized, cooperative breeding programs (species survival plans or SSPs) and taxonomic advisory groups (TAGs) are developing husbandry manuals that include minimum guidelines for environmental enrichment (Wiese and Hutchins 1994). The AZA also has a Mammals Standards Committee that is defining minimum care standards for this taxon; environmental enrichment is among the topics discussed. Similarly, the British Zoo Federation and the Universities Federation for Animal Welfare are collaborating to produce husbandry and welfare guidelines for specific taxa.

Finally, zoos and aquariums must comply with both federal and local animal welfare laws concerning environmental enrichment. It follows that if such standards are not being met, then the species or exhibit should receive high priority for attention. All U.S. exhibitors are regulated by the Animal Welfare Act and its amendments (U.S. Government 1994, 1995), which specify minimum requirements for most mammalian species. The regulations are specific for care, housing, treatment, handling, transportation, and, in some cases, environmental enrichment. Zoos and aquariums must also comply with municipal and state laws, when applicable.

DISCUSSION

There are many reasons why modern, professionally managed zoological institutions should be encouraged to plan and implement environmental enrichment programs. Such plans (1) provide an environment in which captive animals can perform species-typical behaviors, including appropriate reproductive and parental behaviors so critical to successful captive breeding programs; (2) reduce or eliminate stereotypic or other atypical behavior patterns; (3) reduce behavioral stress, thus leading to improved animal health, longevity, and reproduction; (4) provide opportunities to conduct research that is not confounded by behavioral or associated physiological stress; and (5) improve relations with the local community through increased attendance and public appreciation and understanding of animals and their natural behaviors.

In an ideal world, all captive animals would be provided with environments that were maximally conducive to their welfare. Given limited resources, however, prioritization is often necessary and decisions must be made concerning which species or individuals should be the initial or primary focus of more extensive enrichment efforts. We have argued that context is a major determinant of need and can be used as a tool to aid in the decision-making process. Such decisions are often extremely complex in that they involve a number of overlapping contexts and ethical considerations. We have argued that moral pluralism and pragmatism offer an acceptable ethical foundation for making such determinations. This interplay of context and ethics can be illustrated by the following hypothetical example.

A zoo has a lowland gorilla exhibit containing an adult male, three adult females, a juvenile female, and an infant. The gorillas are popular animals that attract many visitors to the zoo, and the species is considered endangered. Although the gorilla exhibit meets the basic regulatory requirements of the Animal Welfare Act, it is old and traditional, with the indoor portion having a concrete floor and tile walls and the outdoor portion being covered only in grass. Nevertheless, the animals are maintained in an appropriate group size and composition, reproduce successfully, and few atypical behaviors have been noted (all the animals, with the exception of one female, were mother raised). The zoo would like to build a larger, more naturalistic habitat for the animals but does not have sufficient funds to begin planning or construction at this time. Furthermore, zoo management has no option for temporarily placing the gorillas elsewhere or transferring them to another enclosure. In the interim, however, it is considering implementation of an environmental enrichment program.

Meanwhile, at the zoo's Bird House, there is an off-exhibit breeding facility for two pairs of adult red birds of paradise *(Paradisaea rubra),* each pair being maintained in a separate enclosure. Like the gorillas, each pair has access to both indoor and outdoor areas. Unlike those of the gorillas, the bird enclosures are relatively naturalistic, containing a soil substrate and dense plantings with multiple perch sites. However, both males were hand-reared and have developed stereotypic flying patterns when moving about their enclosures. In addition, one of the females is overpreening her tail feathers and several are missing. Although viable eggs are laid, hatch rate and chick survival have been low. The most frequent cause of egg and chick mortality is nest destruction and infanticide by the females. The zoo would like to breed these birds, as they are endangered and more consistent reproduction could lead to a cooperative captive breeding program.

Although maintained as pairs, the red bird of paradise is highly polygynous in nature and exhibits a lek mating system. Males gather on a communal display "arena"; females observe the males and presumably use cues such as feather length, color, and behavior to assess male quality and select a mate (Bradbury and Gibson 1983). In addition, once mating has occurred, males do not assist the female in nest-building or parental care (Frith 1976). Even though the area is not open to the public, the zoo would like to modify the interior of the enclosure to promote the formation of leks and create isolated nesting areas for females.

The zoo would like of course to complete both projects as soon as possible, but it has limited funds and personnel time to spend on new exhibit modifications and enrichment efforts. If only one of the projects can be funded, which should be selected? (The two projects would cost about the same in terms of both financial and human resources.) Alternatively, should a compromise be struck, with both species receiving some moderate level of attention?

It has been argued that the species with the greatest neural complexity and cognitive abilities (in this case, the gorillas) should have priority. However, various studies have also suggested that birds have some cognitive abilities and psychological needs (Dawkins 1980, 1992; King 1993), and further research may be necessary to determine the relative weight of this factor. Although the need for enrichment in both species is clear, the two situations (i.e., contexts) are actually quite different. In this case, the interplay of various contexts can help to determine which species should receive the highest priority. For example, the gorillas are housed in an older, traditional zoo enclosure and this would also argue for assigning them the priority. However, the immediate welfare of the primates does not appear to be greatly compromised; no atypical behaviors are currently being exhibited, and the group is healthy and has consistent reproductive success. The welfare of the birds of paradise, however, is questionable because of the frequent occurrence of atypical behaviors (i.e., stereotyped flight patterns and feather plucking in both sexes) and the lack of reproductive success. Possible causative factors include forced proximity of males and females during nest-building and egg laying, as well as the inability of males to display normal, species-typical lekking behavior. Although the birds are housed in a relatively naturalistic, complex enclosure, their group composition is highly species atypical—a factor that would argue for giving them the priority.

The overriding factor in this case could be the intended goal or the use of the animals. Both species are highly endangered in the wild and captive propagation could play a role in their future recovery, but neither is currently involved in reintroduction programs. However, gorillas are reproducing well in captivity and their population is self-sustaining, while birds of paradise are not. If enrichment has the potential of improving the bird of paradise program, thus leading to successful reproduction and an organized, cooperative breeding and conservation program, then perhaps funding should be appropriated to the birds rather than to the apes. Furthermore, even if one of the gorillas in the group displays a stereotypic behavior that does not interfere with group dynamics or with breeding, the bird of paradise enrichment program might still be considered a priority.

This example illustrates the complexity involved in deciding enrichment priorities and there are many other possible variations on this theme. Compromises are sometimes possible and desirable. There may be ways in which the gorilla exhibit could be inexpensively enriched and that simple, low-cost techniques could be used to improve conditions for the birds of paradise at the same time. For example, Hundgen et al. (1991) have used nylon mesh partitions to separate nesting female red birds of paradise from males and to create an artificial lek wherein males could display to one another, as in nature, but could not make physical contact.

We have outlined some significant ethical and practical considerations that can affect the planning and implementation of environmental enrichment programs in zoos and aquariums. From the example presented here, it is clear that there is no one single formula that can be applied to every situation. In an ideal world, animal managers should strive to provide all the animals under their care with sufficient enrichment to prevent undesirable or species-atypical behaviors and to promote their welfare. However, institutional resources are often limited, thus making prioritization a necessity. As we have shown, the decision to provide additional or daily enrichment opportunities for one group or species versus another is based largely on context.

By defining some potentially important factors, it is our hope that these complex decision-making processes can be facilitated. Rather than being arbitrary, such decisions should be carefully thought out and justifiable. We realize, however, that our state of knowledge regarding many species and their psychological and environmental requirements is still in an early stage of development and hope this preliminary model can be further refined.

SUMMARY

In summary, many ethical and practical issues arise in the planning and implementation of environmental enrichment programs in zoos and aquariums, particularly when human and financial resources are limited. No single formula or strategy exists, so enrichment program decisions should be carefully thought out and justifiable. Essential factors to be considered in plan formulation include intended use or goal of maintaining various species in captivity; existing physical and social environments in which species are maintained and alternatives, if any; species-specific needs; variation among individual animals; economic realities; and existing legal and professional standards or guidelines. Prioritization is paramount, although it can be extremely complex. The factors are not mutually exclusive or even hierarchical. To be of use in decision-making, they must be considered within a conceptual framework that accounts for context. Both ethical and practical considerations vary with context, and the influence of each factor can vary with context. Furthermore, interactions between factors can produce various situations, each with its own set of ethical and practical considerations.

REFERENCES

Akers, J. S., and D. S. Schildkraut. 1985. Regurgitation/reingestion and coprophagy in captive gorillas. *Zoo Biology* 4:99–109.

Bayne, K., S. Dexter, and S. Suomi. 1991. Social housing ameliorates behavioral pathology in *Cebus apella. Laboratory Primate Newsletter* 30 (2): 9–12.

Beck, B. B. 1995. Reintroduction, zoos, and conservation. In *Ethics on the Ark: Zoos, Animal Welfare, and Conservation,* ed. B. Norton, M. Hutchins, E. F. Stevens, and T. Maple, 155–163. Washington, D.C.: Smithsonian Institution Press.

Beck, B. B., I. Castro, D. G. Kleiman, J. M. Dietz, and B. Rettberg-Beck. 1988. Preparing captive born primates for reintroduction. *International Journal of Primatology* 8:426.

Berkson, G., J. Goodrich, and I. Kraft. 1966. Abnormal stereotyped movements of marmosets. *Perceptual and Motor Skills* 23:491–498.

Bernstein, I. S. 1989. Breeding colonies and psychological well-being. *American Journal of Primatology (Supplement)* 1:31–36.

Bitgood, S., D. Patterson, and A. Benefield. 1988. Exhibit design and visitor behavior: Empirical relationships. *Environment and Behavior* 20:474–491.

Block, R. 1991. Conservation education in zoos. *Journal of Museum Education* 16:6–7.

Bloomsmith, M. A., P. L. Alford, and T. Maple. 1988. Successful feeding enrichment for captive chimpanzees. *American Journal of Primatology* 16:155–164.

Bloomstrand, K. R., K. A. Riddle, and T. L. Maple. 1986. Objective evaluation of a behavioral enrichment device for captive chimpanzees *(Pan troglodytes). Zoo Biology* 5:293-300.

Boccia, M. L. 1988. Preliminary report on the use of a natural foraging task to reduce aggression and stereotypies in socially housed pigtail macaques. *Laboratory Primate Newsletter* 28:3-4.

Bradbury, J. W., and R. M. Gibson. 1983. Leks and mate choice. In *Mate Choice,* ed. P. Bateson, 109-138. Cambridge: Cambridge University Press.

Carder, B., and K. Berkowitz. 1970. Rats' preference for earned in comparison with free food. *Science* 167:1273-1274.

Carlstead, K., and J. Seidensticker. 1991a. Seasonal variation in stereotypic pacing in an American black bear *Ursus americanus. Behavioural Processes* 25:155-161.

———. 1991b. Environmental enrichment for zoo bears. *Zoo Biology* 10:3-16.

Caro, T. 1993. Behavioral solutions to breeding cheetahs in captivity: Insights from the wild. *Zoo Biology* 12:19-30.

Chamove, A. S., G. R. Hosey, and P. Schaetzel. 1988. Visitors excite primates in zoos. *Zoo Biology* 7:359-369.

Clark, A. S., C. J. Juno, and T. L. Maple. 1982. Behavioral effects of a change in the physical environment: A pilot study of captive chimpanzees. *Zoo Biology* 1:371-380.

Coe, C. L., and J. Scheffler. 1989. Utility of immune measures for evaluating psychological well-being in nonhuman primates. *Zoo Biology (Supplement)* 1:89-99.

Coe, J. C. 1985. Design and perception: Making the zoo experience real. *Zoo Biology* 4:197-208.

———. 1989. Naturalizing environments for captive primates. *Zoo Biology (Supplement)* 1:117-125.

Conway, W. G. 1969. Zoos: Their changing roles. *Science* 163:48-52.

Dawkins, M. S. 1980. *The Science of Animal Welfare.* London: Chapman & Hall.

———. 1992. Behavioural needs in birds. *Animal Welfare* 1:309-312.

Delapa, M. D. 1994. Interpreting hope, selling conservation: Zoos, aquariums, and environmental education. *Museum News* (May/Jun.): 48-49.

Eaton, R. L. 1972. An experimental study of predatory and feeding behaviour in the cheetah *(Acinonyx jubatus). Zeitschrift für Tierpsychologie* 31:270-280.

Ehrlich, P., and A. Ehrlich. 1981. *Extinction: The Causes and Consequences of the Disappearance of Species.* New York: Random House.

Eisenberg, J. F., and D. G. Kleiman. 1975. The usefulness of behaviour studies in developing captive breeding programmes for mammals. *International Zoo Yearbook* 17:81-88.

Erwin, J., and R. Deni. 1979. Strangers in a strange land: Abnormal behaviors or abnormal environments? In *Captivity and Behavior: Primates in Breeding Colonies, Laboratories, and Zoos,* ed. J. Erwin, T. Maple, and G. Mitchell, 148-181. New York: Van Nostrand Reinhold.

Fajz, K., V. Reinhardt, and M. D. Smith. 1989. A review of environmental enrichment strategies for singly caged nonhuman primates. *Lab Animal* 18:23–35.

Finlay, T., L. R. James, and T. L. Maple. 1988. People's perceptions of animals: The influence of the zoo environment. *Environment and Behavior* 20:508–527.

Forthman, D. L., and L. Bakeman. 1992. Environmental and social influences on enclosure use and activity patterns of captive sloth bears *(Ursus ursinus)*. *Zoo Biology* 11:405–415.

Forthman-Quick, D. L. 1984. An integrative approach to environmental engineering in zoos. *Zoo Biology* 3:65–77.

Frith, C. B. 1976. Displays of the red bird of paradise, *Paradisaea rubra,* and their significance, with a discussion on displays and systematics of other Paradisaeidae. *Emu* 76:69–78.

Fritz, J. 1986. Resocialization of asocial chimpanzees. In *Primates: The Road to Self-Sustaining Populations,* ed. K. Bernirschke, 351–359. New York: Springer-Verlag.

Glatston, A., E. Geiloet-Soeteman, E. Hora-Pacek, and J. A. R. A. M. van Hoof. 1984. The influence of the zoo environment on social behavior of groups of cotton topped tamarins, *Saguinus oedipus oedipus*. *Zoo Biology* 3:241–253.

Hancocks, D. 1971. *Animals and Architecture.* New York: Praeger.

———. 1980. Bringing nature into the zoo: Inexpensive solutions for zoo environments. *International Journal for the Study of Animal Problems* 1:170–177.

Hannah, A. C., and B. Brotman. 1990. Procedures for improving maternal behavior in captive chimpanzees. *Zoo Biology* 9:233–240.

Hediger, H. 1964. *Wild Animals in Captivity.* New York: Dover.

———. 1969. *Man and Animal in the Zoo.* New York: Delacorte Press.

Henning, C. W., and W. P. Dunlap. 1978. Tonic immobility in *Anolis carolinensis:* Effects of time and conditions of captivity. *Behavioral Biology* 23:75–86.

Hosey, G. R., and P. L. Druck. 1987. The influence of zoo visitors on the behaviour of captive primates. *Applied Animal Behaviour Science* 18:19–29.

Hundgen, K., M. Hutchins, C. Sheppard, D. Bruning, and W. Worth. 1991. Management and breeding of the red bird of paradise at the New York Zoological Park. *International Zoo Yearbook* 30:192–199.

Hutchins, M. 1988. On the design of zoo research programmes. *International Zoo Yearbook* 27:9–19.

Hutchins, M., and W. Conway. 1995. Beyond Noah's Ark: The evolving role of modern zoological parks and aquariums in field conservation. *International Zoo Yearbook* 34:117–130.

Hutchins, M., and N. Fascione. 1991. Ethical issues facing modern zoos. *Proceedings of the American Association of Zoo Veterinarians* 1991:56–64.

Hutchins, M., D. Hancocks, and T. Calip. 1978–1979. Behavioural engineering in the zoo: A critique. Parts 1–3. *International Zoo News* 25 (7): 18–23; 25 (8): 18–23; 26 (1): 20–27.

Hutchins, M., D. Hancocks, and C. Crockett. 1984. Naturalistic solutions to the behavioral problems of captive animals. *Zoologische Garten* 54:28–42.

Hutchins, M., E. Paul, and J. Bowdoin. 1996. Contributions of zoo and aquarium research to wildlife conservation and science. In *Well-Being of Animals in Zoo- and Aquarium-Sponsored Research,* ed. J. Bielitzki, J. Boyce, G. Burghardt, and D. Schaeffer, 23–39. Greenbelt, Md.: Scientists Center for Animal Welfare.

Hutchins, M., R. Wiese, and K. Willis. 1995. Strategic collection planning: Theory and practice. *Zoo Biology* 14:5–25.

Inglis, I. R., and N. J. K. Fergusson. 1986. Starlings search for food rather than eat freely available, identical food. *Animal Behaviour* 34:614–617.

Keiper, R. R. 1969. Causal factors in the stereotypies of caged birds. *Animal Behaviour* 17:114–119.

King, C. E. 1993. Environmental enrichment: Is it for the birds? *Zoo Biology* 12:509–512.

Kleiman, D. G. 1980. The sociobiology of captive propagation. In *Conservation Biology: An Evolutionary-Ecological Perspective,* ed. M. E. Soulé and B. A. Wilcox, 243-261. Sunderland, Mass.: Sinauer Associates.

———. 1989. Reintroduction of captive mammals for conservation. *BioScience* 39:152–161.

Kreger, M. D., and J. A. Mench. 1995. Visitor-animal interactions at the zoo. *Anthrozoos* 8:143–157.

Laule, G. 1993. Using training to enhance animal care and welfare. *Animal Welfare Information Center Newsletter* (National Agricultural Library) 4:2, 8–9.

Line, S. W., K. N. Morgan, and H. Markowitz. 1991. Simple toys do not alter the behavior of aged rhesus monkeys. *Zoo Biology* 10:473–484.

Maas, T. 1993. Phone books and boxes and balls, oh my! *A to Z* (Philadelphia Zoo) 2:12–15.

Maple, T. L., and T. W. Finlay. 1987. Post-occupancy evaluation in the zoo. *Applied Animal Behavioural Science* 18:5–8.

Maple, T. L., and M. P. Hoff. 1982. *Gorilla Behavior.* New York: Van Nostrand Reinhold.

Maple, T., R. McManamon, and E. Stevens. 1995. Defining the good zoo: Animal care, maintenance, and welfare. In *Ethics on the Ark: Zoos, Animal Welfare, and Conservation,* ed. B. G. Norton, M. Hutchins, E. F. Stevens, and T. L. Maple, 155–163. Washington, D.C.: Smithsonian Institution Press.

Markowitz, H. 1982. *Behavioral Enrichment in the Zoo.* New York: Oxford University Press.

Marmie, W., S. Kuhn, and D. Chizar. 1990. Behavior of captive-raised rattlesnakes *(Crotalus enyo)* as a function of rearing conditions. *Zoo Biology* 9:241–246.

Mason, G. A. 1991. Stereotypies: A critical review. *Animal Behaviour* 41:1015–1037.

Mason, W. A. 1979. Minding, meddling, and muddling through. *Laboratory Primate Newsletter* 18 (1): 1–8.

Menzel, E. W., Jr. 1963. The effects of cumulative experience on responses to novel objects in young, isolation-reared chimpanzees. *Behaviour* 21:1–12.

Moodie, E. M, and A. S. Chamove. 1990. Brief threatening events beneficial for captive tamarins? *Zoo Biology* 9:275–286.

Moor-Jankowski, J., and C. J. Mahoney. 1989. Chimpanzees in captivity: Humane handling and breeding within the confines imposed by medical research and testing. *Journal of Medical Primatology* 18:1–26.

Morris, D. 1964. The response of animals to a restricted environment. *Symposia of the Zoological Society of London* 13:99–118.

Myer-Holzapfel, M. 1968. Abnormal behaviour in zoo animals. In *Abnormal Behaviour in Animals,* ed. M. Fox, 476–503. London: W. B. Saunders.

Neuringer, A. J. 1970. Many responses for food reward with free food present. *Science* 169:503–504.

Norton, B. G. 1991. *Toward Unity among Environmentalists.* New York: Oxford University Press.

Novak, M. A., and S. J. Suomi. 1988. Psychological well-being of primates in captivity. *American Psychologist* 43:765–773.

Ogden, J. J. 1992. A comparative evaluation of natural habitats for captive lowland gorillas *(Gorilla g. gorilla).* Ph.D. dissertation, Georgia Institute of Technology, Atlanta.

Ogden, J. J., T. W. Finlay, and T. L. Maple. 1990. Gorilla adaptations to naturalistic environments. *Zoo Biology* 9:107–121.

Oldberg, F. O. 1987. The influence of cage size and environmental enrichment on the development of stereotypies in bank voles. *Behavioural Processes* 14:155–173.

O'Neil, P. L., M. A. Novak, and S. J. Suomi. 1991. Normalizing laboratory-reared rhesus macaque *(Macaca mulatta)* behavior with exposure to complex outdoor enclosures. *Zoo Biology* 10:237–245.

Poole, T. B. 1992a. The nature and evolution of behavioural needs in mammals. *Animal Welfare* 1:203–220.

———. 1992b. Author's response. *Animal Welfare* 1:30–311.

Popp, J. W. 1984. Interspecific aggression in mixed ungulate species exhibits. *Zoo Biology* 3:211–219.

Renner, M. J. 1987. Experience-dependent changes in exploratory behavior in the adult rat *(Rattus norvegicus):* Overall activity level and interaction with objects. *Journal of Comparative Psychology* 101:94–100.

Sapolsky, R. M. 1990. Stress in the wild. *Scientific American* 262 (1): 116–123.

Shepherdson, D. J. 1988. Environmental enrichment in the zoo. In *Why Zoos?,* 45–53. Potters Bar, U.K.: Universities Federation for Animal Welfare.

———. 1992. Design for behaviour: Designing environments to stimulate natural behaviour patterns in captive animals. In *Proceedings of the Fourth International Symposium on Zoo Design and Construction,* ed. P. Stevens, 156–168. Torquay, U.K.: Whitley Wildlife Conservation Trust.

Shettel-Neuber, J. 1988. Second and third generation zoo exhibits: A comparison of visitor, staff, and animal responses. *Environment and Behavior* 20:396–415.

Snowdon, C. T. 1994. The significance of naturalistic environments for primate behavioral research. In *Naturalistic Environments in Captivity for Animal Behavioral Research,* ed. E. F. Gibbons, E. J. Wyers, E. Waters, and E. W. Menzel, 217–258. Albany: State University of New York Press.

Snyder, R. L. 1975. Behavioral stress in captive animals. In *Research in Zoos and Aquariums,* 41–76. Washington, D.C.: National Academy of Sciences.

Sommer, R. 1972. What do we learn at the zoo? *Natural History* 81:26–27, 84–85.

Stone, C. D. 1987. *Earth and Other Ethics: The Case for Moral Pluralism.* New York: Harper & Row.

Tarpy, C. 1993. New zoos: Taking down the bars. *National Geographic* 184 (1): 2–37.

Tudge, C. 1991. *Last Animals at the Zoo.* London: Hutchinson Radius.

U.S. Government. 1994. Title 7, *U.S. Code* (Animal Welfare Act as Amended), sections 2131 et seq. Riverdale, Md.: APHIS/REAC, U.S. Department of Agriculture.

———. 1997. Title 9, *Code of Federal Regulations* (Animals and Animal Products), Subchapter A (Animal Welfare), Parts 1–3. Riverdale, Md.: APHIS/REAC, U.S. Department of Agriculture.

van Hooff, J. A. R. A. M. 1986. Behavior requirements of self-sustaining primate populations: Some theoretical considerations and a closer look at social behavior. In *Primates: The Road to Self-Sustaining Populations,* ed. K. Benirschke, 307–320. New York: Springer-Verlag.

Visalberghi, E., and J. R. Anderson. 1993. Reasons and risks associated with manipulating captive primates' social environments. *Animal Welfare* 2:3–15.

Watson, D. S. B., B. J. Houston, and G. E. Nacallum. 1989. The use of toys for primate environmental enrichment. *Laboratory Primate Newsletter* 28 (2): 20.

Wechsler, B. 1991. Stereotypies in polar bears. *Zoo Biology* 10:177–188.

Wiese, R., and M. Hutchins. 1994. *Species Survival Plans: Strategies for Wildlife Conservation.* Bethesda, Md.: American Zoo and Aquarium Association.

Wiese, R., K. Willis, and M. Hutchins. 1994. Is genetic and demographic management conservation? *Zoo Biology* 13:297–299.

TREVOR B. POOLE

MEETING A MAMMAL'S PSYCHOLOGICAL NEEDS

Basic Principles

Zoos are commonly criticized for "imprisoning" wild animals and denying them the opportunity to carry out most of their natural behavior patterns. This situation is contrasted with the wild, where animals are perceived as being free and unrestricted. The question is thus raised whether the wild represents an ideal state for an animal, so that the only humane solution when keeping them in captivity is to simulate the natural habitat as closely as possible. Advocates of this view would argue that only natural conditions can meet the behavioral needs of mammals (McKenna et al. 1987). I intend to challenge this point of view by establishing the forms taken by the behavioral needs of mammals and by showing that these needs can also be met in ways that are rather different from situations that the individual is likely to encounter in the wild.

Zoos have been profoundly influenced by the back-to-nature approach, and many have adopted this traditional way of meeting the behavioral needs of mammals. This approach involves attempting to simulate the wild and trying to make up the perceived deficit by environmental enrichment. However, natural features such as predators, disease, hunger, and other life-threatening challenges are omitted from the captive environment. Enclosures of this type are thus sanitized forms of nature, often owing more to human perception, prejudice, and the general ethical beliefs of the staff and visitors than to the real world. For example, in the United Kingdom it is illegal to provide live fish for otters or sea lions even though it is possible that these predators may suffer from an inability to hunt. While some American zoos provide live fish for polar bears *(Ursus maritimus)* or otters, most are reluctant to give carnivores mammal or bird carcasses. Thus, any simulation of nature in a captive environment is restricted because the zoo is responsible for the well-being of its animals and is not prepared either to subject them to natural life-threatening hazards or to antagonize public opinion.

This qualification raises the wider concern as to what behavioral needs a particular mammal may actually experience and whether mammals have specific needs to carry out particular behavior patterns to ensure their well-being; for example, whether predators experience a need to hunt prey. Ethologists have tried to determine whether certain activities represent behavioral needs, but this approach is limited because mammals can exist without any particular behavior that is not essential for survival. Equally, preference tests may show that animals prefer to do A rather than B or work harder to acquire A than B; however, preferences are fluid and particular rewards often have different values in different situations and may also vary with the state of the animal's mind and physiological condition. While the ethologist's approach of examining behavior patterns and their motivation and level of priority has worked well for the harder-wired vertebrates, it has proved less successful in discerning the behavioral needs of mammals.

Simply deciding on the value that a mammal places on a particular behavior is of limited usefulness, because mammals readily substitute one form of action for another depending on the facilities available. For example, the absence of a forest full of interesting foods may be of little concern to a chimpanzee *(Pan troglodytes)* or mandrill *(Mandrillus sphinx)* that has the opportunity to play a computer game (Markowitz 1978; Matsuzawa 1989). Chimps who enjoy working with computers have not been reared in forests, while those whose home is the forest are unlikely to show interest in computer games; the development environment is all important. In considering appropriate forms of environmental enrichment, it is therefore essential to be aware of the rearing environment of the individual. To ensure normal psychological development, a complex and interesting rearing environment must be provided.

If we are to consider meeting a mammal's psychological needs, it is important to define the criteria that should be used to determine whether these needs have been met. Those who are practically involved in assessing well-being in mammals (such as their caregivers) usually rely on, first, the absence of abnormal behaviors; second, whether the individual is busy (has a wide repertoire of behavior); third, whether it is confident (moves around freely showing no fear or aversive behavior); and finally, if it is able to rest in a relaxed manner (without constant signs of vigilance). These criteria are also generally implicit in most scientific studies of mammalian welfare. I therefore try to suggest, in this chapter, ways of meeting the psychological needs of mammals that satisfy, so far as may be practical in the real world, these four criteria.

THE MENTAL WORLD OF MAMMALS

It might be logical to conclude that natural selection has precisely shaped a mammal's behavioral needs to its particular habitat. If this were the case, to satisfy mammals the zoo would have to do everything possible to create an enclosure exactly like the wild. Much evidence conflicts with such a concept, however, because one of the most important mental attributes that mammals possess to promote their survival is intelligence. Intelligence enables individuals to modify their behavior appropriately to suit a wide range of situations. Jerison (1988) defined intelligence as the way one knows the world and uses that knowledge when adapting to changing situations. In other words, intelligence is the ability of an animal to create a model of a situation in its brain derived from multisensory data and experience and to use this model to modify future behavior in an adaptive fashion; for example, by making the right choices and taking appropriate actions to forestall problems or to achieve goals. Thus, mammals can adapt to a range of environments because their behavioral capabilities are sufficiently flexible.

Mammals learn to adapt their behavior to a particular environment during postnatal development when the young are protected by parental vigilance throughout their unique period of childhood. During this period, they learn the attributes of their world through play, curiosity, and imitation. For this reason, the adult mammal is a product of its childhood experience, so that in the wild some chimpanzees fish for termites and others break nuts with hammers and anvils, but in captivity others (in Japan and the United States) play computer games. It also carries the important implication that an individual brought up in a particular situation may find some difficulty in subsequently adapting to a markedly different environment as an adult. This is one of the problems with trying to introduce captive-bred mammals to the wild; their mental model of the world acquired in captivity is largely irrelevant for survival in the natural environment (International Academy of Animal Welfare Sciences 1992).

Although these differing skills are based on the natural attributes of the animals, individuals acquire them when they are young, so that they arise largely from upbringing. In the zoo we must therefore provide a wide range of facilities for youthful learning as well as an adult environment in which the skills that individuals have acquired can be put into practice. Such captive environments may not bear much resemblance to the wild, but the evidence from a wide variety of sources such as psychologists, animal trainers, and keepers supports the view that it is the nature of the challenges that is important rather than what the animal actually does.

ACQUIRING INFORMATION FROM THE ENVIRONMENT

The model of the real world that mammals create in their minds is determined by the physical and perceptual abilities of the species. Thus, evolution provides the mammal with basic physical and sensory skills but much of its behavior, as an individual, is directly related to the world that it has created for itself in its brain (Jerison 1973). This mental world that the mammal acquires in its youth is regularly updated and revised, which is achieved by comparing past experiences with the current situation; individual survival depends on keeping well informed. For example, ignorance of predators that may have taken up residence in their home range could prove fatal to members of a potential prey species. Animals require "food for thought," and I believe that it is the failure to satisfy this need for information that leads to boredom (Wemelsfelder 1984; see also Mench, Chapter 3, this volume) and abnormal behavior in captivity.

Mammals have evolved a capacity to expect change and challenges and a need to seek information. Thus, they are unsuited to a dependent lifestyle in captivity where all their material needs can be met without any effort on their part. Mammals may also spend time improving their skills, for example, by developing new foraging methods, but in addition if time and energy allow they also carry out leisure activities (Poole 1992). I have defined a leisure activity as one that does not provide a reward in the form of some immediate benefit in terms of survival. This can be contrasted with work, which consists of actions of immediate relevance to the survival of the animal or its genes. Leisure activities have a low priority and take place only when a mammal's immediate needs have been satisfied, as Baldwin and Baldwin (1972, 1974, 1976) demonstrated for play in wild and captive squirrel monkeys *(Saimiri oerstidii)*. Leisure activities such as play and so-called idle curiosity provide information about the nature of the individual's environment and the physical skills required to manipulate it. This information, acquired from seemingly frivolous activity, more often than not may prove to be of relevance in some future, unforeseen situation and this is why it has survival value.

BEHAVIOR IN THE WILD

Having considered the biological background of the behavioral needs of mammals, we first of all need to know what the particular mammal does in its natural habitat, which varies considerably from species to species. A day in the life of a small carnivore, such as a mink *(Mustela vison)* (Dunstone 1993), goes something like this: wake up, leave burrow, defecate, urinate, travel through a portion of the home range scent marking and seeking information about the presence of larger carnivores, conspecifics, potential prey, etc. Scent marks provide information about species, identifying how long ago the individual was present and whether it has made its home in the area. For conspecifics, the personal signature, sex, status, and reproductive condition may also be inferred from scent marks, urine, and feces (Gosling 1981). Any changes in topography are noted, especially those that impinge on the animal's pathways.

While patrolling its home range, the animal may stalk and capture small rodents and insects or visit the river and catch some fish. Any surplus prey may be cached; the mink then returns to its home base or to a convenient bolt-hole for a rest. Although it spends eighteen to twenty hours a day in its den, the four to six hours spent moving around outside are very energetic and the mink typically travels 5 to 6 kilometers during its active period. In the wild, therefore, the mink requires a secure base, a complex environment that includes land, water, and vegetation, opportunities to achieve objectives such as catching prey, and an element of novelty of which it must be aware.

TRANSLATION TO CAPTIVITY

The aim in providing an appropriate captive environment should be to create facilities that enable the animal to carry out a program of activity similar in complexity to that which it undertakes in the wild. We need to determine the kinds of things that the animal ought to be able to do and what facilities may be required for it to carry out a program offering a similar quality of life. It is important also to appreciate that carnivores, which take large meals of high-quality food, may spend very little time active, so we must create an environment that also allows the animals good facilities for resting (see Mellen et al., Chapter 12, this volume). To provide a suitable captive environment, we must examine the features of the animal's natural daily life that we could simulate in the zoo.

Security

The animal must have a safe haven in which to rest and feel secure. This may be a den, an elevated resting place such as the fork of a tree, sufficient space in the enclosure to exceed the animal's flight distance (Hediger 1955), or, for some species, companions that help to protect it and warn of danger. For mammals that make nests, we should provide appropriate bedding material, while social species should have the company of compatible conspecifics. In practice, it may not be possible to provide the same kind of refuge in captivity as in the wild, but substitutes can be devised. For example, a nest box with a tunnel may act as a satisfactory substitute for an earth burrow for foxes. Security is an important requirement; although sometimes neglected, it should be regarded as an essential aspect of environmental enrichment. It is easy to concentrate on giving animals something to do while neglecting to provide them somewhere to do nothing.

Complexity

The captive environment should also be sufficiently complex to allow a full range of locomotor activities, including walking, climbing, swimming, or burrowing as appropriate to the species concerned. In the wild, a mammal chooses a living area that offers suitable facilities for its needs, so the zoo manager should do the same for those in his care. When considering the quality of an enclosure, the activities for which facilities are provided and those that are omitted can be listed and thus related to the animal's abilities and natural lifestyle. When providing adequate complexity for a mammal within a restricted enclosure, it must be borne in mind that individuals may range over several kilometers under natural conditions. The quality of an environment may be compared with the wild by considering how much time the animal spends actively carrying out different normal behaviors. For most species, the size of enclosure will be a fraction of the natural home range, but it must be appreciated that the distance traveled in nature is partly determined by the availability of food and the individual's metabolic needs, so that there may be no necessity for a well-fed zoo animal to cover a similar distance in captivity. The time that the animal gains through living in a smaller range can be spent in acquiring survival skills and also information for its own sake (play and curiosity), in other words, in leisure activities.

Achievement

Although security and unvarying complexity allow mammals to devise games, it is also important to ensure that they can achieve objectives in their environment, or what is commonly referred to as "control." The animal should be able to obtain rewards for appropriate courses of action and be in a situation where it can make demands on the environment that have a probability of being satisfied. In planning these facilities, the zoo tends to prefer to create an environment for the animal that looks natural, but this must not be achieved at the expense of its welfare. Clearly the inclusion of artifacts in the enclosure should be explained to the public and their importance to the welfare of the animal emphasized. A chimpanzee's intelligence may be much better employed in a highly artificial situation where it can play a computer and learn tasks from a human trainer rather than simply being placed in an outdoor enclosure that is little more than a field surrounded by a moat. The latter may seem more natural to the human observer, but it offers far less interest to the animal. Likewise, a mink that spends an hour or so every day carrying out discrimination tests (Dunstone and Sinclair 1978) may enjoy a much better quality of life than one that lives out its life in a grassy enclosure with little to occupy its intelligence. To use a human analogy, in urban societies we live full and interesting lives in a highly artificial environment that bears little resemblance to the natural habitat of our ancestors.

The early work of Markowitz (1982) clearly showed that technology can enrich the life of mammals by providing challenges and opportunities for achievement. While this approach may be too expensive for the average zoo, it is quite wrong to dismiss it as "artificial," as Forthman-Quick (1984) pointed out. Not only does this application satisfy the intellectual needs of the animal, but it also demonstrates to the public the plasticity and adaptability of mammalian behavior.

Novelty

Captive environments should include an element of unpredictability because change and novelty provide the individual with the anticipated input of new information and also satisfy curiosity. Some degree of novelty can be introduced by adding materials or artifacts for short periods of time and changing them at intervals.

Table 6.1 illustrates how facilities can be provided in an enclosure to meet a mammal's needs for security, complexity, opportunities for achievement, and novelty, and the work and leisure activities associated with each practical suggestion (see also Mellen, Chapter 12, this volume).

THE HUMAN FACTOR

The relationship between the mammal and its caregiver is of critical importance for many species. Mammals in zoos are usually aware of their dependency on their keepers, and their security and well-being are greatly enhanced if there is a trusting, friendly relationship between human and animal. This feeling can even

Table 6.1
Features of Facilities Designed to Accommodate Activity Programs with Different Emphases

Program emphasis and facility feature	Type of opportunity	
	Work[a]	Leisure[a]
Security		
Burrow, bolt-holes	Sleep, rest	Hide-and-seek
Nest materials	Nest-building	Investigation, manipulation
Social group	Affiliative, sexual, agonistic, cooperative behaviors	Social play
Perches	Safety, rest	Acrobatics
Cliffs	Sleep, rest, seeking safety	Acrobatics
Space	Seeking safety	Locomotor play
Complexity		
Varied substrate	Foraging, scent-marking, digging, hiding	Social play and games utilizing substrate
Varied topography	Hiding	Social play and games utilizing topography
Vegetation	Ambushing, hiding, manipulation, feeding	Social play and games utilizing vegetation
Climbing frame, trees	Running, jumping, chasing	Social play and games utilizing climbing apparatus
Pond	Bathing, swimming, diving	Social play and games involving water
Opportunities for achievement		
Hidden food, dispensers	Foraging, learning, developing skills, searching and finding	None
Tools, operant devices	Learning properties, learning, cooperation	Curiosity, object play
Destructible objects	Learning properties	Curiosity, object play
Training by humans	Developing skills	Object or social play with trainer
Unpredictability and novelty		
Food	Foraging	None
Toys	Using as tools	Curiosity, play
Animal intruders (e.g., mice, birds, insects)	Chasing and catching, predation	None
Varied climate	Sunbathing, showering	None
Seasonal changes in vegetation	Manipulating and feeding on vegetation or visiting insects	Curiosity

Note: The features listed here are somewhat generalized; in practice, all features of the facility must be appropriate for the species.

[a] *Work* is any type of activity immediately relevant to the survival of the animal or its genes, whereas *leisure* is the kind of activity that an animal engages in only when its immediate needs have been satisfied.

be reflected in reproductive success in captivity, as Mellen (1991) has shown. This trust can be further reinforced by training animals to commands and encouraging them to learn simple routines using positive reinforcement (Poole and Kastelein 1990; see also Laule and Desmond, Chapter 17, this volume). Zoos sometimes fear that they will be accused of not being serious (the chimps' tea party) or of turning into circuses, but training undoubtedly improves the welfare of the animal and also makes routine examination and veterinary treatment much easier and less stressful (Kirkwood et al. 1990). Regular contact and good relations between caregivers and the mammals for which they are responsible can benefit the animal by providing security and greater complexity in its life, and humane training can add opportunities for both achievement and novelty.

SPECIES DIFFERENCES IN BEHAVIORAL NEEDS

It is well known that some species, such as bears and primates, are more prone to abnormal behavior in bare captive environments than others, such as deer. Different species therefore appear to have different needs for mental stimulation, needs that vary according to longevity, complexity of foraging techniques, vulnerability to predators, social environment, and nature of the terrain utilized. I suggest that each of these factors is likely to contribute to the overall complexity of the species behavioral needs.

Longevity

Long-lived animals can build up experience over many years and thus may be able to rely on this experience for decision-making by recalling distant past events. Some long-lived species, for example, elephants, may have a prodigious memory and the intellectual ability to utilize this knowledge and revise it by taking into account new experiences (Moss 1988). However, the possibility should also be borne in mind that some short-lived mammals such as shrews, which have a very high rate of metabolism, may be capable of acquiring experience at a higher rate than say a chimpanzee or an elephant.

Foraging

Complexity of foraging technique is also an important consideration. Omnivores, which depend on a wide variety of foods with different acquisition and handling techniques, require a high degree of skill and knowledge, as do predators that hunt highly active prey; in contrast, grazers require a minimum of skill in seeking and handling their food.

Vulnerability to Predation

Vulnerability to predators depends primarily on body size, but also on the defensive weapons, such as spines or stink glands, that the animal may possess. Large animals, such as elephants or the African buffalo *(Syncerus caffer),* have a lesser need to seek information about potential dangers from other animals, and their need for vigilance is also much less than that of a smaller animal such as a hyrax.

Complexity of Social Life

Elephants, monkeys, and mongooses, which live in large groups with friends, relatives, and rivals, also require a high degree of intelligence to maintain their positions in society (Byrne and Whiten 1988; de Waal 1991). A social group provides frequent changes and challenges, and aggression or submission alone may not solve many of the problems presented by group living. Success may depend more on alliances and deception than on confrontation and reconciliation.

Topography

Mammals that live in complex three-dimensional habitats need to practice their skills and judgment in an environment that has a varied topography. Thus they devise chasing games to maintain and improve these skills, which require a high degree of physical and mental agility (Fagen 1981).

Animal Senses

In planning environmental enrichment we should make provision for the fact that the degree of sensitivity of perception for most mammals is arranged in the following order, from greatest to least importance: smell, hearing, touch, and sight. The particular species should be assessed with respect to the four senses and facilities provided accordingly. A zoo thus should design suitable environments to meet the species need for varied sensory stimulation, enabling mammals to use their intelligence in interpreting a variety of different combinations of stimuli. It is particularly important that the intelligence of species relying on senses that human observers are unable to appreciate (tactile vibrissae, ultrasonic hearing, and sense of smell) should not be underestimated. For example, it is surprising that comparative psychologists found rats to be so intelligent in spite of the fact that they tested the rat's poorest sensory modality, vision, at a time when the animal would normally have been asleep. Rats might have proved even more clever if smell or hearing had been used in experiments and these had been conducted during their waking hours.

CONCLUSIONS

Mammals rely on their intelligence for survival and on their ability to keep well informed. Zoos should ensure that mammals live in environments that meet these psychological needs. To do this, the natural lifestyle and perceptual abilities of the species should be scored in relation to the likely complexity of the associated psychological needs. When this has been done, the environment can be planned to enable the particular species of mammal to carry out a program of activity that meets its need for security, provides adequate complexity for a full range of normal behaviors, and gives adequate mental stimulation by incorporating opportunities to achieve goals and some degree of novelty. These facilities should also offer opportunities for both work and leisure activities. If this is done, there seems little doubt that the psychological needs of the species can be met.

Although enclosures that look natural to the human eye are more aesthetically pleasing, zoos must not be afraid to provide artificial features if these meet a mammal's psychological needs. The reasons for doing so can be explained to the visiting public.

REFERENCES

Baldwin, J. D., and J. I. Baldwin. 1972. The ecology and behavior of squirrel monkeys *(Saimiri oerstedii)* in a natural forest in western Panama. *Folia Primatologica* 18:161–184.

———. 1974. Exploration and social play in squirrel monkeys *(Saimiri oerstedii)*. *American Zoologist* 14:303–315.

———. 1976. Effects of food ecology on social play: A laboratory simulation. *Zeitschrift für Tierpsychologie* 40:1–14.

Byrne, R. W., and A. Whiten. 1988. *Machiavellian Intelligence: Social Expertise and the Evolution of Intellect in Monkeys, Apes, and Humans.* Oxford, U.K.: Clarendon Press.

de Waal, F. B. M. 1991. The chimpanzee's sense of social regularity and its relation to the human sense of justice. *American Behavioral Scientist* 34:335–349.

Dunstone, N. 1993. *The Mink*. London: Poyser.

Dunstone, N., and W. Sinclair. 1978. Comparative aerial and underwater visual acuity of the mink (*Mustela vison* Schreber). *Animal Behaviour* 26:14–21.

Fagen, R. 1981. *Animal Play Behavior.* New York: Oxford University Press.

Forthman-Quick, D. L. 1984. An integrative approach to environmental engineering in zoos. *Zoo Biology* 3:65–77.

Gosling, L. M. 1981. Demarcation in a gerenuk territory: An economic approach. *Zeitschrift für Tierpsychologie* 56:305–322.

Hediger, H. 1955. *Studies of the Psychology and Behaviour of Captive Animals in Zoos and Circuses.* London: Butterworths.

International Academy of Animal Welfare Sciences. 1992. *Welfare Guidelines for the Reintroduction of Captive Bred Mammals to the Wild.* Potters Bar, U.K.: Universities Federation for Animal Welfare.

Jerison, H. J. 1973. *Evolution of the Brain and Intelligence.* New York: Academic Press.

———. 1988. Evolutionary biology of intelligence: The nature of the problem. In *Intelligence and Evolutionary Biology,* ed. H. J. Jerison and I. Jerison, 1–10. Berlin: Springer.

Kirkwood, J. K., C. Kichenside, and W. A. James. 1990. Training zoo animals. In *Animal Training: A Review and Commentary on Current Practice,* 93–99. Potters Bar, U.K.: Universities Federation for Animal Welfare.

Markowitz, H. 1978. Engineering environments for behavioral opportunities in the zoo. *Behavior Analyst* 1:34–47.

———. 1982. *Behavioral Enrichment in the Zoo.* New York: Van Nostrand Reinhold.

Matsuzawa, T. 1989. *The Perceptual World of the Chimpanzee.* Kyoto, Japan: Kyoto University Primate Research Institute.

McKenna, V., W. Travers, and J. Wray, eds. 1987. *Beyond the Bars: The Zoo Dilemma.* Wellingborough, U.K.: Thorsons.

Mellen, J. D. 1991. Factors influencing reproductive success in small captive exotic felids (*Felis* spp.): A multiple regression analysis. *Zoo Biology* 10:95–110.

Moss, C. 1988. *Elephant Memories.* London: Elm Tree Books.

Poole, T. B. 1992. The nature and evolution of behavioural needs in mammals. *Animal Welfare* 1 (3): 203–220.

Poole, T. B., and R. A. Kastelein. 1990. The role of training in the welfare of zoo mammals. *Ratel* 17:108–115.

Wemelsfelder, F. 1984. Animal boredom: Is a scientific study of the subjective experiences of animals possible? In *Animal Welfare Science,* ed. M. W. Fox and L. D. Mickley, 1–115. Washington, D.C.: Humane Society of the United States.

Part Two

ENVIRONMENTAL ENRICHMENT IN ANIMAL CONSERVATION AND WELFARE

BRIAN MILLER, DEAN BIGGINS, ASTRID VARGAS,
MICHAEL HUTCHINS, LOUIS HANEBURY,
JERRY GODBEY, STAN ANDERSON, CHRIS WEMMER,
AND JOHN OLDEMEIER

7

THE CAPTIVE ENVIRONMENT AND REINTRODUCTION

The Black-Footed Ferret as a Case Study with Comments on Other Taxa

Reintroduction of captive-raised animals has become a popular tool in the effort to conserve biodiversity, but the effects of captivity on behaviors—particularly those related to survival and reproduction in the wild—are not well defined, and there is much variability among species (Kleiman 1989; Derrickson and Snyder 1992; also see Castro et al., Chapter 8, this volume). Reintroduction, therefore, may not be as easy as it is seems (Hutchins et al. 1995a,b; Beck 1995). To increase the chances of success, we recommend that the possible effects of the captive environment on both behavior and survival be tested systematically. Many factors can influence postrelease survival and even minor details should not be taken for granted. For example, a failure to provide learning opportunities at critical times in an animal's development can result in reduced probability of survival (Beck 1995). In addition, without proper care, unconscious artificial selection or the lack of natural selection may erode traits necessary for survival in the wild, particularly in very long term breeding programs (Derrickson and Snyder 1992; Hutchins et al. 1995b). Mellen (1991) reported that small exotic felids reproduced more successfully when habituated to their keepers. However, habituation to humans is often maladaptive in the wild (Derrickson and Snyder 1992).

The captive environment may also erode morphological, behavioral, or physiological traits necessary for survival if those traits are expensive to maintain genetically (Derrickson and Snyder 1992). In addition, if the genetically determined range of expression for a particular trait is no longer modified during a critical period of development, one effect of captivity may be to increase individual behavioral variability to the point where adaptive traits (or the ability to perform those traits with the efficiency necessary for survival) are underrepresented in the reintroduced population. For example, our studies have noted considerable individual behavioral variation within captive-raised Siberian ferrets *(Mustela eversmannii)* even though the entire stock originated from only three females and two males.

The captive environment, therefore, can have a profound effect on development of behavioral traits, and the constraints and objectives of a captive breeding program may also differ from the objectives of reintroduction (Kreger et al., Chapter 5, this volume). In a captive breeding program, animals are raised with great care and reduced risk to ensure continued reproduction and high survivability. After those captive-raised animals are released into a natural habitat, however, they must overcome many dangers and hardships, including hunger, predation, disease, parasites, and inclement weather, to survive.

Successful wild animals have developed an array of finely honed behavioral and physiological responses to adapt to these conditions. The effective development of these adaptive responses requires the natural expression of genetically influenced traits, and it can also require exposure to the appropriate stimuli at a critical time (Gossow 1970). Indeed, the development of any complex behavior pattern is the result of extensive interaction between genetic heritage and experience (Polsky 1975). Some behaviors may require repeated cues throughout juvenile development if they are to be performed efficiently as adults (Gossow 1970). The failure of captive-raised reintroduced animals to survive compared to wild-born, translocated animals of the same species (Schladweiler and Tester 1972; Griffith et al. 1989; Beck et al. 1991; Biggins et al. 1991) emphasizes the importance of providing an appropriate captive environment (Shepherdson 1994). The reasons for decreased survival could include a host of variables, including one as simple as lack of physical conditioning (see Castro et al., Chapter 8, this volume).

On the basis of these reasons, we recommend that at least a sample of young animals targeted for release be raised in an environment as close to natural as is feasible, particularly if little is known about the species' normal developmental processes (Miller et al. 1990a). Indeed, prerelease exposure to the enriched environment of an enclosed prairie dog arena has been beneficial for the development of prey-killing and movement behaviors of captive-raised Siberian ferrets that were subsequently released into the wild (Biggins et al. 1991), and prerelease exposure to a similar type of arena increased the survival of captive-raised black-footed ferrets *(Mustela nigripes)* after reintroduction (Vargas 1994). Similarly, the program for reintroduction of the golden lion tamarin *(Leontopithecus rosalia)* used prerelease conditioning strategies successfully (Kleiman et al. 1986; Bronikowski et al. 1989; see Castro et al., Chapter 8, this volume). Survival of the masked bobwhite *(Colinus virginianus ridgwayi)* also improved with a prerelease conditioning program (Ellis et al. 1977), and large enclosures were used to establish social group structure before the successful reintroduction of the Arabian oryx *(Oryx leucoryx)* in Oman (Stanley-Price 1989).

In some cases (e.g., domestic dogs, *Canis familiarus*), wild behavioral repertoires can be restored after generations in captivity by recreating the correct environment during critical periods of development (Freedman et al. 1961, cited in Scott and Fuller 1965). This requirement varies from species to species, however, particularly for behaviors that are culturally transmitted. Indeed, learned behaviors can erode much faster than genetic diversity (May 1991; Shepherdson 1994). The ability to restore learned behaviors and the expense of restoring them will vary among species and traits. If captive breeding is augmenting a fragile wild population, the wild individuals may be able to transmit a certain amount of learned knowledge back to reintroduced individuals. If the species has critical traits that are culturally transmitted and there are no individuals left in the wild, an enriched environment which mimics nature may be crucial for a sample of the population throughout the entire captive program. However, even an enriched environment may not be sufficient for the transmission of some knowledge (e.g., migration routes).

Shepherdson (1994) listed two interrelated factors that are important to survival of captive-raised animals slated for release: (1) proficiency gained from earlier experience and (2) the ability to learn new skills after reintroduction and to adjust those skills as needed in the dynamic natural environment. Both these factors can be enhanced by enriching the prerelease environment (Shepherdson 1994). Shepherdson cautioned, however, that reproducing natural cues in a beneficial manner is not always easily accomplished. Golden lion tamarins that ranged free in a large wooded area of the National Zoological Park (Washington, D.C.) demonstrated beneficial behavioral adaptations after reintroduction, but previous attempts to train animals in enriched enclosures failed to bestow quantifiable advantages over those animals that were not trained (Bronikowski et al. 1989).

In this chapter, we discuss critical periods, recognition of home sites and movement, early neural development, avoidance of predators, capture of prey, and interaction of behavioral traits as these factors relate to prerelease conditioning of captive-raised animals scheduled for reintroduction. The discussion focuses on our experience with the black-footed ferret and its experimental analog, the more common Siberian ferret. However, experiences with other taxa are also summarized as necessary and relevant. We advocate systematic testing of the factors that could affect survival after release. We also discuss the need to integrate captive studies and field research for a successful and efficient recovery program.

BEHAVIOR AND THE EARLY ENVIRONMENT: CRITICAL PERIODS

Imprinting is a phase-specific learning process that requires specific stimuli during a sensitive period in the animal's development. The length of the sensitive period varies from species to species and behavior to behavior, and it is related to the speed of development (Immelmann 1975). Imprinting during the sensitive period is permanent and ensures that critical information is available to the individual before its first application. Animals that do not receive the specific stimuli during a sensitive period may still be able to develop particular behavioral traits later in life but will do so less efficiently (Hasler 1966: Gossow 1970; Immelmann 1975; Caro 1979).

Most of the evidence for a sensitive period has been reported in the avian literature, but there is information for some carnivores and other mammals. Dogs, wolves *(Canis lupus)*, domestic cats *(Felis silvestris)*, and European ferrets *(Mustela putorius)* have a sensitive period for socialization early in life (Scott and Fuller 1965; Leyhausen 1979; Poole 1972). Similarly, some species, such as domestic cats, weasels *(Mustela erminea)*, and European ferrets must be exposed to prey during juvenile development or they do not kill efficiently as adults (Gossow 1970; Caro 1979, 1980). Olfactory imprinting at two to three months of age has been correlated with neural development in European ferrets (Apfelbach 1986), and it influenced later food preferences in European ferrets (Apfelbach 1978, 1986) and black-footed ferrets (Vargas and Anderson, n.d.).

Raising animals in an enriched environment should reduce the risk of missing specific stimuli during an early sensitive period. If a sensitive period is missed, the animals may be forced to learn a critical behavioral trait at a time when efficiency cannot be maximized. If the stimuli are first experienced after the animal is reintroduced to the wild, that animal not only may have to learn at an inefficient time, but it may have to learn at a time when it is confronted by the complexity of the natural environment.

RECOGNITION OF HOME SITES AND MOVEMENT PATTERNS

One of the ecological implications of sensitive periods can be entrainment on a home site. In one case, captive mallard ducks *(Anas platyrhynchos)* raised in elevated wooden nest boxes chose elevated wooden nest boxes after reintroduction, as did their wild-born offspring. Conversely, mallard ducks raised in ground nests chose ground nests after attaining sexual maturity (Hess 1972). Cage-raised golden lion tamarins were released from a familiar nest box fastened to a tree without a release cage; they typically remained close to the nest box for periods averaging from twelve to eighteen months, but eventually moved farther from the site of release (B. Beck, personal communication). Cage-raised Siberian and black-footed ferrets that were released onto a prairie dog colony via release cages both used burrows and returned to the boxes (Biggins et al. 1991; Hnilica and Luce 1992).

When postrelease movements of cage-raised and pen-raised black-footed ferrets were compared (cumulative distance traveled each night and dispersal distance from the release location), cage-raised ferrets moved further each night and dispersed over larger areas, which made them more susceptible to surface predators such as coyotes *(Canis latrans)* (Biggins et al. 1992, 1993).

EARLY NEURAL DEVELOPMENT

For many mammals, the number of cranial neurons and synapses is not complete at birth, and the division of neurons and formation of synapses can continue for weeks or months (or for larger mammals, even years) post partum (Immelmann 1975). An enriched early environment can beneficially alter brain morphology (as well as increase brain weight); increase the number, pattern, and quality of synaptic connections; and enhance other cerebral measures that affect behavior later in life (Greenough and Juraska 1979; Rosenzweigh 1979).

Animals experiencing environmental complexity early in life are able to to better employ cues in problem solving. Henderson (1970) demonstrated that genetic potential of food-finding ability was reduced considerably by raising animals in a typical laboratory cage. In contrast, rats *(Rattus norvegicus)* from an enriched environment were more adept at maneuvering through an unfamiliar maze (Greenough and Juraska 1979). Renner (1988) showed that rats raised in an enriched environment could avoid predator models faster than rats raised in an impoverished environment. Juvenile Siberian ferrets familiar with underground burrow systems in an enriched environment spent less time on the surface than cage-raised Siberian ferrets when introduced to an unfamiliar arena, and excessive surface time increases the risk of predation (Miller et al. 1992). Siberian ferrets (Miller et al. 1992) and black-footed ferrets (Vargas 1994) raised in enriched environments were better able to secure prey. Therefore, when animals scheduled for reintroduction are raised in an enriched environment, they can acquire beneficial neural changes that can increase problem-solving abilities later in life. This process could enhance survival for a captive-raised individual that is introduced into a novel and dynamic wild environment.

AVOIDING PREDATORS

Successful predator avoidance behavior involves both recognition of a potential predator and the correct escape response performed in an efficient manner. Bolles (1970) speculated that innate, species-specific defense reactions occur in wild animals when the animals encounter certain stimuli. The responses are always near the threshold, so that the reaction (such as fleeing, freezing, and fighting) occurs whenever the correct stimulus is present. Evolutionarily, it makes sense that a trait as urgent as escape from predators be developmentally fixed (Alcock 1979; Coss and Owings 1985; Magurran 1989).

A number of studies (reviewed by Shalter 1984) have indicated that prey recognize a few key characteristics of classes of predators. Shalter stated that young animals might possess the ability to respond to a wide variety of predator-related stimuli; through a process of habituation, they narrow the responses to only those producing a real and present threat. If appropriate antipredator responses are solely the result of spatial and temporal novelty, then any reinforcement is unnecessary (Shalter 1984). The mother probably plays an important early role in shaping the infant's responses to novel and familiar experiences (Bronson 1968). This habituation process not only determines which species are a threat but can be fine-tuned to determine whether an individual predator is hunting or satiated. Along these lines, our experience with wild-born black-footed ferrets indicated that juveniles were more wary and difficult to trap shortly after they began to appear above ground and gradually became more tolerant of people, cattle, and pronghorn antelope (Miller and Anderson 1993).

Although some species maintain stereotypic antipredator responses whether in captivity or the wild (Smith 1975; Shalter 1984; Kleiman et al. 1986), a depauperate environment may reduce even relatively innate abilities over time (Derrickson and Snyder 1992). Schaller and Emlin (1962), Price (1972), and Smith (1972) demonstrated different responses between domesticated animals and their wild counterparts. Captive-raised masked bobwhites showed inefficient antipredator responses until they were exposed to dogs, humans, and a trained Harris hawk *(Parabuteo unctinctus)* before reintroduction. This prerelease exposure resulted in greater mobility, covey coordination, and predator avoidance skills (Ellis et al. 1977). It may be a matter not of keeping or losing a trait but of maintaining it at maximum efficiency. When predator pressures were reduced on some wild populations, those animals responded correctly but less efficiently (Curio 1969, cited in Shalter 1984; Morse 1980; Coss and Owings 1985; Loughry 1988).

When naive, captive-raised Siberian ferrets were exposed to avian and terrestrial models, they showed an ineffective escape response at two months of age, but naive Siberian ferrets increased alertness and escape time at three and four months of age; they also demonstrated the ability to improve antipredator reactions at three and four months of age after a single mild aversive stimulus (Miller et al. 1990a). Exposure to a live dog further improved antipredator responses (Miller et al. 1992), but that exposure did not translate into a significant difference in survival time after the animals were released into the wild (Biggins et al. 1991).

The lack of significant difference in survival time when Siberian ferrets were (or were not) exposed to potential predators may be the result of several factors. On one hand, detecting survival differences usually requires large sample sizes. Survival is affected by a number of variables, and it is often possible to demonstrate differences in behavioral traits before survival differences become detectable. We could not however analyze discrete predator avoidance behaviors in the wild as we did in captivity. Indeed, in other experiments in which we could measure behavioral responses in the wild, there were significant differences in behavior without significant differences in survival.

On the other hand, exposure to mock or real predators may not always be safe or practical. The stimulus may have to be presented in such an unnatural context that animals may not recognize the cue in a natural setting. In addition, if exposure is frequent without sufficient aversive conditioning, there is a danger of habituation to the predator (which we saw with Siberian ferrets). The development of antipredator behavior in Siberian ferrets may have resulted as much from the neural stimulation, enhanced problem-solving ability, familiarity with the natural setting, development of natural movement and activity patterns, and increased physical conditioning as it did from exposure to the mock predator.

LOCATING AND SECURING PREY

Black-footed ferrets kill prey that are equal to their own size. Only in highly specialized (anatomically and behaviorally) Viverridae, Mustelidae, Felidae, and Canidae is this predatory pattern observed (Eisenberg and Leyhausen 1972). Captive-raised mustelids have at least a rudimentary capacity to kill prey. Captive-raised ermine (Gossow 1970), European ferrets (Wustehusse 1960), long-tailed weasels *(Mustela frenata)* (Powell 1982), fishers *(Martes pennanti)* (Powell 1982), mink *(Mustela vison)* (Powell 1982), black-footed ferrets (Vargas 1994), and Siberian ferrets (Miller et al. 1992) all have killed during their first exposure to prey.

Even if the black-footed ferret neck bite is genetically predetermined, proficiency has been observed to increase with experience (Miller and Anderson 1993; Vargas 1994) and with exposure to an enriched environment (Vargas 1994). Similarly, captive Siberian ferrets killed more effectively with experience (Miller et al. 1992), and the only Siberian ferrets verified as taking prey after release to the wild had experience with prey and burrow systems in an enclosed prairie dog arena (Biggins et al. 1991). Killing a prey item outside its natural setting (especially one as unnatural as a cage) may be easier than killing a prey item in its own home range. A ferret that has gone without food for a time may become weakened, and each tiring and unsuccessful attempt at killing may make subsequent predatory opportunities more difficult.

Killing prey, however, is only one part of the predatory sequence. For example, two nine-month-old, captive-raised fishers killed porcupines *(Erithizon dorsatum)* on the first opportunity in captivity but starved to death after release in the wild because they were unable to successfully search for food (Kelly 1977). Thus, other facets of predation include learning to search in a particular place, altering direction to increase contact with the prey item, search image, learning the appropriate time to attack, and specialized hunting techniques (Krebs 1973; Lawrence and Allen 1983).

Wild black-footed ferrets use a zigzag hunting pattern that is typical of Mustelinae (Powell 1982; Richardson et al. 1987). They probably alter direction to increase probability of contact with desired prey items (Krebs 1973). Verbeek (1985) cited many instances of immature raptors pursuing inappropriate prey, and Griffiths (1975) hypothesized that immature predators attack prey as they find it while adults select their victims. Naive Siberian ferrets were capable of locating prey in a mock prairie dog arena, and they located food progressively faster between two and four months of age.

During winter, white-tailed prairie dogs *(Cynomys leucurus)* hibernate, and black-footed ferrets must locate these prairie dogs when burrows are plugged with soil. Naive Siberian ferrets located food in a burrow behind a 0.67-meter-diameter dirt plug during winter (Miller et al. 1990b), and reintroduced black-footed ferrets were observed digging in the typical manner of wild black-footed ferrets. Both maturation and experience probably play a role in predatory ability (Gossow 1970; Polsky 1975; Caro 1979, 1980; Tan and Counsilman 1985; Langley 1986; Miller et al. 1990b, 1992; Vargas 1994). The enriched environment of an arena provides ferrets with an opportunity to gain valuable experience with prey in a natural burrow system before release to the wild. It may also allow the young ferrets to contact prey at the appropriate time of development that maximizes killing efficiency during adulthood.

INTERACTION OF TRAITS

Behavioral traits do not exist in isolation. An animal must react to an array of stresses, risks, and conditions in its environment, and the strategy necessary to survive and reproduce depends on the circumstances of the moment or the season. Individual behaviors must be performed in the context of a host of other simultaneous behaviors necessary for survival; the expression of an individual behavior may be adjusted according to each situation that the animal faces.

For example, predation pressures determine the time of breeding in the tropical clay-colored robin *(Turdus grayi)* (Morton 1971). Solitary cervids flee quicker than individuals in a herd (Altmann 1958). Yellow-eyed juncos *(Junco phaeonotus),* ostriches *(Struthio camelus),* and coati *(Nasua narica)* alter feeding and drinking strategies when threatened with predation (Bertram 1980; Caraco et al. 1980; Burger and Gochfeld 1992). Faneslow and Lester (1988) showed that domestic rats at risk of predation decreased their foraging time with a compensatory increase in meal size that maintained their total daily intake and body weight. Even in moderately large carnivores, hunting behavior, socialization, and predator avoidance behaviors influence each other (Caro 1989).

Behavioral traits of wild, free-ranging animals must be performed efficiently under a variety of situations. The seminatural prerelease environment allows animals to develop responses (by whatever processes) to a variety of ecologically relevant stimuli. The approach also seems to be safest when there is incomplete understanding of developmental processes, and it provides the most holistic prerelease preparation for captive-raised animals.

MEASURING SUCCESS: THE IMPORTANCE OF INTEGRATING CAPTIVE STUDIES AND FIELD WORK

Many constraints are imposed by the small numbers and legal status of threatened populations. Some of our research used the closely related but common Siberian ferret to test hypotheses concerning behavioral development, prerelease conditioning, and release techniques before those ideas were applied to the endangered black-footed ferret. To prevent establishment of an exotic, Siberian ferrets used to test release techniques were first sterilized (by methods that left the steroid-producing organs intact), and all released animals were monitored by radiotelemetry.

The primary goal of reintroduction is to establish a wild population, and that aim may preclude the rigorous scientific schedules applied to laboratory studies. That does not mean, however, that the principles of science can be ignored when designing reintroduction techniques. A well-designed and monitored program will increase efficiency and provide accurate insight for other programs. The results will be far more credible if they are published in peer-reviewed journals, and those results should be accessible to the public. This approach will stimulate the creative input necessary for solving the complex problems of reintroduction.

A solid, scientifically designed base of data from the field and laboratory is essential to the conservation of declining species. Such data can help establish if or when captive breeding is necessary, the potential viability of a captive breeding program, the viability of reintroduction, and the captive conditions necessary to successfully reintroduce captive-raised species into the wild. The degree of culturally transmitted behaviors (e.g., parenting, migration routes), the duration of time necessary in captivity (shorter durations may result in fewer genetic and behavioral changes), and the causes of population decline will have a major impact on conservation decisions that involve captive propagation and reintroduction.

When animals are being bred in captivity for reintroduction, the captive program should analyze the effects of different captive environments on behavioral development, including an environment as nearly natural as possible. We urge that all options be considered and each species be tested individually. If data on behavioral development or behavioral traits critical to survival are unavailable from the field, they should be collected from a sample of wild-caught individuals as soon as they enter captivity. Comparison of trait expression among captive generations can provide valuable information about long-term effects of captive environments on behavior. Data could also be gathered on the possibilities of restoring eroded behaviors by using naturalistic enclosures during developmental periods, or by capturing a sample of reintroduced animals for behavioral comparison after a period of time in the wild.

It is important that reintroductions be monitored intensively to document causes of mortality, movements, and other attributes of reintroduced animals (Miller et al. 1993), including keeping accurate records on offspring born in the wild after reintroduction. These data are crucial to improving future reintroduction attempts. Reintroduction should test several release strategies including a "hard release" that serves as a control (U.S. Fish and Wildlife Service 1988; Miller et al. 1993). If different reintroduction techniques are compared early, the program has a higher probability of finding the most effective road to recovery quickly. In the long term, this will be a more efficient use of animals, time, and money.

In contrast, lack of comparison can be very costly if the sole technique fails. The financial expense is obvious, but extra time in captivity can also increase behavioral and genetic erosion. In general, the faster a captive species can be returned to the wild, the easier a reintroduction may be. Similarly, lack of comparison can cause a partially successful technique to be uniformly adopted when there actually may be a more efficient and effective method. When faced with known (but modest) success, some people may be prevented by fear of failure from looking for a better technique. The program will still make progress toward recovery but at a less efficient rate. In these times of shrinking budgets and escalating conservation problems, efficiency and economics must be prime considerations. Finally, unless work in captivity is coordinated with work in the field after reintroduction, a technique that looked promising in captivity but actually provided no benefit to survival after release may be adopted. There may also be situations in which a technique that produces a slight benefit to individual animals is so costly that this disadvantage outweighs any advantages to the population.

There are various measures for defining success as it relates to reintroduction. Ultimately, the number of animals surviving to successfully reproduce is the best comparative measure, but mortality is usually high in early reintroduction attempts. Postrelease analysis of behavioral traits may, therefore, provide important information about release strategies and contribute clues to increase survival in future attempts. Selection of behavioral attributes to be analyzed is influenced primarily by two considerations: (1) attributes that can be reliably recorded (via observation, telemetry, etc.) in free-ranging animals, and (2) traits which seem most important to the animal's overall success. To evaluate the first black-footed ferret reintroductions, we analyzed causes of mortality, daily survival rate, fidelity to release site, litter effects, diet, and movements (Biggins et al. 1992, 1993). In the first few releases of captive-raised animals, it may be desirable to consider knowledge gained toward an effective release strategy as the highest priority goal (Miller et al. 1993).

CONCLUSIONS

Because the captive environment can differentially influence species as well as individuals of the same species, we advocate the systematic testing of factors that could influence the success of reintroduction. We have outlined some possibilities and how they could potentially be affected by an enriched environment. Careful consideration of these factors and integration of captive studies and field research should enhance the probability of successfully reintroducing captive-raised animals to the wild.

ACKNOWLEDGMENTS

The Conservation and Research Center of the National Zoological Park, the Wyoming Game and Fish Department, and the U.S. Army Pueblo Depot supplied facilities and donated time and materials. The Wyoming Game and Fish Department allowed access to the black-footed ferrets at their captive breeding center. Funding sources included Wildlife Preservation Trust International, the U.S. Fish and Wildlife Service National Ecology Research Center, the U.S. Fish and Wildlife Service Wyoming Cooperative Research Unit, the U.S. Fish and Wildlife Service Region 6 Enhancement, Smithsonian Institution, the National Zoological Park, Friends of the National Zoo, Chevron USA, National Fish and Wildlife Foundation, Wyoming Game and Fish Department, Ferret Fanciers of Louden County, and Brookfield Zoo. Many biologists and technicians contributed to the field research, captive research, and raising black-footed ferrets and Siberian ferrets in captivity. Fritz Knopf and Constantino Macio Garcia provided constructive comments on the manuscript. We thank Ben Beck, Scott Derrickson, and Devra Kleiman for valuable discussions on the captive environment and reintroduction.

REFERENCES

Alcock, J. 1979. *Animal Behavior: An Evolutionary Approach.* Sunderland, Mass.: Sinauer Associates.

Altmann, S. A. 1958. Avian mobbing behavior and predator recognition. *Condor* 58:241-258.

Apfelbach, R. 1978. A sensitive phase for the development of olfactory preference in ferrets (*Mustela putorius f. furo* L.). *Zeitschrift für Saeugetierkunde* 43:289-295.

———. 1986. Imprinting on prey odours in ferrets (*Mustela putorius F. furo* L.) and its neural correlates. *Behavioural Processes* 12:363-381.

Beck, B. 1995. Reintroduction, zoos, conservation, and animal welfare. In *Ethics on the Ark: Zoos, Animal Welfare, and Wildlife Conservation,* ed. B. G. Norton, M. Hutchins, E. F. Stevens, and T. L. Maple, 155-163. Washington, D.C.: Smithsonian Institution Press.

Beck, B. B., D. G. Kleiman, J. M. Dietz, I. Castro, C. Carvalho, A. Martins, and B. Rettberg-Beck. 1991. Losses and reproduction in the reintroduced golden lion tamarins *(Leontopithecus rosalia). Dodo, Journal of the Jersey Wildlife Preservation Trust* 27:50-61.

Bertram, B. R. C. 1980. Vigilance and group sizes in ostriches. *Animal Behaviour* 28:278-286.

Biggins, D. E., J. Godbey, and A. Vargas. 1993. Influence of pre-release experience on reintroduced black-footed ferrets *(Mustela nigripes).* U.S. Fish and Wildlife Service Report. Fort Collins, Colo.: U.S. Fish and Wildlife Service, National Ecology Research Center.

Biggins, D. E., L. Hanebury, B. J. Miller, R. A. Powell, and C. Wemmer. 1991. Release of Siberian ferrets *(Mustela eversmanni)* to facilitate reintroduction of black-footed ferrets. U.S. Fish and Wildlife Service Report. Fort Collins, Colo.: U.S. Fish and Wildlife Service, National Ecology Research Center.

Biggins, D. E., B. J. Miller, and L. Hanebury. 1992. First reintroduction of the black-footed ferret. U.S. Fish and Wildlife Service Report. Fort Collins, Colo.: U.S. Fish and Wildlife Service, National Ecology Research Center.

Bolles, R.C. 1970. Species-specific defense reactions and avoidance learning. *Psychological Review* 77:32–48.

Bronikowski, J., B. Beck, and M. Power. 1989. Innovation, exhibition, and conservation: Free-ranging tamarins at the National Zoological Park. In *Proceedings of American Association of Zoological Parks and Aquariums Annual Conference,* 540–546. Wheeling, W.Va.: AAZPA.

Bronson, G. W. 1968. The fear of novelty. *Psychological Bulletin* 69:350–358.

Burger, J., and M. Gochfeld. 1992. Effect of group size on vigilance while drinking in the coati, *Nasua narica,* in Costa Rica. *Animal Behaviour* 44:1053–1057.

Caraco, T., S. Martindale, and H. R. Pulliam. 1980. Flocking: Advantages and disadvantages. *Nature* 285:400–401.

Caro, T. M. 1979. Relations of kitten behavior and adult predation. *Zeitschrift für Tierpsychologie* 51:158–168.

———. 1980. Effects of mother, object play, and adult experience on predation in cats. *Behavioral and Neurological Biology* 29:29–51.

———. 1989. Missing links in predator and anti-predator behavior. *Trends in Ecology and Evolution* 4:333–334.

Coss, R. G., and D. H. Owings. 1985. Restraints on ground squirrel anti-predator behavior: Adjustment over multiple time scales. In *Issues in the Ecological Study of Learning,* ed. T. D. Johnston and A. T. Pietrewicz, 167–200. Hillsdale, N.J.: Lawrence Erlbaum.

Derrickson, S. R., and N. F. R. Snyder. 1992. Potentials and limits of captive breeding in parrot conservation. In *New World Parrots in Crisis: Solutions from Conservation Biology,* ed. S. R. Beissinger and N. F. R. Snyder, 133–163. Washington, D.C.: Smithsonian Institution Press.

Eisenberg, J. E., and P. Leyhausen. 1972. The phylogenesis of predatory behavior in mammals. *Zeitschrift für Tierpsychologie* 30:59–72.

Ellis, D. H., S. J. Dobrott, and J. G. Goodwin, Jr. 1977. Reintroduction techniques for masked bob-whites. In *Endangered Birds,* ed. S. A. Temple, 345–354. Madison: University of Wisconsin Press.

Faneslow, M. S., and L. S. Lester. 1988. A functional behavioristic approach to aversively motivated behavior: Predatory imminence as a determinant of the topography of defensive behavior. In *Evolution and Learning,* ed. R. C. Bolles and M. D. Beecher, 185–212. Hillsdale, N.J.: Lawrence Erlbaum.

Gossow, H. 1970. Vergleichende verhaltensstudien an Marderartigen. I. Über Lautausserungen und zum Beuteurhalten. *Zeitschrift für Tierpsychologie* 27:405–480.

Greenough, W. T., and J. M. Juraska. 1979. Experience induced changes in brain fine structure: Their behavioral implications. In *Development and Evolution of Brain Size: Behavioral Implications,* ed. M. E. Hahen, C. Jensen, and B. C. Dudek, 263-294. New York: Academic Press.

Griffith, B., J. M. Scott, J. W. Carpenter, and C. Reed. 1989. Translocation as a species conservation tool: Status and strategy. *Science* 345:447-480.

Griffiths, D. 1975. Prey availability and the food of predators. *Ecology* 56:1209-1214.

Hasler, A. D. 1966. *Underwater Guideposts: Homing of Salmon.* Madison: University of Wisconsin Press.

Henderson, N. D. 1970. Genetic influences on the behavior of mice can be obscured by laboratory rearing. *Journal of Comparative and Physiological Psychology* 72:505-511.

Hess, H. H. 1972. The natural history of imprinting. *Annals of the New York Academy of Science* 193:124-136.

Hnilica, P., and B. Luce. 1992. Post-release surveys of free-ranging black-footed ferrets in Shirley Basin during the fall and winter of 1991. In *Black-Footed Ferret Reintroduction in Shirley Basin, Wyoming: 1991 Annual Completion Report,* ed. B. Oakleaf, B. Luce, E. T. Thorne, and S. Torbit, 172-195. Cheyenne: Wyoming Game and Fish Department.

Hutchins, M., K. Willis, and R. J. Wiese. 1995a. Strategic collection planning: Theory and practice. *Zoo Biology* 14:5-25.

———. 1995b. Author's response. *Zoo Biology* 14:67-80.

Immelmann, K. 1975. Ecological significance of imprinting and early learning. *Annual Review of Ecology and Systematics* 6:15-37.

Kelly, G. M. 1977. Fisher *(Martes pennanti)* biology in the White Mountain National Forest and adjacent areas. Ph.D. dissertation, University of Massachusetts, Amherst.

Kleiman, D. G. 1989. Reintroduction of captive mammals for conservation. *Bioscience* 39:152-161.

Kleiman, D. G., B. B. Beck, J. M. Dietz, L. A. Dietz, J. D. Ballou, and A. F. Coimbra-Filho. 1986. Conservation program for the golden lion tamarin: Captive research and management, ecological studies, educational strategies, and reintroduction. In *Primates: The Road to Self-Sustaining Populations,* ed. K. Benirschke, 959-979. New York: Springer-Verlag.

Krebs, J. R. 1973. Behavioral aspects of predation. In *Perspectives in Ethology,* ed. P. P. G. Bateson and P. H. Klopfer, 73-111. New York: Plenum Press.

Langley, W. M. 1986. Development of predatory behaviour in the southern grasshopper mouse *(Onychomys torridus). Behaviour* 99:275-295.

Lawrence, E. S., and J. A. Allen. 1983. On the search image. *Oikos* 40:313-314.

Leyhausen, P. 1979. *Cat Behavior: Predatory and Social Behavior of Domestic and Wild Cats* (trans. B. A. Tonkin). New York: Garland Press.

Loughry, W. J. 1988. Population differences in how black-tailed prairie dogs deal with snakes. *Behavioral Ecology and Sociobiology* 22:61-67.

Magurran, A. E. 1989. Acquired recognition of predator odour in the European minnow *(Phoxinus phoxinus). Ethology* 82:216-223.

May, R. 1991. The role of ecological theory in planning the reintroduction of endangered species. *Symposia of the Zoological Society of London* 62:145–163.

Mellen, J. D. 1991. Factors influencing reproductive success in small exotic felids (*Felis* spp.): A multiple regression analysis. *Zoo Biology* 10:95–110.

Miller, B. J., and S. H. Anderson. 1993. *Descriptive Ethology of the Black-Footed Ferret*. Advances in Ethology, Vol. 37. Berlin: Paul Parey.

Miller, B., D. Biggins, L. Hanebury, C. Conway, and C. Wemmer. 1992. Rehabilitation of a species: The black-footed ferret *(Mustela nigripes)*. *Wildlife Rehabilitation* 9:183–192.

Miller, B., D. Biggins, L. Hanebury, and A. Vargas. 1993. Reintroduction of the black-footed ferret. In *Creative Conservation: Interactive Management of Wild and Captive Animals,* ed. G. Mace, P. Olney, and A. Feistner, 455–463. London: Chapman & Hall.

Miller, B., D. Biggins, C. Wemmer, R. Powell, L. Calvo, L. Hanebury, and T. Wharton. 1990a. Development of survival skills in captive-raised Siberian polecats. II. Predator avoidance. *Journal of Ethology* 8:95–104.

Miller, B., D. Biggins, C. Wemmer, R. Powell, L. Hanebury, D. Horn, and A. Vargas. 1990b. Development of survival characteristics in Siberian ferrets *(Mustela eversmanni)*. I. Locating prey. *Journal of Ethology* 8:89–94.

Morton, E. S. 1971. Nest predation affecting the breeding season of the clay-colored robin, a tropical songbird. *Science* 171:920–921.

Morse, D. H. 1980. *Behavioral Mechanisms in Ecology.* Cambridge: Harvard University Press.

Polsky, R. H. 1975. Developmental factors in mammalian predation. *Behavioral Biology* 15:353–382.

Poole, T. 1972. Some behavioral differences between the European polecat, *M. putorius,* the ferret, *M. furo,* and their hybrids. *Journal of Zoology* (London) 66:25–35.

Powell, R. A. 1982. *The Fisher.* Minneapolis: University of Minnesota Press.

Price, E. O. 1972. Novelty-induced self-food deprivation in wild and semi-domestic deer mice *(Peromyscus maniculatus bairdii)*. *Behaviour* 41:91–104.

Renner, M. J. 1988. Learning during exploration: The role of behavioral topography during exploration in determining subsequent adaptive behavior. *International Journal of Comparative Psychology* 2:43–56.

Richardson, L., T. W. Clark, S. C. Forrest, and T. M. Campbell. 1987. Winter ecology of the black-footed ferret at Meeteetse, Wyoming. *American Midland Naturalist* 117:225–239.

Rosenzweigh, M. R. 1979. Responsiveness of brain size to individual experience: Behavioral and evolutionary implications. In *Development and Evolution of Brain Size: Behavioral Implications,* ed. M. E. Hahn, C. Jensen, and B. C. Dudek, 263–294. New York: Academic Press.

Schaller, G. B., and J. T. Emlen. 1962. The ontogeny of avoidance behaviour in some precocial birds. *Animal Behaviour* 10:370–381.

Schladweiler, J. L., and J. R. Tester. 1972. Survival and behaviour of hand-reared mallard released into the wild. *Journal of Wildlife Management* 36:1118–1127.

Scott, J. P., and J. L. Fuller. 1965. The critical period. In *Genetics and Social Behavior of the Dog,* ed. J. P. Scott and J. L. Fuller, 117–150. Chicago: University of Chicago Press.

Shalter, M. D. 1984. Predator-prey behavior and habituation. In *Habituation, Sensitization, and Behavior,* ed. H. V. S. Peeke and L. Petrinovich, 349-391. New York: Academic Press.

Shepherdson, D. 1994. The role of environmental enrichment in captive breeding and reintroduction of endangered species. In *Creative Conservation: Interactive Management of Wild and Captive Animals,* ed. G. Mace, P. Olney, and A. Feistner, 167–177. London: Chapman & Hall.

Smith, R. H. 1972. Wildness and domestication in *Mus musculus:* A behavioral analysis. *Journal of Comparative and Physiological Psychology* 79:22–29.

Smith, S. M. 1975. Innate recognition of coral snake pattern by a possible avian predator. *Science* 187:759–760.

Stanley-Price, M. R. 1989. *Animal Reintroduction: The Arabian Oryx in Oman.* Cambridge: Cambridge University Press.

Tan, P. L., and J. J. Counsilman. 1985. The influence of weaning on prey-catching behavior in kittens. *Zeitschrift für Tierpsychologie* 70:148–164.

U.S. Fish and Wildlife Service. 1988. *Black-Footed Ferret Recovery Plan.* Denver, Colo.: U.S. Fish and Wildlife Service.

Vargas, A. 1994. Ontogeny of the endangered black-footed ferret *(Mustela nigripes)* and the effects of captive upbringing on predatory behavior and post-release survival for reintroduction. Ph.D. dissertation, University of Wyoming, Laramie.

Vargas, A., and S. H. Anderson. n.d. The effects of diet on black-footed ferret *(Mustela nigripes)* food preferences. *Zoo Biology* (in press).

Verbeek, N. A. M. 1985. Behavioral interactions between avian predators and their avian prey: Play behavior or mobbing? *Zeitschrift für Tierpsychologie* 67:204–214.

Wustehusse, C. 1960. Beitrage zur Kenntnis besonders des Spiel und Beutefangverhaltens einheimisher Musteliden. *Zeitschrift für Tierpsychologie* 17:579–613.

M. INÊS CASTRO, BENJAMIN B. BECK,
DEVRA G. KLEIMAN, CARLOS R. RUIZ-MIRANDA,
AND ALFRED L. ROSENBERGER

8 ENVIRONMENTAL ENRICHMENT IN A REINTRODUCTION PROGRAM FOR GOLDEN LION TAMARINS (*LEONTOPITHECUS ROSALIA*)

Animals that lack the necessary skills for survival in a wild habitat are less likely to be successfully introduced. If a captive population is managed genetically, the potential for expression of species-typical behaviors will probably persist (Ballou 1992). However, even though we strive to preserve genetic diversity, we may not be maintaining the original behavioral repertoire because the "typical" captive environment may not encourage the development of species-appropriate behaviors that promote survival after reintroduction (but see Miller et al., Chapter 7, this volume). We know little about the effects of zoo environments on the development and expression of natural behaviors. For example, we do not know if individuals born in captivity need to experience adequate environmental determinants of species-typical behavior before they can survive in the wild. In this chapter, we regard training as a type of enrichment aimed at preserving or acquiring a species-specific behavior in form, frequency, and function.

We examine here whether and how environmental enrichment has any effect on individual behavior relevant to survival after reintroduction. Using golden lion tamarins *(Leontopithecus rosalia)* as an example, we evaluate two null hypotheses: (1) providing captive individuals with environmental enrichment before reintroduction does not promote longer survival after release into the wild, and (2) there is no difference between the postrelease behavior of captive-born individuals and that of wild individuals.

Golden lion tamarins are small endangered primates endemic to a restricted portion of the Atlantic rain forest of Brazil. The Golden Lion Tamarin Conservation Program (GLTCP), initiated about two decades ago by the National Zoological Park in Washington, D.C., has been active in preserving the species through research and captive breeding and through field studies and conservation efforts, such as reintroduction of captive-born individuals to the native habitat of the species in and around the Poço das Antas Biological Reserve (state of Rio de Janeiro, southeastern Brazil) (Kleiman et al. 1986). Through the GLTCP research efforts, we have had a rare opportunity to study the effect of the captive environment on behavioral characteristics, for the following reasons. (1) The long-term management strategy for the captive population is to maintain the genetic diversity of the original founders, therefore maximizing the potential preservation of species-typical behaviors and their variability; (2) the in-depth studies on captive groups provide us with solid knowledge of the species biology and behavior; (3) golden lion tamarin groups are housed in environmental conditions in zoos ranging from "standard" zoo enclosures to "enriched" free-ranging exhibits; (4) field studies on the behavioral ecology of wild golden lion tamarins provide data on the natural behavioral repertoire of the species; and (5) reintroduction of golden lion tamarins results in differential postrelease performance and survival, which can be correlated with different prerelease environmental conditions and experience.

BACKGROUND ON THE GOLDEN LION TAMARIN CONSERVATION PROGRAM

Captive Population

When wild golden lion tamarins were brought into captivity in the early to mid-1960s, breeding and survival were poor. By the early 1970s, it was evident that both wild and captive populations would become extinct if no measures were taken to reverse the trend in population decline. Researchers at the National Zoological Park and other holding institutions started to investigate several aspects of their behavior, nutrition, genetics, physiology, and pathology. Application of research findings (Kleiman 1981; Kleiman et al. 1988) to captive breeding resulted in rapid growth of the golden lion tamarin population in the following decade. By the early 1980s, the captive breeding strategy began to emphasize reproduction of genotypes from founder lineages that were less well represented. In 1995, there were about 470 golden lion tamarins in more than 138 institutions around the world. These animals represent a stable population that is into its eighth generation in captivity (Kleiman et al. 1986). With a secure captive population, reintroducing golden lion tamarins into their native habitat became feasible. Before implementing the actual reintroduction program, however, in 1983 we initiated censuses and research on the behavioral ecology of the wild population to provide baseline data (Kleiman et al. 1986). We also began other activities to enhance the success potential of the conservation program, such as conservation education for the local community and professional education and training for Brazilians (Dietz and Nagagata 1986, 1995).

Selection for Reintroduction

To prevent losses of individuals genetically valuable to the captive population, the first golden lion tamarins to be reintroduced were chosen from lineages overrepresented in captivity. As the reintroduction program and the breeding of less well represented lineages progressed, we chose reintroductees from a wider range of descendants of the original founders. Ideally, future reintroductions will include individuals from all lineages for good representation of the genetic diversity of the captive population (Ballou 1992).

Release, Provisioning, and Monitoring

Golden lion tamarins initially reintroduced into Brazil were released in Poço das Antas Biological Reserve. Subsequently, they have been released in private tracts of protected forests in areas adjacent to the reserve. Reintroduced golden lion tamarins are observed and monitored daily and provided with food and water; such provisioning is gradually reduced during the eighteen months following the release. In general, observation hours decrease as provisioning is reduced, but monitoring of birth and losses is continuous. The first individuals released in 1984, for example, are still surveyed at least twice a year. Some individuals have required rescuing (e.g., when they became lost or wounded) and may or may not be returned to the program. Even when a rescued individual is returned to the program, the rescue date is considered as the "date of death," or loss, when calculating survivorship estimates (Beck et al. 1991).

From 1984 until 1994, 136 golden lion tamarins, of which 129 were captive born and 7 were wild born, were reintroduced into their natural habitat in and around Poço das Antas Biological Reserve. Of these, 32 (23.5 percent) survived until 1994, 30 (23 percent) captive-born and 2 (29 percent) wild-born individuals. About 143 offspring have been born to reintroduced parents (or their offspring) in the wild; of these progeny, 92 (64 percent) survived until 1994. Most wild-born young of reintroduced parentage appear to be nearly independent of human support. Therefore, for these ten years, the net result of 136 reintroductees is 124 individuals, constituting 28 groups with breeding potential, living in the Poço das Antas Biological Reserve or on eleven privately owned farms. To participate in the GLTCP and receive reintroduced golden lion tamarins on their land, the farmers have pledged to protect their forests, representing an area of 2,300 hectares (43 percent of all protected habitat for golden lion tamarins).

Family groups containing young individuals seem to fare better after reintroduction than pairs of adults; also, wild-born offspring of reintroduced tamarins exhibit better survivorship and earlier independence than their parents (Kleiman et al. 1986, 1991; Beck et al. 1991). Pairing, sexual behavior, and parental care behavior seem not to be negatively affected by captivity as captive-born animals reproduce successfully after release. The major causes of loss (i.e., death, disappearance, and rescues) (Beck et al. 1991) include theft by humans, starvation, exposure, bee stings, disease, wounds resulting from social conflict, consumption of toxic fruits, and snakebite.

NULL HYPOTHESIS 1: ENVIRONMENTAL ENRICHMENT BEFORE REINTRODUCTION DOES NOT PROMOTE LONGER SURVIVAL AFTER RELEASE INTO THE WILD

Decisions such as when and how to implement training or whether to implement it at all have been an important part of reintroduction, because different methods of training captive-born individuals may have varying effects on enhancing behavioral skills for long-term survival after release in their natural habitat. Since the first golden lion tamarin reintroduction took place, the GLTCP has modified release protocols to improve the chances of individuals for survival in the wild. We discuss the release conditions briefly here (also see Table 8.1).

In 1984, the first year of the reintroduction project, one family group and three pairs (one male was a confiscated wild individual), totaling fifteen golden lion tamarins, were released in Poço das Antas Biological Reserve. Before the release, all fourteen captive-born animals received ten months of in-cage training that concentrated on improving foraging skills. The animals had no training after release but received food and water provisioning for seven to ten days. In 1985, two groups were reintroduced outside Poço das Antas Reserve: Group "NO" experienced three months of training in a large enclosure (112 cubic meters) before release and for one month after release, focusing on foraging and locomotion; the control group, "WITCH," experienced no training before release, but did receive three months of postrelease training. Both groups received postrelease support, that is, provisioning, for six to twelve months. In 1987, reintroduced golden lion tamarins had no training before and six months of training after

Table 8.1
Release Conditions of Reintroduced Golden Lion Tamarins

Year of release	Months of training provided: Prerelease[a]	Months of training provided: Postrelease	Veteran in release group?[b]	Release provisioning: Duration (months)	Release provisioning: Special feeders used?[c]
1984	10	0	Yes	0.2–0.3	No
1985					
"NO" group	3	1	No	6–12	No
"WITCH" group	0	3	No	6–12	No
1987	0	6	No	12–18	No
1988	0 (or FR)	12–18	No	12–18	Yes
1989, 1990–1991, 1992–1993	0 (or FR)	12–18	Yes	12–18	Yes

[a]The parenthetical "or FR" indicates that some of the animals were free-ranging before release.

[b]A veteran is an experienced (wild or previously introduced) individual.

[c]The special feeders were trays specifically designed to encourage foraging behavior.

release; individuals were provisioned for twelve to eighteen months (Kleiman et al. 1986; Beck et al. 1991).

During the fourth year, 1988, groups destined for reintroduction either received no training before release or had a few months of free-ranging experience in wooded areas on the grounds of the National Zoological Park or other zoos in the United States. (For more details on the free-ranging training process, see Bronikowski et al. 1989.) Groups were then released in suitable forest habitat on private farms near the Poço das Antas Reserve, where provisioning, which continued for twelve to eighteen months, was combined with training by supplying food in a feeder designed to encourage natural foraging behaviors as well as distributing provided food in space and time. From 1989 on, some groups of tamarins have been released with a previously reintroduced golden lion tamarin, a wild-born offspring of reintroduced parents, or a wild individual that had been in captivity for some time (usually as a pet) and confiscated by authorities. Each type of "veteran" (i.e., a golden lion tamarin with experience in the wild) can potentially help the newcomers adapt more readily to the wild.

Results from the 1985 reintroduction supported hypothesis 1: There was no statistical difference in rates of survival, indicating that the three months of foraging and locomotion training for the "NO" group before release did not confer any long-term advantage over the "WITCH" group, which received only postrelease training. (For more details on training procedures, see Beck et al. 1991.) Despite the lack of significant differences, it is perhaps worth noting that observers on the ground, on the basis of subjective impressions, generally agreed that the trained group showed more behavior resembling that of wild groups than did the untrained group.

Numbers of days of postrelease survival were scored for a total of fifty-nine golden lion tamarins reintroduced between 1987 and 1991. The sample consisted of nineteen individuals that received free-ranging experience at a National Zoological Park wooded site and forty individuals without such experience (matched for age, sex, and social grouping). Excluding animals that were still alive, the average number of days survived by the group with previous free-ranging experience was 572.6 (SD = 692.7) and by the inexperienced group 585.9 (SD = 484.4) (t-test: $t = -0.068$; $n = 49$; $p = .947$, 95% CI), again showing no effect of previous training.

While working on this hypothesis, however, we have identified many confounding variables related to each individual's background, and thus it is more difficult to arrive at an unambiguous answer to our question. Experiences differ, throughout all phases of life, with regard to physical characteristics of the enclosures during the first months of life, appropriate social structure and number of siblings the individual helped raise, exposure to different types of live food (including vermin), levels of habituation to people and interactions with keepers, previous exposures to live vegetation and number of changes in lifestyle (e.g., movements among enclosures, composition of groups). These are just some of the experiential elements that can influence an individual's behavior, much as can a conscious program of environmental enrichment. Therefore, most of the results presented here are trends that need to be further examined in the light of individual experiences before the reintroduction.

NULL HYPOTHESIS 2: THERE IS NO DIFFERENCE IN THE BEHAVIORS OF CAPTIVE-BORN AND WILD GOLDEN LION TAMARINS

We compared some aspects of behavior in wild and captive-born golden lion tamarins to determine if reintroduced tamarins are at a disadvantage after release and to clarify reasons for a rate of loss of 41 percent among reintroduced golden lion tamarins during their first year in the wild (Beck et al. 1991). Experimental studies on the use of space, locomotion, predator avoidance, vocal communication, and foraging are being conducted to compare the behavior of individuals with different backgrounds (i.e., wild and captive-born reintroduced golden lion tamarins) now living in the same environment. Some preliminary results of ongoing studies follow.

Use of Space

The territory size of recently reintroduced groups is much smaller than that of wild groups. Wild golden lion tamarin groups studied at Poço das Antas Biological Reserve have territories averaging 41.4 (SD = 21.1) hectares ($n = 47$; Dietz and Baker 1993). This may be conservative, because the reserve appears to be at its carrying capacity and suitable forested areas are at a premium for golden lion tamarins. After release, captive-born tamarins initially use an area with a diameter of 50 meters. However, after two to four years, when wild-born offspring have reached maturity, groups enlarge their home ranges to about 40 hectares.

Communication

Auditory communication probably initiates and guides most social interactions, travel, and group cohesion of golden lion tamarins (Boinski et al. 1994), thus influencing survival and reproductive success. It is important, therefore, to understand how the captive environment affects auditory communication abilities. We have investigated this issue by two methods: through playbacks of long calls to wild-born and captive-born reintroduced animals, and by naturalistic recording of the vocalizations and concomitant behavior of both captive and free-ranging golden lion tamarins at the National Zoological Park and reintroduced and wild individuals in Brazil. Golden lion tamarin long calls are loud, long, and conspicuous vocalizations composed of two or three phrases, each phrase with a different number of syllables (Halloy and Kleiman 1994). The long-call playbacks were conducted on thirteen groups of wild tamarins and six groups of reintroduced tamarins (Kleiman, unpublished data). One set of recordings of long calls derives from twenty-eight wild individuals (sixteen males and twelve females) and twenty-one captive-born reintroduced individuals (seven males and fourteen females). Other vocalization data come from the studies by Green (1979) on twenty-one captive animals and on eight free-ranging zoo, twenty-three reintroduced, and twenty-one wild animals (Ruiz-Miranda, unpublished data).

Captive-born golden lion tamarins express the entire species-typical vocal repertoire. The main differences between captive-born and wild golden lion tamarins are in (1) the pitch of specific calls; (2) the context in which calls are given; (3) the way some vocalizations are combined; (4) the rate of emission of some calls;

and (5) the responses to playbacks of recorded vocalizations. Specifically, the long calls of captive-born reintroduced tamarins have a significantly higher frequency range than those of wild animals; calls of females also have a higher frequency range than those of males (Ruiz-Miranda, unpublished data; Kleiman, unpublished data). A factor analysis performed on sixteen variables from the first and second phrases of long calls of forty-nine golden lion tamarins (averaging a minimum of five and a maximum of ten long calls per individual) showed that the six measures contributing significantly to factor 1 were all frequency-related variables. This difference in frequency between captive-born reintroduced and wild golden lion tamarins may result from differences in perception of the sound of long calls in more enclosed captive environments during the learning phase for this call. This result contrasts with other studies of primates showing that rearing environment did not affect the form of expressive behaviors (Mason 1985) or isolation vocalizations (Newman and Symmes 1982), and with the consensus that mammalian vocalizations are not affected by experience.

The "tsick" vocalization of wild golden lion tamarins is given almost exclusively while foraging (Ruiz-Miranda, unpublished data), while animals in captivity also emit this sound when excited and under alarm (Green 1979). In three groups of adults in the National Zoo's free-ranging exhibit, 50 percent of the "trill" vocalizations were emitted while the animals were stationary within visual contact of others (Ruiz-Miranda, unpublished data). In contrast, wild animals emit most trills when locomoting or foraging on the periphery of the group (Boinski et al. 1994). We are currently conducting an analysis of the context in which intragroup vocalizations are emitted.

Another measure of vocal communication is the structure of the bouts of vocalizations. Vocal communication changes with the transition from the zoo environment to the natural habitat, creating a difference between the reintroduced animals and the other two experimental groups. Wild and free-ranging zoo adult golden lion tamarins combine vocalizations (i.e., they have "multitypic bouts") in 34 percent of their bouts, whereas captive-reared reintroduced adults combine them in 44 percent of their bouts (Ruiz-Miranda, unpublished data). A possible explanation of this difference, which is statistically significant (*t*-test; $p < .05$), is that captive-born animals are presented with new sets of contingencies and problems when exposed to natural environments, and thus the change in communication style reflects ambiguous perceptions of the animals or changes in the referents of the vocalizations.

This change in behavior associated with a change in the physical environment is also seen in the overall rate of vocalization (vocalizations per minute), as well as in the rate of emission of each different type of vocalization (whine, trill, and cluck, as defined by Green 1979). Overall vocalization rates of captive individuals are significantly lower than those of reintroduced and wild golden lion tamarins (Ruiz-Miranda, unpublished data). This could be explained by lower rates of vocalizations, for example, a trill, related to foraging while traveling, to alarm, and to intergroup encounters in captive individuals. The trill is a call used by wild tamarins while foraging and locomoting (Boinski et al. 1994; Ruiz-Miranda et al., unpublished data) and thus is thought to promote group cohesion. Captive and free-ranging zoo tamarins do not engage in foraging while traveling because they are fed in specific places (although free-ranging tamarins have foraging site choices, they do not travel far) and are almost always within visual contact of each other, while reintroduced and wild tamarins forage often while traveling. Captive tamarins also emit fewer clucks than reintroduced and wild individuals, which is not surprising because clucks are vocalizations used by wild golden lion tamarins during foraging, mobbing, and territorial encounters.

These three contexts are present in the environments of free-ranging (foraging and mobbing), reintroduced (foraging, mobbing, and territorial), and wild (foraging, mobbing, and territorial) golden lion tamarins. The captive animals cluck during feeding time and when excitement builds in the colony after people walk into the rooms. Similarly, it is not surprising for captive animals to utter few whines, a mild alarm vocalization used by wild animals, while mobbing in interspecific interactions or when startled. It seems that changes in rate of emission of vocalizations are related to differences in the physical characteristics of different environments. Especially, changes in vocalization rates are more evident from individuals in the captive environment to those free-ranging, reintroduced, and wild. These results differ from studies on rhesus monkeys *(Macaca mulatta)* (Mason 1985) and baboons *(Papio hamadryas)* (Kummer and Kurt 1965), which found that the rate of use of expressive behaviors was similar among captive-born and wild individuals. To evaluate our preliminary findings further, we are examining whether reintroduced adults change their vocal behavior over time in the wild and whether the wild-born young of reintroduced animals vocalize like their parents or like wild animals.

Finally, captive-born and wild animals differ in their responses to playbacks of long calls. Kleiman (1990) found that although the responses of reintroduced and wild adults were similar in form (traveling in the direction of the playback), the responses of wild tamarins were more sustained over time. In addition, responses of wild tamarins were consistently different to each type of long-call vocalization, suggesting that they clearly differentiated the call types; they also altered their direction of travel and calling rates more significantly than captive-born, reintroduced tamarins in response to playbacks of long calls. Similarly, in a pilot study, Ruiz-Miranda (unpublished data) found that to obtain a response from reintroduced adults to playbacks of infant vocalizations ("rasps" and "trills"), the recording had to contain three to four times more calls and had to be played longer than playbacks used for wild adults. Reintroduced adults also returned more quickly to previous activities than wild ones. In other words, reintroduced tamarins needed more stimulation to show a response that was more delayed and less intense than that of their wild counterparts.

In summary, variations between wild and captive-born individuals appear to be related to the environment and are evident some time after reintroduction. Some differences in the vocalizations may result from the physical characteristics of the environment while others seem to be correlated with the opportunities for different experiences offered by different environments. As an example, wild tamarins have frequent energy-consuming encounters with bordering groups (Peres 1989) that help maintain territorial integrity and monitor activities in neighboring groups. These encounters are ritualized, mostly through long calls and other vocalizations, and include occasional chases. The wild tamarins seem to need to learn to ritualize their intergroup aggressive behaviors. Zoo animals usually do not have the opportunity to meet other groups and may show more aggressive behavior in their first encounters.

Dealing with Danger

Beck et al. (1991) reported that twelve of thirty-six (33 percent) losses of reintroduced tamarins of known cause involved encounters with dangerous animals and toxic plants. It is particularly difficult to introduce enrichment in the area of recognition and avoidance of danger because of the potential for injury or habituation. In an experimental study conducted with captive-born and wild-born lion tamarins in captivity, Castro (1990) noted that all individuals, regardless of origin, responded similarly to the presentation of an aerial predator model (stuffed hawk). The immediate reaction to the hawk and its control (plastic ball) suggested that these responses are innate in *Leontopithecus.* In the same study, tamarins mobbed a model of an arboreal-terrestrial predator (rubber snake) but did not mob the control item (piece of bamboo). Further observations suggested that mobbing behavior in lion tamarins may also be inborn, but the ontogeny of predator-recognition mechanisms is still unclear.

Of all behaviors recorded in this study, only two types differed between wild and captive-born individuals: Captive-born animals scanned the ground more often after the hawk and went to the ground less often after the snake. These differences may result from the different perceptions of safety areas that tamarins possibly have depending on the environments in which they were raised. For instance, captive-born individuals may perceive the open "understory" area of their enclosures as more dangerous than the protected "canopy" area, therefore becoming more vigilant and avoiding the lower part of the enclosure when under threat. These results again suggest that the expression of some golden lion tamarin behaviors may depend on the environments in which they were raised. If the differences in vigilance behaviors seen between captive-born and wild individuals actually reflect different perceptions of safety areas, then captive-born lion tamarins were not presenting the "wrong" antipredator behavior for the particular circumstance in which they were raised. However, they still need to learn the appropriate context for their behavior in the forest after reintroduction.

Locomotion

Golden lion tamarins that have been recently reintroduced position themselves at a significantly lower level in the forest canopy than wild golden lion tamarins (Kleiman 1990). It is still unclear if this difference is the expression of locomotor difficulties or preference. Additionally, studies conducted by Rosenberger and Stafford (1994) using video analyses and footprint tracks showed that captive-born individuals more often use a quadrupedal-type locomotion called transaxial bounding, which they continue to exhibit in the wild after reintroduction. Captive-born golden lion tamarins living in unenriched zoo enclosures show higher frequencies of this locomotor pattern than those of free-ranging animals living in zoo woods (Stafford et al. 1994) and higher than those of wild individuals (Stafford and Rosenberger, unpublished data). These results suggest that transaxial bounding is controlled by environmental variables such as substrate type and availability (Rosenberger and Stafford 1994; Stafford et al. 1994). We do not yet know whether the frequency of this locomotor pattern decreases after time in the wild or whether it is present (and to what degree) in wild-born offspring of reintroduced parents. We also do not know to what degree it inhibits ease of movement through the midcanopy level where golden lion tamarins spend most of their time.

DISCUSSION

Null Hypothesis 1: Environmental Enrichment before Reintroduction Does Not Promote Longer Survival after Release into the Wild

We found no evidence that the form of prerelease training and free-ranging experiences described here had any effect on postrelease survival. It is possible that the many confounding variables, as mentioned earlier, can function as environmental enrichment and affect individual survival. As an example, when studying squirrel monkeys *(Saimiri sciureus),* Masataka (1993) found that even naive, captive-born individuals developed a fear of snakes and responded to snakes appropriately on first encounters if they were given the opportunity to experience live insects in their diet. Masataka concluded that individual squirrel monkeys were sensitized to this particular stimulus (i.e., snakes) after experiencing other small live animals. An alternative explanation for our results may be that the form of training offered by the reintroduction component of the GLTCP does not provide individuals with the appropriate level or type of experience for survival in the wild.

Stimulation designed for enrichment may sometimes be stressful (Beck 1991, 1995; Beck and Castro 1994). Moodie and Chamove (1990) found that captive cotton-top tamarins *(Saguinus oedipus),* after brief periods of arousal, behaved similarly to individuals who experienced more enriched conditions. The authors proposed that short-term threatening events are beneficial to individuals in captivity. Shepherdson (1994) suggested that although elimination of all stress is not required to maximize captive animal well-being, the optimal level of stimulation should be provided. Similarly, behavioral training for reintroduction may also provide more adequate results when certain stress components are added. If so, training procedures and zoo environments should reflect even more closely the animal's natural habitat, including stressful experiences (Beck 1991, 1995; Beck and Castro 1994; see also Shepherdson 1994; and Kreger et al., Chapter 5, this volume).

In summary, hypothesis 1, regarding training before reintroduction, should be investigated further, taking into consideration both the background experiences of each individual and the type of enrichment offered to them before reintroduction.

Null Hypothesis 2: There Is No Difference in the Behaviors of Captive-Born and Wild Golden Lion Tamarins

Results from these studies, although preliminary, support the idea that even when reared in a traditional zoo environment golden lion tamarins show the complete behavior repertoire of their species; that is, there has been no loss of behavior patterns per se. Although captive-born golden lion tamarins are able to express the same behaviors as wild individuals, they may do so under different contexts or at different frequencies. Marler et al. (1980) suggested that the timing and orientation of a behavior can be as flexible as its motor pattern and that each may be controlled by different mechanisms. Our observations suggest that golden lion tamarins raised traditionally in standard zoo enclosures may not be able to acquire the experience needed to fine-tune their behaviors fully and always express them adequately in the appropriate situations. Therefore, keeping golden lion tamarins in free-ranging conditions may help them develop those behavioral skills that are dependent on physical characteristics of the environment (e.g., some vocalizations). As a side benefit, because zoo visitors are notably attracted to free-ranging exhibits, this type of presentation may make it easier to convey conservation messages about the species and its habitat.

Although the GLTCP is continually exploring ways to improve reintroduction techniques, reintroduction of golden lion tamarins has been successful in that it increased the size of the in situ golden lion tamarin population by 20 percent and the amount of available protected habitat by more than 40 percent in the original area of occurrence. By providing captive-reared golden lion tamarins with food, a nest site, and some care after release in the wild (see Beck et al. 1991), individuals are more likely to survive to experience their new environment, to learn adequate responses to new situations, and to reproduce. This combination of provisioning and postrelease training is still proving to be the most cost-effective and time-effective procedure for golden lion tamarin reintroduction.

SUMMARY

Current evidence suggests that golden lion tamarins reared in traditional captive environments are able to express the full behavior repertoire (i.e., there has been no loss of behavior per se), although when compared to wild golden lion tamarins, captive-reared individuals may express some behaviors under functionally different circumstances or at different frequencies. Traditional captive conditions may not provide adequate stimuli or opportunities for the expression of certain behaviors (e.g., some vocalizations) that seem to be more properly performed in more naturalistic environments such as free-ranging exhibits.

Environmental enrichment in the form of prerelease training (i.e., foraging and locomoting in enclosures or free-ranging experiences) appears not to confer any advantage in survivorship of golden lion tamarins after reintroduction. However, the effect of prerelease training on postrelease survival needs to be studied further in the light of individual background experiences and the type of enrichment offered before reintroduction. Providing captive-reared golden lion tamarins with time in the wild to experience their new environment seems to be a more cost-effective and time-effective way of enabling individuals to learn appropriate contexts for responses and to fine-tune their behavior adequately.

ACKNOWLEDGMENTS

Funding for M.I.C. to attend the First Environmental Enrichment Conference, Portland, Oregon, was provided by Metro Washington Park Zoo, Portland, and to attend the Conference on Indoor Naturalistic Facilities, Stony Brook, New York, by Wildlife Preservation Trust International. The graduate study of M.I.C. was supported by a scholarship from CNPq, Brazil, and by Friends of the National Zoo (FONZ). M.I.C. and C.R.R.-M. were supported by grants from FONZ while writing this manuscript. A.L.R. was supported by a Research Development Award (National Zoological Park) and the Chicago Zoological Society. The Golden Lion Tamarin Conservation Program is supported by the National Zoological Park and has been funded by grants from the Smithsonian Institution (International Environmental Sciences Program [IESP] and Scholarly Studies), Friends of the National Zoo, World Wide Fund for Nature-WWF, National Science Foundation (grant DBS9008161), National Geographic Society, Frankfurt Zoological Society-Help for Threatened Wildlife, Jersey Wildlife Preservation Trust, Wildlife Preservation Trust International, and TransBrasil Airline.

REFERENCES

Ballou, J. D. 1992. Genetic and demographic considerations in endangered species captive breeding and reintroduction programs. In *Wildlife 2001: Populations,* ed. D. McCullough and R. Barrett, 262-275. Barking, U.K.: Elsevier.

Beck, B. B. 1991. Managing zoo environments for reintroduction. In *Proceedings of the American Association of Zoological Parks and Aquariums Annual Conference,* 436-440. Wheeling, W.Va.: AAZPA.

———. 1995. Reintroduction, zoos, conservation, and animal welfare. In *Ethics on the Ark: Zoos, Animal Welfare, and Wildlife Conservation,* ed. B. G. Norton, M. Hutchins, E. F. Stevens, and T. L. Maple, 155-163. Washington, D.C.: Smithsonian Institution Press.

Beck, B. B., and M. I. Castro. 1994. Environments for endangered primates. In *Naturalistic Environments in Captivity for Animal Behavior Research,* ed. E. F. Gibbons, E. Wyers, E. Waters, and E. Menzel, 259-270. Albany: State University of New York Press.

Beck, B. B., D. G. Kleiman, J. M. Dietz, M. I. Castro, C. Carvalho, A. Martins, and B. Retteberg-Beck. 1991. Losses and reproduction in reintroduced golden lion tamarin, *Leontopithecus rosalia. Dodo, Journal of the Jersey Wildlife Preservation Trust* 27:50-61.

Boinski, S., E. Moraes, D. G. Kleiman, J. M. Dietz, and A. J. Baker. 1994. Intragroup vocal behaviour in wild golden lion tamarins, *Leontopithecus rosalia:* Honest communication of individual activity. *Behaviour* 130:53-75.

Bronikowski, E. J., B. B. Beck, and M. Power. 1989. Innovation, exhibition, and conservation: Free-ranging tamarins at the National Zoological Park. In *Proceedings of the*

American Association of Zoological Parks and Aquariums Annual Conference, 540–546. Wheeling, W.Va.: AAZPA.

Castro, M. I. 1990. A comparative study of anti-predator behavior in the three species of lion tamarins *(Leontopithecus)* in captivity. Master's thesis, University of Maryland, College Park.

Dietz, J. M., and A. Baker. 1993. Polygyny and female reproductive success in golden lion tamarin, *Leontopithecus rosalia. Animal Behaviour* 46:1067–1078.

Dietz, L. A., and E. Y. Nagagata. 1986. Projeto mico-leão: Programa de educação comunitária para a conservação do mico-leão-dourado *Leontopithecus rosalia* (Linnaeus, 1766). Desenvolvimento e avaliação de educação como tecnologia para a conservação de uma espécie em extinção. In *A Primatologia no Brasil, 2,* ed. M. Thiago de Mello, 249–256. Brasília: Sociedade Brasileira de Primatologia.

———. 1995. Golden lion tamarin program: A community education effort for forest conservation in Rio de Janeiro State, Brazil. In *Conserving Wild Life: International Education and Communication Approaches,* ed. S. K. Jacobson, 64–86. New York: Columbia University Press.

Green, K. M. 1979. Vocalizations, behavior, and ontogeny of the golden lion tamarin *Leontopithecus rosalia rosalia.* Ph.D. dissertation, Johns Hopkins University, Baltimore.

Halloy, M., and D. G. Kleiman. 1994. Acoustic structure of long calls in free-ranging groups of golden lion tamarins, *Leontopithecus rosalia. American Journal of Primatology* 32:303–310.

Kleiman, D. G. 1981. *Leontopithecus rosalia. Mammalian Species* 148:1–7.

———. 1990. Responses to long call playbacks: Differences between wild and reintroduced golden lion tamarins *(Leontopithecus rosalia).* Paper presented at the Animal Behavior Society Annual Meeting, State University of New York, Binghamton, June 10–15, 1990.

Kleiman, D. G., B. B. Beck, J. M. Dietz, and L. A. Dietz. 1991. Costs of a re-introduction and criteria for success: Accounting and accountability in the golden lion tamarin conservation program. *Symposia of the Zoological Society of London* 62:125–142.

Kleiman, D. G., B. B. Beck, J. M. Dietz, L. A. Dietz, J. D. Ballou, and A. Coimbra-Filho. 1986. Conservation program for the golden lion tamarin: Captive research and management, ecological studies, educational strategies, and reintroduction. In *Primates: The Road to Self-Sustaining Populations,* ed. K. Benirschke, 959–979. New York: Springer-Verlag.

Kleiman, D. G., B. J. Hoage, and K. M. Green. 1988. The lion tamarins, genus *Leontopithecus. Ecology and Behavior of Neotropical Primates* 2:299–347.

Kummer, H., and F. Kurt. 1965. A comparison of social behavior in captive and wild hamadryas baboons. In *The Baboon in Medical Research,* ed. H. Vagtborg, 65–80. Austin: University of Texas Press.

Marler, P. R., R. J. Dooling, and S. Zoloth. 1980. Comparative perspectives on ethology and behavioral development. In *Comparative Methods in Psychology,* ed. M. Burnstein, 189–230. Hillsdale, N.J.: Lawrence Erlbaum.

Masataka, N. 1993. Effects of experience with live insects on the development of fear of snakes in squirrel monkeys, *Saimiri sciureus. Animal Behaviour* 46:741–746.

Mason, W. A. 1985. Experiential influences on the development of expressive behaviors in rhesus monkeys. In *The Development of Expressive Behaviors: Biology-Environment Interactions,* ed. G. Zivin, 117–152. San Diego: Academic Press.

Moodie, E. M., and A. S. Chamove. 1990. Brief threatening events beneficial for captive tamarins? *Zoo Biology* 9:275–286.

Newman, J. D., and M. Symmes. 1982. Inheritance and experience in the acquisition of primate acoustic behavior. In *Primate Communication,* ed. C. T. Snowdon, C. H. Brown, and M. R. Petersen, 259–278. New York: Cambridge University Press.

Peres, C. 1989. Costs and benefits of territorial defense in wild golden lion tamarins, *Leontopithecus rosalia. Behavioral Ecology and Sociobiology* 25:227–233.

Rosenberger, A. L., and B. J. Stafford. 1994. Locomotion in captive *Leontopithecus* and *Callimico:* A multi-media study. *Zoo Biology* 94:379–394.

Shepherdson, D. 1994. The role of environmental enrichment in the captive breeding and reintroduction of endangered species. In *Creative Conservation: Interactive Management of Wild and Captive Animals,* ed. G. Mace, P. Olney, and A. Feistner, 167–177. New York: Chapman & Hall.

Stafford, B. J., A. L. Rosenberger, and B. B. Beck. 1994. Locomotion of free ranging golden lion tamarins *(Leontopithecus rosalia)* at the National Zoological Park. *Zoo Biology* 13:333–344.

CAROLYN M. CROCKETT

PSYCHOLOGICAL WELL-BEING OF CAPTIVE NONHUMAN PRIMATES

Lessons from Laboratory Studies

Old-style zoo enclosures lacked furnishings and bedding materials, and many laboratory primate cages still do. However, in 1985 the concept of psychological well-being was introduced in an amendment to the 1966 Animal Welfare Act. The attention to psychological well-being derived from ethical concerns for the welfare of our fellow creatures and from a need to ensure a healthy exhibit animal or research subject. As stated in the 1991 Animal Welfare Standards:

> Dealers, exhibitors, and research facilities must develop, document, and follow an appropriate plan for environment enhancement adequate to promote the psychological well-being of nonhuman primates. The plan must be in accordance with the currently accepted professional standards as cited in appropriate professional journals or reference guides, and as directed by the attending veterinarian. . . . The . . . plan must include specific provisions to address the social needs of nonhuman primates of species known to exist in social groups in nature. . . . The physical environment in the primary enclosures must be enriched by providing means of expressing noninjurious species-typical activities. Species differences should be considered when determining the type or methods of enrichment. Examples of environmental enrichment include providing perches, swings, mirrors, and other cage complexities; providing objects to manipulate; varied food items; using foraging or task-oriented feeding methods; and providing interaction with the care giver or other familiar and knowledgeable person consistent with personnel safety precautions (U.S. Department of Agriculture 1991, 6499–6500).

According to this regulation, zoos and research institutions must formulate plans to enrich the physical environment and meet the social needs of captive primates. The requirement for social housing can be waived for various reasons, including overly aggressive behavior, physical debilitation or disease, or, in the case of laboratory primates, an approved research protocol that precludes social contact. There is a growing body of literature on techniques for nonhuman primate environmental enrichment that addresses the concept of psychological well-being and how to measure it (Moberg 1985; Novak and Suomi 1988; Segal 1989; Thomas and Lorden 1989; Bayne 1991; Mendoza 1991b; Novak and Petto 1991; U.S. Department of Agriculture et al. 1992).

Although animal rights groups challenged the 1991 regulations in court as being inadequate to promote well-being, I believe that rigid rules of the sort these groups demand are premature (Crockett 1993). Rules based on untested assumptions may not have the desired outcome, which is improving the lives of captive primates. Given the still-limited data on many aspects of the promotion of psychological well-being in nonhuman primates, professional judgment continues to play an important role in the care of captive animals. Animals in enhanced environments do not necessarily experience improved psychological well-being, and we must decide how to measure well-being so as to evaluate the effectiveness of enrichment. We cannot simply embellish enclosures with furnishings and toys and hope for the best. Environmental enrichment has a direct cost measured in dollars and personnel time that must be weighed against measured benefit. Furthermore, presumed enrichments may not always be beneficial; some may actually cause harm by entangling the animals or acting as bacterial vectors (Bielitzki 1992; Bayne et al. 1993a; Murchison 1993).

Animals on exhibit in zoos are presumed to benefit from the environmental complexity provided by naturalistic habitat enclosures (Hutchins et al. 1984). If the naturalistic exhibit also provides educational benefits for zoo visitors, then—so long as the enclosure does not compromise the residents' health—the time and effort may be justified. For laboratory enclosures or off-exhibit areas of zoos where every effort to enrich has a monetary cost, this may not be true.

I focus here on an ongoing research program at the University of Washington's Regional Primate Research Center for which I serve as project coordinator. In 1988 we began a series of studies, funded by the National Institutes of Health, to investigate the psychological well-being of monkeys, primarily macaques (*Macaca* spp.), in a variety of environmental conditions typically experienced by laboratory primates. Although environmental enrichment of compound-housed primates has more obvious relevance to zoos, in studies of singly caged monkeys we can test measures of psychological well-being in a more controlled environment. A few of our studies have evaluated particular environmental enhancements, but the major studies are concerned with meeting the social needs of laboratory monkeys and with responses to features of captivity such as the sizes of individual cages, transfer to a new room, and sedation for routine procedures. We are finding that some preconceived notions about what is or is not good for captive monkeys are wrong (Crockett and Bowden 1994). These studies, some published and others still in progress, involve the efforts of many people, including scientists, students, and support staff.

MEASURES OF PSYCHOLOGICAL WELL-BEING

To evaluate psychological well-being, we use a variety of measures because single measures may not provide a complete picture (Broom 1988; Novak and Suomi 1988; International Primatological Society 1993; Mason and Mendl 1993). We record behavior: normal behavior, abnormal or stereotypic behavior, activity cycle data, and, if relevant, the use of enrichment objects. Abnormal or stereotypic behavior in particular is often thought to reflect deficiencies in welfare (Goosen 1981; Capitanio 1986; Dantzer 1986; Mason 1991; Lawrence and Rushen 1993; also see Carlstead, Chapter 11, this volume), but normal behaviors that occur at atypically high or low rates, or altered activity cycles, may also reflect poor welfare (International Primatological Society 1993).

Our studies are conducted in special observation rooms where we can videotape behavior via computer-controlled videocassette tape recorders and infrared-sensitive cameras. The behavior is coded later with the aid of a computer program (Crockett and Bowden 1994; Crockett et al. 1994a, 1995). This setup allows us to obtain detailed samples of behavior at all times of day, even at night. The coding system allows abnormal behavior to be summed separately or to be combined with the behavior from which it was derived (for example, normal locomotion versus stereotyped locomotion versus total locomotion). For studies of enrichment devices provided to colony animals—those not specifically assigned to our major projects—we simply use paper-and-pencil data sheets (Crockett et al. 1989; Crockett 1990; Heath et al. 1992).

In several studies we have collected urine for cortisol assay as a physiological indicator of stress and thereby of negative psychological well-being (Crockett et al. 1993a,b, 1994a). Urinary corticosteroids have been used to assess stress in humans for many years (Fishman et al. 1962; Hamburg 1962; Friedman et al. 1963; Bunney et al. 1965; Lundberg 1980). Urine can be collected unobtrusively, and the ratios of free cortisol to creatinine levels in the urine have been shown to reflect adrenopituitary activity, and hence a presumed stress response, in macaques, domestic and nondomestic felids, and bighorn sheep *(Ovis canadensis canadensis)* (Miller et al. 1991; Carlstead et al. 1992; Crockett et al. 1993a).

Although the levels of cortisol in the blood (serum or plasma component) have been used as an indicator of stress (Line et al. 1987; Mendoza 1991b; Reinhardt et al. 1991a), blood collection cannot be unobtrusive and, unless animals are specially trained, may result in stress-related elevation of cortisol (Reinhardt et al. 1991b). Urine collection is easiest with individually housed animals, but is also feasible with group-living and even wild primates (Kelley and Bramblett 1981; Bond 1991; Byrne and Suomi 1991; van Schaik et al. 1991).

We have also used appetite as an indicator of psychological well-being (Crockett et al. 1990, 1993b). The monkeys receive a precounted number of biscuits (more than they are expected to consume), and the next day we count the remaining biscuits before daily cleaning. Appetite suppression is calculated as the decline in biscuit consumption (grams of biscuit per kilogram of body weight) relative to average intake. We have used other measures of well-being in some of our studies, such as heart rate and other cardiac parameters (Bowers et al. 1993), but behavior, urinary cortisol, and appetite are most suitable for use in a zoo setting or wherever unobtrusive measurements are desired.

The next sections summarize some of our findings to date and include a brief review of findings from other laboratories, for a broader perspective.

MINIMAL EFFECTS OF CAGE SIZE IN THE LABORATORY SETTING

Animal Welfare rules rigidly regulate the minimum cage size for nonhuman primates (U.S. Department of Agriculture 1991). Required cage size is directly tied to the animal's weight; if more than one primate is housed in the same enclosure, the total area must be at least of the sum of the floor area required for each animal. Although meeting cage size minimums is not likely to be a common problem for zoos, many research institutions are faced with making expensive changes to increase cage size by a few inches. Cage size minimums that were "recommendations" by the National Institutes of Health (U.S. Public Health Service 1985) became "regulations" in the Animal Welfare Standards, Final Rule (U.S. Department of Agriculture 1991). The cage size minimums are based on arbitrary judgments as to how much space is required to allow nonhuman primates to make normal postural adjustments, for example, "minimum floor space equal to an area at least three times the area occupied by the primate when standing on four feet" (U.S. Department of Agriculture 1991, 6499).

Cage size minimums as dictated by these arbitrary standards were long viewed by veterinarians as adequate but by certain sectors of the public as entirely too

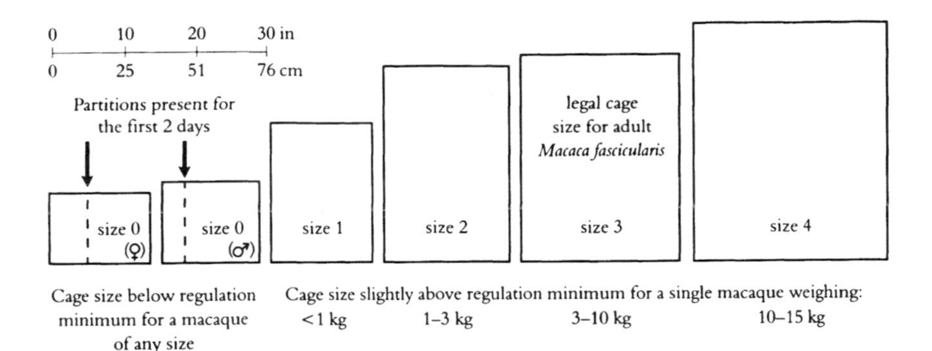

Figure 9.1. Sizes of cages in which adult longtailed macaques (each weighing 3–8 kilograms) were singly housed in a well-controlled study done to evaluate the benefits of increased cage size. The weight classes of macaques that regulations (U.S. Department of Agriculture 1991) would allow to be housed singly in such cages are given for sake of contrast. (Adapted from Crockett et al. 1993a.)

small. Initially, neither group had objective data on which to base their opinions. The widespread opinion that small space produces unhappy inhabitants, however, has not been borne out by recent quantitative research. For group-living primates, aggression is not directly related to area available per primate but varies as a function of structural complexity and social stability (Erwin 1979; de Waal 1989; Erwin and Sackett 1990; Bercovitch and Lebron 1991; Estep and Baker 1991). For singly housed monkeys, the increased incidence of abnormal behavior in smaller cages found in earlier studies (Draper and Bernstein 1963; Paulk et al. 1977) has not been substantiated by recent studies (Bayne and McCully 1989; Line et al. 1989, 1990b, 1991a). In the earlier studies, monkeys were moved to novel cages for testing, whereas in the more recent studies monkeys were observed in home cages of different sizes. Macaques moved to novel cages of the same size also displayed increased abnormal behavior (Mitchell and Gomber 1976).

Our own studies added substantially to the emerging evidence that modest variations in cage size have little measurable effect on the psychological well-being of monkeys. We evaluated cage size in a well-controlled study in which ten adult female longtailed macaques *(Macaca fascicularis)* weighing 3-4 kilograms and, subsequently, ten adult males of the same species weighing 5-8 kilograms, spent two weeks singly housed in each of five cage sizes ranging from 20 percent to 148 percent of the floor area designated for primates of their weight class (3-10 kilograms) (Figure 9.1). These twenty monkeys were all wild caught as adults, which until recently was typical of research subjects of this species.

Neither urinary cortisol, appetite suppression, nor abnormal behavior varied significantly as a function of cage size (Crockett et al. 1990, 1993a, 1995). Our singly housed monkeys spent only 5 percent of their time engaged in activities that were scored as "abnormal," and none were self-injurious. (Group-housed zoo macaques and baboons [*Papio* spp.] average 3 percent of the time in stereotypies [Marriner and Drickamer 1994].) The only behavioral effect we found was that time spent in locomotor behavior (and its correlate, frequency of all behaviors per minute) was significantly reduced in the two smallest cages (cage sizes 0 and 1 in Figure 9.1), as one might expect given the small area they provided. The monkeys showed no differences on any measure in the other three sizes tested: the regulation-size cage (cage size 3), one size larger, and one size smaller. To counter the hypothesis that the monkeys might be stressed in any size cage, we verified that urinary cortisol levels in response to cage size changes were low compared with elevated levels we found after expected stressors such as sedation and surgery and after experimental stimulation of the adrenocortical system with adrenocorticotropic hormone (Crockett et al. 1993a). When we replicated this study with eight females of another species, the pigtailed macaque *(M. nemestrina),* cage size again had no significant effect on urinary cortisol, appetite, or behavior (Crockett et al. 1993b, 1994b).

From these results, we have concluded that cage size has only minimal effects on the psychological well-being of singly housed macaques. Certainly, within the range of cage sizes regulated by the Animal Welfare rules (U.S. Department of Agriculture 1991), there is no evidence that increasing a cage of, say, 4.0 square feet (0.37 square meter) to 4.3 square feet (0.40 square meter; cage size 3, the minimum for adult-size longtailed macaques) will measurably improve the psychological well-being of the cage's inhabitant. In our studies, even increasing cage size substantially, to 6.4 square feet (0.59 square meter), did not produce significant changes in psychological well-being, as measured by behavior, cortisol, or appetite. Preliminary analysis of time spent "exercising" in an even larger cage, 12.5 square feet (1.16 square meters), revealed only a small increase (Leu et al. 1993); locomotion primarily increased when the animals moved to different positions to engage in visual contact ("social interest") with other monkeys (Crockett et al., manuscript in preparation). Thus, simply increasing cage size does not constitute environmental enrichment for a singly housed macaque. I am not advocating that longtailed macaques should be kept in size-2 cages (3.0 square feet; 0.28 square meter) rather than the regulation-size cage (4.3 square feet; 0.40 square meter) just because our measures revealed no significant difference between the two; the smaller size might compromise the installation of meaningful enrichment devices such as perches (see following). Nor do I mean to discourage facilities from keeping their monkeys in larger cages than the minimums required, if such cages and space are available. Rather, I want to dispel the illusions that increasing cage size, within the range likely to be possible in a research lab or behind-the-scenes zoo setting, will provide meaningful enrichment to macaques.

Further, our data revealed nothing about the appropriateness of extrapolating floor area per monkey in single housing or in small groups (of five or fewer) to larger groups in compounds. However, even in group housing it is clear that well-being is not a linear function of living space. For example, the rate of aggression in large groups (more than fifty individuals) of rhesus macaques *(M. mulatta)* was nearly identical whether their housing allowed 15.5 square meters per monkey or 1.25 square meters per monkey (de Waal 1989). In comparison, the cage size minimums established by the Animal Welfare regulations are 0.40–0.56 square meter per adult macaque, depending on the size of the animal.

MONKEYS AND ENVIRONMENTAL CHANGE: BENEFIT OR DISTURBANCE?

Captive animals are often thought to be bored (Wemelsfelder 1993). One might expect that an obvious solution is to stimulate them by changing their environment frequently, but we have found that environmental changes disturb the monkeys temporarily and can cause minor distress. Macaques however also demonstrate habituation over time as the novelty of the situation declines. Novelty can stimulate exploratory behavior but can also elicit fear and disturbance (Mitchell and Gomber 1976; Clarke et al. 1988a; Mench 1994, and Chapter 3, this volume). Defining stress, and distinguishing adaptive stress responses from disruptive chronic distress, compounds our difficulties in measuring psychological well-being (Moberg 1985, 1987). One study even suggested that brief threatening events were beneficial for captive tamarins *(Saguinus oedipus)* because behavioral responses resembled those following environmental enrichment (Moodie and Chamove 1990); however, stress hormones were not measured.

The cage size study described in the previous section was preceded by a room change and a move to glass-fronted observation cages of the size designated for the weight class of the monkeys (cage size 3 in Figure 9.1). During the first twenty-four hours in the new room, the monkeys were somewhat distressed, as demonstrated by significantly elevated cortisol (Crockett et al. 1993a), by significant sleep disruption, decreased self-grooming, and increased inactivity (Crockett et al. 1995), and by appetite suppression, which lasted for several days

(Crockett et al. 1990). The appetite suppression and behavioral disruption occurred to a lesser extent after every change to a new, clean cage (regardless of size), but no elevation in cortisol occurred after cage change. Appetite suppression decreased with repeated cage changes, suggesting habituation to the novelty of being moved within the room. These responses seem to indicate that monkeys experience mild, transitory distress when faced with environmental changes but then adapt rather quickly. Such changes are unavoidable at some level because cages must be washed and might even provide "beneficial" arousal for some monkeys. However, I believe that our cortisol and behavioral data show that moving monkeys to a strange room is sufficiently distressing that it should not be done more often than necessary.

INANIMATE ENRICHMENT

The types of environmental enrichment mentioned in the 1991 Animal Welfare regulations (U.S. Department of Agriculture 1991, 6499–6500) are not equally effective. Nor, for that matter, has the effectiveness of all types, singly or in combination, been tested for a wide variety of primate species. The impact of an enrichment device may be measured by time spent using it, by associated decreased abnormal behavior and increased desirable behavior, or by stress hormone levels.

Inanimate enrichments for laboratory-housed nonhuman primates are primarily of four categories: structures (e.g., perches), manipulable objects with no food reward (e.g., dog toys), devices that require the animal to "work" for food (e.g., foraging boards), and external sensory stimuli, with or without choice or control (e.g., music, TV, murals) (Schapiro et al. 1991). The devices most consistently used by monkeys housed individually or in pairs are perches, foraging apparatus, and fleece-covered boards that stimulate grooming behavior (Fajzi et al. 1989; Reinhardt 1989b, 1990b; Bayne et al. 1991; Kopecky and Reinhardt 1991; Bayne et al. 1992b; Reinhardt and Reinhardt 1992).

The amount of time spent on perches varies widely, ranging from 7 percent to 48 percent of the daytime (Reinhardt 1989b, 1990b; Kopecky and Reinhardt 1991; Bayne et al. 1992b; Reinhardt and Reinhardt 1992; Shimoji et al. 1993). Monkeys in lower-level cages may spend more time on perches than upper-level monkeys (Reinhardt 1989b). We found that monkeys spent more time on the front half of the perch when the cages were oriented so that the perch provided a view of the door into the room (Crockett et al. 1996). As further evidence of monkey hesitance in response to environmental change, we found that six of twenty wild-born adult longtailed macaques would not use a perch during the first three days after it was installed, although all twenty eventually did (Shimoji et al. 1993).

Only devices that stimulate foraging and grooming have been shown to reduce abnormal behavior consistently (Bayne et al. 1991; Lam et al. 1991; Line et al. 1991a; Bayne et al. 1992a,b). However, one series of studies found sustained reductions in self-directed abnormal behavior when *M. fascicularis* were presented with pet toys (Weld and Erwin 1990; Weld et al. 1991). Researchers differ as to the effectiveness of pieces of natural wood versus rubber, plastic or nylon toys in eliciting manipulation (Line et al. 1991b; Reinhardt and Reinhardt 1992). One study singled out a translucent flexible plastic as being the most effective toy substance (Weld et al. 1991). Many simple toys, such as the rubber Kong developed for dogs, are initially attractive but primates rapidly habituate to them. In many studies, the percentage of daytime spent using or contacting most enrichment objects dropped rapidly to a remarkably similar 10 percent or less (Crockett et al. 1989; Maki and Bloomsmith 1989; Line and Morgan 1991; Line et al. 1991b). In at least one study, toy use did not change significantly over time (Bayne et al. 1992b). Although not occupying as much time as most reported perch use, 10 percent of the day is not an insignificant portion of the activity budget for captive monkeys to spend interacting with objects, whether or not abnormal behavior is changed.

Although simple toys are ineffective in consistently altering monkey behavior, changing objects periodically may increase their use (Line and Morgan 1991; Line et al. 1991b; Reinhardt and Reinhardt 1992; Weld 1992). Further, monkeys have been found to increase their use of simpler food puzzles and foraging or grooming boards over time (Bayne et al. 1991, 1992a; Murchison and Nolte 1992). Such devices certainly require more time and money than the provision of simple toys and tend to be used only when containing food. An alternative foraging device is a simple food box that requires some manipulation to extract a biscuit. Such devices have been found to engage macaques longer than regular food boxes even though they contain the same kind of food (Reinhardt 1994a). In addition, whole fruits and vegetables are natural foraging devices that are much appreciated by nonhuman primates (Smith et al. 1989; Crockett, personal observation).

Several studies have found significant differences in use of enrichment devices as a function of subject species, sex, age, and origin (wild- or captive-born) (Crockett et al. 1989; Maki and Bloomsmith 1989; Line et al. 1991b; Murchison and Nolte 1992; Weld 1992). For example, patas monkeys *(Erythrocebus patas)* did not respond to manipulable objects (Weld 1992). We found, in two different studies, that early responses to enrichment objects predicted long-term interest; only the macaques that showed immediate interest in Kong toys ever used them to any extent (Crockett et al. 1989). In a study to further evaluate a relatively inexpensive food puzzle developed at our Primate Field Station (Murchison 1991), we found that juvenile macaques differed at an early stage in their ability to solve the puzzle (Heath et al. 1992). Three of nine monkeys (one rhesus and two pigtailed macaques) quickly learned to shove a peanut from one level of the PVC (polyvinyl chloride) tube puzzle to another until it reached the lowest of four levels and a hole large enough to remove the peanut. The other monkeys seemed to lose interest when they could not obtain the food reward and seldom manipulated the puzzle.

A study on preference of enriched versus unenriched cages is worth special mention (Bayne et al. 1992b). Eight subadult and adult male rhesus macaques, all with stereotypies or other behavioral abnormalities, were housed singly in two connected single cages. All demonstrated a preference for one side even though no consistent difference with respect to lighting, door into colony room, or social relationship with neighbors was identified. Then the nonpreferred cage side was enriched with a perch, a grooming board, and two toys anchored with chains so they could not be moved to the other side. This clever design tested whether the monkey's initial preference for one side could be changed by embellishing the other side and whether any reductions in abnormal behavior would result. Only half the monkeys switched cage side preference to the enriched side. However, the subjects spent an average of 27 percent of the time interacting with the enrichments, and all showed reductions in behavioral abnormalities. When the furnishings were removed at the end of the enrichment phase, behavioral abnormalities returned to preenrichment levels. Thus, although enrichment may not always entice a monkey to change its housing preference, it may attract enough use to reduce the animal's abnormal behavior.

Exposure to substantial physical, feeding, and sensory (videotape) enrichment compared with no enrichment had no significant effect on abnormal behavior or cortisol levels in yearling rhesus macaques (Schapiro et al. 1993; Schapiro and Bloomsmith 1994). Both physical and feeding enrichment led to changes in other behavior (increased play and decreased self-grooming), but sensory enrichment was of little benefit (Schapiro and Bloomsmith 1994).

This brief summary of research on inanimate enrichment indicates that provision of inanimate environmental enrichment to captive primates does not predictably reduce abnormal behavior. However, the evidence supports provision of some inanimate enrichment because most animals use it at least some of the time. Our institutional environmental enhancement plan for laboratory primates includes a minimum of a perch, a manipulable object (usually Kongs or other rubber or plastic toys), and food treats (fruits or vegetables) at least two days a week.

SOCIAL ENRICHMENT

The mixed effectiveness of inanimate enrichment in measurably improving the psychological well-being of captive primates has led several authors to propose social contact as the best solution to meet the intent of the Animal Welfare Act (Crockett 1990; Line et al. 1991b; Reinhardt and Reinhardt 1992). The regulations call for meeting primate social needs, preferably by some sort of group housing, although physical contact is not absolutely required under all circumstances (Bayne 1991; U S. Department of Agriculture 1991). Contact with a human caregiver bearing food treats may be an alternative to social contact with a conspecific (Johnson-Delaney and White 1992; Bayne et al. 1993b).

One of the most obvious means of meeting the social needs of primates in a laboratory setting is pair housing, which has been tried with success for both sexes of rhesus and stumptailed *(M. arctoides)* macaques (Reinhardt et al. 1988, 1989; Reinhardt 1989a, 1990a,b). Pair housing, however, raises important questions for colony management: What kind of partner is likely to promote a given animal's psychological well-being? Does a pairing that promotes one partner's psychological well-being automatically promote that of the other? Is it possible to screen for mutually satisfactory partners in a way that does not risk trauma to either animal?

To evaluate the effects of same-sex pair housing on the psychological well-being of *M. fascicularis* adults, we studied behavioral compatibility and urinary cortisol excretion in fifteen female–female and fifteen male–male pairs (Crockett et al. 1994a). One side of each cage was a removable panel that could be replaced with a sheet-metal sliding door to form a pair with the adjacent cage. Insertion of various combinations of opaque and clear plastic panels allowed either no contact, visual-only contact, or physical contact with entry into the adjoining cage. Each subject experienced two weeks with each partner. On day 1 of the physical contact condition, each pair was monitored continuously for ninety minutes and then separated. For the remainder of the physical contact phase, pairs were housed together for seven hours during the day. The results showed that same-sex pair housing improved the psychological well-being of female *M. fascicularis* but not of most males. Every aspect of data analysis indicated that the success of pairing was strongly related to gender. All the female pairs were compatible, but nearly half the male pairs had to be separated prematurely, and only a third of the male pairs showed as much compatibility as females. Females spent more than a third of the day in social grooming and readily accepted strangers as cagemates, but paired adult males had much lower interaction rates. Furthermore, males were initially somewhat distressed by introduction to a cagemate, as indicated by increased excretion of urinary cortisol. Females, despite their eagerness to groom their cagemate, which I assume to be a behavioral indicator of social arousal (Moodie and Chamove 1990), displayed the same urinary cortisol levels when housed singly as when housed with other females. This result further suggests that elevated cortisol indicates distress, not just beneficial excitement.

We concluded, for adult wild-born female *M. fascicularis* on research protocols that permit social contact, that the social grouping requirement of the U.S. Department of Agriculture can usually be met by same-sex pair housing. For males of this species, same-sex pairing in completely interconnected cages was often unsuccessful. At least 20 percent of males could not be pair-housed with other males because of fighting. A few male pairs were quite compatible, but we were unable to predict the successful pairs based on prior noncontact preference testing (Crockett et al. 1994a). The marked differences between the sexes and the low rate of affiliative social interaction of male pairs suggested that same-sex pairing does not promote the psychological well-being of most male longtailed macaques. This outcome is consistent with macaque society because male-male competition and low rates of affiliative behavior are features of most polygynous, nonhuman primate societies (Melnick and Pearl 1987; Lindburg 1991; Crockett et al. 1994a). However, given the success of male pairings of other macaque species, our study may represent another example of species differences in response to potential enrichment, perhaps related to underlying differences in temperament and behavior (Thierry 1986; Clarke and Mason 1988; Clarke et al. 1988a,b; Mendoza 1991a,b; Clarke and Lindburg 1993). Such species and gender differences, as well as individual differences, certainly present additional challenges to those who want to provide social enrichment to laboratory primates.

It was not clear from our initial study whether same-sex housing failed to meet a social need of adult male longtailed macaques or whether adult males simply have a low need for social contact (Crockett et al. 1994a). To investigate this question, we introduced the element of choice so that the monkeys could indicate whether they preferred social contact or not. Giving animals control over their captive environment and allowing them to make choices and exhibit preferences has been advocated as a way to select appropriate enrichments to improve welfare (Chamove and Anderson 1989; Dantzer 1989; Dawkins 1989; Line et al. 1991a; Bayne et al. 1992b; Mench 1994, and Chapter 3, this volume). The monkeys were housed in adjoining cages equipped with sliding mesh doors and widely spaced bars (Figure 9.2) that prevented aggressive pursuit when one monkey moved out of reach of the other (Crockett et al. 1995). If both monkeys opened their doors, grooming contact was possible through the widely spaced bars from 10 A.M. to 3 P.M. At hourly intervals from 11 A.M. to 2 P.M., we closed any doors that were open, thus beginning a new trial (five per day). Each pair of cages was videotaped for one minute, four times per hour. At 3 P.M., a visual barrier was inserted and the doors locked, ending the day's session. The same female-female and male-male pairs from the first pairing study were tested with this option for choice, four days per week, two weeks per pair (familiar pairs). Subsequently, each of the ten males was paired with two of the females to see whether male-female pairing could provide suitable social enrichment for both sexes. The third part of this study involved pairing the remaining possible male-male pairs (each male paired with six unfamiliar males).

Behavioral data are still being coded and analyzed. However, preliminary results were estimated from notes taken on daily data sheets, recording how many animals opened their doors at 10 A.M. Given that both monkeys must open their doors for physical contact to be possible, we can estimate the percentage of time in potential contact by squaring the mean percentage of time that each individual's door was open. Both members of fifteen familiar male pairs opened their doors an estimated 49 percent of the time compared with 61 percent for thirty unfamiliar pairs (we do not yet know whether this difference is statistically significant). With the choice of access to the widely spaced bars, none of the familiar male pairs had to be separated, even though 47 percent of these same pairs had been separated in the original study when no bars prevented aggressive pursuit. Of the unfamiliar male pairs, 13 percent had to be separated before the end of the eight days of potential contact. Overall, four (9 percent) of the forty-five male-male pairings with widely spaced bars had to be separated, which compares with the 12 percent incompatibility found in continuous male-male pair housing of *M. mulatta* (Reinhardt 1994b). Although some *M. fascicularis* male pairs have been aggressive or reluctant to interact, others have been compatible, engaging in grooming or occasional playing. The widely spaced grooming-contact bars seem to be a solution for providing social enrichment for some males.

The male-female pairing showed that all males were interested in affiliative social interaction with females, even though some males appeared disinterested in males. Females appeared to be as interested in affiliative contact with males as with females. Male-female pairs both opened their doors an estimated 95 percent of the time, very similar to the 93 percent for female-female pairs. Male-female pairs engaged in considerable social grooming, but we cannot statistically compare the time spent grooming among the various categories of pairings until the videotapes have been behavior-coded. Unfortunately, from a management point of view, the widely spaced bars allowed copulation, resulting in one unplanned infant (in twenty pairings). However, we have now tested modifications to the

Figure 9.2. Social grooming between longtailed macaque pairs through widely spaced bars. Each monkey must open a mesh door in its cage to make such contact. *Top,* male grooming male; *bottom,* male grooming female. (Photos by C. Crockett.)

bar design that sucessfully prevented copulation while allowing safe contact (i.e., appendages did not get caught in the bars). Nine male-female pairs have been housed twenty-four hours per day in adjoining grooming-contact cages for many months with no resulting pregnancies or injuries (Crockett et al., n.d.).

Our preliminary conclusions are that most adult longtailed macaques do have social needs and that widely spaced bars provide the animals with the means to control the amount of contact and to minimize injury. Most male longtailed macaques, although generally less sociable than females, do seem to experience improved welfare in paired caging that allows them to control the amount of contact when appropriate partners are presented. The problem remains of identifying compatible male pairs while avoiding any injuries. Some injuries and incompatibility are inevitable, even among female pairs (Line et al. 1990a). The monetary cost and risk of injury in identifying compatible partners may be too high for many laboratories to attempt male pairing on a widespread basis. Male-female pairing seems to be a solution now that we have developed a bar design that prevents copulation and pregnancy.

IMPLICATIONS FOR ZOO ENRICHMENT

As zoos attempt to comply with Species Survival Plans, they may increasingly need to house excess adult males of various nonhuman primate species (Bound et al. 1988). Thus, laboratory studies concerned with meeting the potential social needs of adult males are relevant because zoos are equally subject to the Animal Welfare regulations (U.S. Department of Agriculture 1991). Zoos may sometimes be faced with keeping adult males in single cages, and lessons from the laboratory suggest that in such environments perches and foraging devices are the most beneficial inanimate enrichment. Foraging litter such as wood chips or hay, while shown to be beneficial (Chamove et al. 1982; Bryant et al. 1988; Byrne and Suomi 1991), is more practical for zoo enclosures and some laboratory compounds than for standard laboratory cages where drainage problems may preclude its use. Laboratory studies also indicate that individual, species, age, sex, and rearing differences have marked effects on use and effectiveness of environmental enrichments. Our studies show that small changes in cage floor area have little impact on psychological well-being of laboratory-housed macaques. On the other hand, changing of rooms was moderately distressful, suggesting that animal moves should not be made unnecessarily.

Effective promotion of the psychological well-being of laboratory primates depends on accurate knowledge of the animal's needs and preferences and a commitment to apply that knowledge in day-to-day management. By systematically testing the conventional wisdom, we can begin to assemble a substantial body of scientifically based facts, thereby obtaining a more accurate picture of enrichment strategies that effectively promote laboratory primate welfare (Crockett and Bowden 1994).

CONCLUSIONS

Zoos as well as research laboratories must comply with the 1991 Animal Welfare regulations that stipulate the implementation of environmental enhancement plans to promote the psychological well-being of nonhuman primates. Controlled laboratory studies can identify which factors of the captive environment actually impinge on primate psychological well-being. Cage size alone is not an important factor. Merely increasing cage size has not been associated with meaningful improvements in psychological well-being among singly housed macaques. Environmental change, such as being moved into a new room, disturbs monkeys, as demonstrated by appetite suppression, disruption of the normal activity cycle, and mildly elevated stress hormones, and thus animals should not be moved without good reason.

Inanimate environmental enrichment devices do not invariably improve psychological well-being, and the effectiveness of enrichment devices in reducing abnormal behaviors has been mixed. Perches and foraging devices appear to be used most consistently; toys of some shapes and materials appear to be favored over others. There is sufficient evidence, based on time spent using them, that enrichment objects should be included in cages but that dramatic changes in behavior should not be expected.

Adult female longtailed macaques benefit from social enrichment through pairing with other females. Adult males also have social needs, although they are more likely to express them toward females. Many males ignore or behave aggressively toward other males, although some male pairs are highly compatible. Housing longtailed macaque males in paired caging with widely spaced grooming-contact bars prevents aggressive pursuit and increases the success rate of male pairing. Housing males and females in the same type of caging is highly successful, and we have designed widely spaced bar panels that permit grooming but prevent unplanned pregnancies. Zoos will benefit from the knowledge gained from laboratory studies when they are required with increasing frequency to house excess males in separate caging.

A PERSONAL PERSPECTIVE

My earlier research focused on the behavior of primates in natural habitats and zoo settings (Crockett and Wilson 1980; Gaspari and Crockett 1984; Hutchins et al. 1984; Crockett and Eisenberg 1987; Crockett 1996). This background compels me to discover what aspects of the captive environment really matter to nonhuman primates. There are too many unsubstantiated opinions about what is good or not good for promoting the psychological well-being of captive primates. Furthermore, time and money to implement environmental enrichment are in limited supply, and I want them to be spent constructively.

Although I prefer to observe macaques in the rain forest, I have learned fascinating things about their psyches that could only be revealed in captivity. Many of my own preconceived notions simply were not borne out by systematic study. This leads me to the conviction that we should seek knowledge derived from objective studies to meaningfully enrich the captive environment of these intelligent animals who entertain and teach us.

ACKNOWLEDGMENTS

Douglas Bowden, principal investigator of the grant funding this research, has been integral in designing the experiments and preparing the manuscripts. Charles Bowers, Mika Shimoji, Matthias Leu, and Rita Bellanca actively participated in data collection, with the assistance of many student helpers acknowledged by name in the cited publications. Other collaborators include G. P. Sackett, O. A. Smith, F. A. Spelman, J. T. Bielitzki, C. Emerson, C. Johnson-Delaney, and W. R. Morton. This research would have not been possible without the assistance of the Colony and Bioengineering Divisions of the Regional Primate Research Center. The manuscript has been improved by constructive comments from K. Elias and M. Hutchins and by the invaluable bibliographic services of the Primate Information Center. This research was supported by National Institutes of Health (NIH) grants RR00166 and RR04515.

REFERENCES

Bayne, K. A. L. 1991. Alternatives to continuous social housing. *Laboratory Animal Science* 41:355–369.

Bayne, K. A. L., S. L. Dexter, J. K. Hurst, G. M. Strange, and E. E. Hill. 1993a. Kong toys for laboratory primates: Are they really an enrichment or just fomites? *Laboratory Animal Science* 43:78–85.

Bayne, K. A. L., S. Dexter, H. Mainzer, C. McCully, G. Campbell, and F. Yamada.

1992a. The use of artificial turf as a foraging substrate for individually housed rhesus monkeys *(Macaca mulatta). Animal Welfare* 1:39–53.

Bayne, K. A. L., S. L. Dexter, and G. M. Strange. 1993b. The effects of food treat provisioning and human interaction on the behavioral well-being of rhesus monkeys *(Macaca mulatta). Laboratory Animal Science* 32 (2): 6–9.

Bayne, K. A. L., J. K. Hurst, and S. L. Dexter. 1992b. Evaluation of the preference to and behavioral effects of an enriched environment on male rhesus monkeys. *Laboratory Animal Science* 42:38–45.

Bayne, K. A. L., H. Mainzer, S. Dexter, G. Campbell, F. Yamada, and S. Suomi. 1991. The reduction of abnormal behaviors in individually housed rhesus monkeys *(Macaca mulatta)* with a foraging/grooming board. *American Journal of Primatology* 23:23–35.

Bayne, K. A. L., and C. McCully. 1989. The effect of cage size on the behavior of individually housed rhesus monkeys. *Lab Animal* 18 (7): 25–28.

Bercovitch, F. B., and M. R. Lebron. 1991. Impact of artificial fissioning and social networks on levels of aggression and affiliation in primates. *Aggressive Behavior* 17:17–25.

Bielitzki, J. T. 1992. Letter to the editor: Enrichment hazards. *Laboratory Primate Newsletter* 31 (3): 36.

Bond, M. 1991. How to collect urine from a gorilla. *Gorilla Gazette* 5 (3): 12–13.

Bound, V., H. Shewman, and J. Sievert. 1988. The successful introduction of five male lion-tailed macaques *(Macaca silenus)* at Woodland Park Zoo. In *Proceedings of the American Association of Zoological Parks and Aquariums Regional Conference,* 122–133. Wheeling, W.Va.: AAZPA.

Bowers, C. L., C. M. Crockett, M. Shimoji, R. Bellanca, and D. M. Bowden. 1993. Heart rate variability and psychological well-being. *American Journal of Primatology* 30 (4): 22.

Broom, D. M. 1988. The scientific assessment of animal welfare. *Applied Animal Behaviour Science* 20:5–19.

Bryant, C. E., N. M. J. Rupniak, and S. D. Iversen. 1988. Effects of different environmental enrichment devices on cage stereotypies and autoaggression in captive cynomolgus monkeys. *Journal of Medical Primatology* 17:257–269.

Bunney, W. E., J. W. Mason, and D. A. Hamburg. 1965. Correlations between behavioral variables and urinary 17-hydroxycorticosteroids in depressed patients. *Psychosomatic Medicine* 27:299–308.

Byrne, G. D., and S. J. Suomi. 1991. Effects of woodchips and buried food on behavior patterns and psychological well-being of captive rhesus monkeys. *American Journal of Primatology* 23:141–151.

Capitanio, J. P. 1986. Behavioral pathology. In *Comparative Primate Biology.* Vol. 2A, *Behavior, Conservation, and Ecology,* ed. G. Mitchell and J. Erwin, 411–454. New York: Alan R. Liss.

Carlstead, K., J. L. Brown, S. L. Monfort, R. Killens, and D. E. Wildt. 1992. Urinary monitoring of adrenal responses to psychological stressors in domestic and nondomestic felids. *Zoo Biology* 11:165–176.

Chamove, A. S., and J. R. Anderson. 1989. Examining environmental enrichment. In *Housing, Care, and Psychological Well-Being of Captive and Laboratory Primates,* ed. E. F. Segal, 183–202. Park Ridge, N.J.: Noyes Publications.

Chamove, A. S., J. R. Anderson, S. C. Morgan-Jones, and S. P. Jones. 1982. Deep woodchip litter: Hygiene, feeding, and behavioral enhancement in eight primate species. *International Journal for the Study of Animal Problems* 3:308–318.

Clarke, A. S., and D. G. Lindburg. 1993. Behavioral contrasts between male cynomolgus and lion-tailed macaques. *American Journal of Primatology* 29:49–59.

Clarke, A. S., and W. A. Mason. 1988. Differences among three macaque species in responsiveness to an observer. *International Journal of Primatology* 9:347–364.

Clarke, A. S., W. A. Mason, and G. P. Moberg. 1988a. Differential behavioral and adrenocortical responses to stress among three macaque species. *American Journal of Primatology* 14:37–52.

———. 1988b. Interspecific contrasts in responses of macaques to transport cage training. *Laboratory Animal Science* 38:305–309.

Crockett, C. M. 1990. Psychological well-being and enrichment workshop held at Primate Centers' Directors' Meeting. *Laboratory Primate Newsletter* 29 (3): 3–6.

———. 1993. Rigid rules for promoting psychological well-being are premature. *American Journal of Primatology* 30:177–179.

———. 1996. Data collection in the zoo setting, with emphasis on behavior. In *Wild Mammals in Captivity,* ed. D. G. Kleiman, M. E. Allen, K. V. Thompson, S. Lumpkin, and H. Harris, 545–565. Chicago: University of Chicago Press.

Crockett, C. M., R. U. Bellanca, C. L. Bowers, and D. M. Bowden. n.d. Grooming-contact bars provide social contact for individually caged laboratory macaques. *Contemporary Topics in Laboratory Animal Science* (in press).

Crockett, C. M., J. Bielitzki, A. Carey, and A. Velez. 1989. Kong toys as enrichment devices for singly-caged macaques. *Laboratory Primate Newsletter* 28 (2): 21–22.

Crockett, C. M., and D. M. Bowden. 1994. Challenging conventional wisdom for housing monkeys. *Lab Animal* 23 (2): 29–33.

Crockett, C. M., C. L. Bowers, R. Bellanca, M. Shimoji, and D. M. Bowden. 1995a. How often do singly housed longtailed macaques choose grooming contact with a neighbor? *American Journal of Primatology* 36:118.

Crockett, C. M., C. L. Bowers, D. M. Bowden, and G. P. Sackett. 1994a. Sex differences in compatibility of pair-housed adult longtailed macaques. *American Journal of Primatology* 32:73–94.

Crockett, C. M., C. L. Bowers, G. P. Sackett, and D. M. Bowden. 1990. Appetite suppression and urinary cortisol responses to different cage sizes and tethering procedures in longtailed macaques. *American Journal of Primatology* 20:184–185.

———. 1993a. Urinary cortisol responses of longtailed macaques to five cage sizes, tethering, sedation, and room change. *American Journal of Primatology* 30:55–74.

Crockett, C. M., C. L. Bowers, M. Shimoji, R. Bellanca, and D. M. Bowden. 1994b. Behavioral responses to four sizes of home cage by adult female pigtailed macaques. *American Journal of Primatology* 33:203-204.

Crockett, C. M., C. L. Bowers, M. Shimoji, M. Leu, R. Bellanca, and D. M. Bowden. 1993b. Appetite and urinary cortisol responses to different cage sizes in female pigtailed macaques. *American Journal of Primatology* 30:305.

Crockett, C. M., C. L. Bowers, M. Shimoji, M. Leu, D. M. Bowden, and G. P. Sackett. 1995b. Behavioral responses of longtailed macaques to different cage sizes and common laboratory experiences. *Journal of Comparative Psychology* 109:368-383.

Crockett, C. M., and J. F. Eisenberg. 1987. Howlers: Variations in group size and demography. In *Primate Societies,* ed. B. B. Smuts, D. L. Cheney, R. M. Seyfarth, R. W. Wrangham, and T. T. Struhsaker, 54-68. Chicago: University of Chicago Press.

Crockett, C. M., and W. L. Wilson. 1980. The ecological separation of *Macaca nemestrina* and *M. fascicularis* in Sumatra. In *The Macaques: Studies in Ecology, Behavior, and Evolution,* ed. D. G. Lindburg, 148-181. New York: Van Nostrand Reinhold.

Crockett, C. M., J. Yamashiro, S. DeMers, and C. Emerson. 1996. Engineering a rational approach to primate space requirements. *Lab Animal* 25 (9): 44-47.

Dantzer, R. 1986. Behavioral, physiological, and functional aspects of stereotyped behavior: A review and a re-interpretation. *Journal of Animal Science* 62:1776-1786.

———. 1989. Assessment of psychological well-being in animals: Lessons from farm animal studies. *American Journal of Primatology Supplement* 1:5-7.

Dawkins, M. S. 1989. From an animal's point of view: Consumer demand theory and animal welfare. *Behavioral and Brain Sciences* 13:1-61.

de Waal, F. B. M. 1989. The myth of a simple relation between space and aggression in captive primates. *Zoo Biology Supplement* 1:141-148.

Draper, W. A., and I. S. Bernstein. 1963. Stereotyped behavior and cage size. *Perceptual and Motor Skills* 16:231-234.

Erwin, J. 1979. Aggression in captive macaques: Interaction of social and spatial factors. In *Captivity and Behavior: Primates in Breeding Colonies, Laboratories, and Zoos,* ed. J. Erwin, T. L. Maple, and G. Mitchell, 139-171. New York: Van Nostrand Reinhold.

Erwin, J., and G. P. Sackett. 1990. Effects of management methods, social organization, and physical space on primate behavior and health. *American Journal of Primatology* 20:23-30.

Estep, D. Q., and S. C. Baker. 1991. The effects of temporary cover on the behavior of socially housed stumptailed macaques *(Macaca arctoides). Zoo Biology* 10:465-472.

Fajzi, K., V. Reinhardt, and M. D. Smith. 1989. A review of environmental enrichment strategies for singly caged nonhuman primates. *Lab Animal* 18 (3): 23-35.

Fishman, J. R., D. A. Hamburg, J. H. Handlon, J. W. Mason, and E. Sachar. 1962. Emotional and adrenal cortical responses to a new experience: Effect of social environment. *Archives of General Psychiatry* 6 (2): 29-36.

Friedman, S. B., J. W. Mason, and D. A. Hamburg. 1963. Urinary 17-hydroxycorticosteroid levels in parents of children with neoplastic disease: A study of chronic psychological stress. *Psychosomatic Medicine* 25:364–376.

Gaspari, M. K., and C. M. Crockett. 1984. The role of scent marking in *Lemur catta* agonistic behavior. *Zoo Biology* 3:123–132.

Goosen, C. 1981. Abnormal behavior patterns in rhesus monkeys: Symptoms of mental disease? *Biological Psychiatry* 16:697–716.

Hamburg, D. A. 1962. Plasma and urinary corticosteroid levels in naturally occurring psychologic stresses. *Proceedings of the Association for Research on Nervous Mental Disease* 40:406–413.

Heath, S., M. Shimoji, J. Tumanguil, and C. Crockett. 1992. Peanut puzzle solvers quickly demonstrate aptitude. *Laboratory Primate Newsletter* 31 (1): 12–13.

Hutchins, M., D. Hancocks, and C. Crockett. 1984. Naturalistic solutions to the behavioral problems of captive animals. *Zoologische Garten* 54:28–42.

International Primatological Society. 1993. IPS international guidelines: IPS code of practice. 1. Housing and environmental enrichment. *Primate Report* 35 (1): 8–16.

Johnson-Delaney, C. A., and J. White. 1992. Human primate/non-human primate social interaction program [abstract]. Regional Proceedings, American Association of Laboratory Animal Science.

Kelley, T. M., and C. A. Bramblett. 1981. Urine collection from vervet monkeys by instrumental conditioning. *American Journal of Primatology* 1:95–97.

Kopecky, J., and V. Reinhardt. 1991. Comparing the effectiveness of PVC swings versus PVC perches as environmental enrichment objects for caged female rhesus macaques *(Macaca mulatta)*. *Laboratory Primate Newsletter* 30 (2): 5–6.

Lam, K., N. M. J. Rupniak, and S. D. Iversen. 1991. Use of a grooming and foraging substrate to reduce cage stereotypies in macaques. *Journal of Medical Primatology* 20:104–109.

Lawrence, A. B., and J. Rushen, eds. 1993. *Stereotypic Animal Behaviour: Fundamentals and Applications to Welfare.* Wallingford, U.K.: CAB International.

Leu, M., C. M. Crockett, C. L. Bowers, and D. M. Bowden. 1993. Changes in activity levels of singly housed longtailed macaques when given the opportunity to exercise in a larger cage. *American Journal of Primatology* 30:327.

Lindburg, D. G. 1991. Ecological requirements of macaques. *Laboratory Animal Science* 41:315–322.

Line, S. W., A. S. Clarke, and H. Markowitz. 1987. Plasma cortisol of female rhesus monkeys in response to acute restraint. *Laboratory Primate Newsletter* 26 (4): 1–4.

Line, S. W., H. Markowitz, K. N. Morgan, and S. Strong. 1991a. Effects of cage size and environmental enrichment on behavioral and physiological responses of rhesus macaques to the stress of daily events. In *Through the Looking Glass: Issues of Psychological Well-Being in Captive Non-human Primates,* ed. M. A. Novak and A. J. Petto, 160–179. Washington, D.C.: American Psychological Association.

Line, S. W., and K. N. Morgan. 1991. The effects of two novel objects on the behavior of singly caged adult rhesus macaques. *Laboratory Animal Science* 41:365–369.

Line, S. W., K. N. Morgan, and H. Markowitz. 1991b. Simple toys do not alter the behavior of aged rhesus monkeys. *Zoo Biology* 10:473-484.

Line, S. W., K. N. Morgan, H. Markowitz, J. A. Roberts, and M. Riddell. 1990a. Behavioral responses of female long-tailed macaques *(Macaca fascicularis)* to pair formation. *Laboratory Primate Newsletter* 29 (4): 1-5.

Line, S. W., K. N. Morgan, H. Markowitz, and S. Strong. 1989. Influence of cage size on heart rate and behavior in rhesus monkeys. *American Journal of Veterinary Research* 50:1523-1526.

———. 1990b. Increased cage size does not alter heart rate or behavior in female rhesus monkeys. *American Journal of Primatology* 20:107-113.

Lundberg, U. 1980. Catecholamine and cortisol excretion under psychologically different laboratory conditions. In *Catecholamines and Stress: Recent Advances,* ed. E. Usdin, S. Kvetnansky, and I. J. Kopin, 455-460. Amsterdam: Elsevier-North Holland.

Maki, S., and M. A. Bloomsmith. 1989. Uprooted trees facilitate the psychological well-being of captive chimpanzees. *Zoo Biology* 8:79-87.

Marriner, L. M., and L. C. Drickamer. 1994. Factors influencing stereotyped behavior of primates in a zoo. *Zoo Biology* 13:267-275.

Mason, G. J. 1991. Stereotypies: A critical review. *Animal Behaviour* 41:1015-1037.

Mason, G. [J.], and M. Mendl. 1993. Why is there no simple way of measuring animal welfare? *Animal Welfare* 2:301-319.

Melnick, D. J., and M. J. Pearl. 1987. Cercopithecines in multimale groups: Genetic diversity and population structure. In *Primate Societies,* ed. B. B. Smuts, D. L. Cheney, R. M. Seyfarth, R. W. Wrangham, and T. T. Struhsaker, 121-134. Chicago: University of Chicago Press.

Mench, J. 1994. Environmental enrichment and exploration. *Lab Animal* 23 (2): 38-41.

Mendoza, S. P. 1991a. Behavioural and physiological indices of social relationships: Comparative studies of New World monkeys. In *Primate Responses to Environmental Change,* ed. H. O. Box, 311-335. London: Chapman & Hall.

———. 1991b. Sociophysiology of well-being in nonhuman primates. *Laboratory Animal Science* 41:344-349.

Miller, M. W., N. T. Hobbs, and M. C. Sousa. 1991. Detecting stress responses in Rocky Mountain bighorn sheep *(Ovis canadensis canadensis):* Reliability of cortisol concentrations in urine and feces. *Canadian Journal of Zoology* 69:15-24.

Mitchell, G., and J. Gomber. 1976. Moving laboratory rhesus monkeys *(Macaca mulatta)* to unfamiliar home cages. *Primates* 17:543-546.

Moberg, G. P. 1985. Biological response to stress: Key to assessment of animal well-being? In *Animal Stress,* ed. G. P. Moberg, 27-49. Bethesda, Md.: American Physiological Society.

———. 1987. Problems in defining stress and distress in animals. *Journal of the American Veterinary Medical Association* 191:1207-1211.

Moodie, E. M., and A. S. Chamove. 1990. Brief threatening events beneficial for captive tamarins? *Zoo Biology* 9:275-286.

Murchison, M. A. 1991. PVC-pipe food puzzle for singly caged primates. *Laboratory Primate Newsletter* 30 (3): 12–14.

———. 1993. Potential animal hazard with ring toys. *Laboratory Primate Newsletter* 32 (1): 7.

Murchison, M. A., and R. E. Nolte. 1992. Food puzzle for singly caged primates. *American Journal of Primatology* 27:285–292.

Novak, M. A., and A. J. Petto, eds. 1991. *Through the Looking Glass: Issues of Psychological Well-Being in Captive Non-human Primates.* Washington, D.C.: American Psychological Association.

Novak, M. A., and S. J. Suomi. 1988. Psychological well-being of primates in captivity. *American Psychologist* 43:765–773.

Paulk, H. H., H. Dienske, and L. G. Ribbens. 1977. Abnormal behavior in relation to cage size in rhesus monkeys. *Journal of Abnormal Psychology* 86:87–92.

Reinhardt, V. 1989a. Behavioral responses of unrelated adult male rhesus monkeys familiarized and paired for the purpose of environmental enrichment. *American Journal of Primatology* 17:243–248.

———. 1989b. Evaluation of the long-term effectiveness of two environmental enrichment objects for singly caged rhesus macaques. *Lab Animal* 18 (6): 31–33.

———. 1990a. Environmental enrichment program for caged stump-tailed macaques *(Macaca arctoides). Laboratory Primate Newsletter* 29 (2): 10–11.

———. 1990b. Time budget of caged rhesus monkeys exposed to a companion, a PVC perch, and a piece of wood for an extended time. *American Journal of Primatology* 20:51–56.

———. 1994a. Caged rhesus macaques voluntarily work for ordinary food. *Primates* 35:95–98.

———. 1994b. Continuous pair-housing of caged *Macaca mulatta:* Risk evaluation. *Laboratory Primate Newsletter* 33 (1): 1–4.

Reinhardt, V., D. Cowley, and S. Eisele. 1991a. Serum cortisol concentrations of single-housed and isosexually pair-housed adult rhesus macaques. *Journal of Experimental Animal Science* 34:73–76.

Reinhardt, V., D. Cowley, S. Eisele, and J. Scheffler. 1991b. Avoiding undue cortisol responses to venipuncture in adult male rhesus macaques. *Animal Technology* 42 (2): 83–86.

Reinhardt, V., D. Houser, D. Cowley, S. Eisele, and R. Vertein. 1989. Alternatives to single caging of rhesus monkeys *(Macaca mulatta)* used in research. *Zeitschrift für Versuchstierkunde* 32:275–279.

Reinhardt, V., D. Houser, S. Eisele, D. Cowley, and R. Vertein. 1988. Behavioral responses of unrelated rhesus monkey females paired for the purpose of environmental enrichment. *American Journal of Primatology* 14:135–140.

Reinhardt, V., and A. Reinhardt. 1992. Quantitatively tested environmental enrichment options for singly-caged nonhuman primates: A review. *Humane Innovations and Alternatives* 6:374–384.

Schapiro, S. J., and M. A. Bloomsmith. 1994. Behavioral effects of enrichment on singly-housed, yearling rhesus monkeys: An analysis including three enrichment conditions and a control group. *American Journal of Primatology* 35:89–101.

Schapiro, S. J., M. A. Bloomsmith, A. L. Kessel, and C. A. Shively. 1993. Effects of enrichment and housing on cortisol response in juvenile rhesus monkeys. *Applied Animal Behaviour Science* 37:251–263.

Schapiro, S. J., L. Brent, M. A. Bloomsmith, and W. C. Satterfield. 1991. Enrichment devices for nonhuman primates. *Lab Animal* 20 (6): 22–28.

Segal, E. F., ed. 1989. *Housing, Care, and Psychological Well-Being of Captive and Laboratory Primates.* Park Ridge, N.J.: Noyes Publications.

Shimoji, M., C. L. Bowers, and C. M. Crockett. 1993. Initial response to introduction of a PVC perch by singly caged *Macaca fascicularis. Laboratory Primate Newsletter* 32 (4): 8–11.

Smith, A., D. G. Lindburg, and S. Vehrencamp. 1989. Effect of food preparation on feeding behavior of lion-tailed macaques. *Zoo Biology* 8:57–65.

Thierry, B. 1986. A comparative study of aggression and response to aggression in three species of macaque. In *Primate Ontogeny, Cognition, and Social Behaviour,* ed. J. G. Else and P. C. Lee, 307–313. New York: Cambridge University Press.

Thomas, R. K., and R. B. Lorden. 1989. What is psychological well-being? Can we know if primates have it? In *Housing, Care, and Psychological Well-Being of Captive and Laboratory Primates,* ed. E. F. Segal, 12–26. Park Ridge, N.J.: Noyes Publications.

U.S. Department of Agriculture. 1991. Animal welfare, standards, final rule (part 3, subpart D: Specifications for the humane handling, care, treatment, and transportation of nonhuman primates). *Federal Register* 56 (32): 6495–6505.

U.S. Department of Agriculture, U.S. Public Health Service, and Primate Information Center. 1992. *Environmental Enrichment Information Resources for Nonhuman Primates: 1987–1992.* Beltsville, Md.: National Agricultural Library, Animal Welfare Information Center.

U.S. Public Health Service. 1985. *Guide for the Care and Use of Laboratory Animals.* NIH 85-23. Bethesda, Md.: National Institutes of Health.

van Schaik, C. P., M. A. van Noordwijk, T. van Bragt, and M. A. Blankenstein. 1991. A pilot study of the social correlates of levels of urinary cortisol, prolactin, and testosterone in wild long-tailed macaques *(Macaca fascicularis). Primates* 32:345–356.

Weld, K. P. 1992. Environmental enrichment of laboratory-housed nonhuman primates. Master's thesis, University of Maryland, College Park.

Weld, K., and J. Erwin. 1990. Provision of manipulable objects to cynomolgus macaques promotes species-typical behavior. *American Journal of Primatology* 20:243.

Weld, K., B. Metz, and J. Erwin. 1991. Environmental enrichment for *Macaca fascicularis:* Effects of shape and substance of manipulable objects. *American Journal of Primatology* 24:139.

Wemelsfelder, F. 1993. The concept of animal boredom and its relationship to stereotyped behaviour. In *Stereotypic Animal Behaviour: Fundamentals and Applications to Welfare,* ed. A. B. Lawrence and J. Rushen, 65–95. Wallingford, U.K.: CAB International.

KATHLEEN N. MORGAN, SCOTT W. LINE,
AND HAL MARKOWITZ

10

ZOOS, ENRICHMENT, AND THE SKEPTICAL OBSERVER

The Practical Value of Assessment

Public sentiment, the law, and our own sense of morality dictate that we accelerate our efforts to improve the well-being of our animal charges. Research institutions that maintain live animal collections are coming under increasing pressure from diverse animal rights groups. Zoological parks have not escaped the attention of animal rights activists either, and in the United States zoos also have a legal responsibility for meeting the demands of the Animal Welfare Act. In addition to this legal obligation, zoos may recognize other incentives for improving and enriching the lives of the animals in their care.

For example, one incentive for zoo enrichment efforts is increased public interest. Zoos are largely public institutions, dependent in part on private donation and membership fees for their revenue. Visitors are frequently concerned about what they consider to be unnatural behavior in zoo animals housed in older or more traditional barren exhibits. Enriching these habitats in ways that stimulate species-typical behaviors may excite the public interest, increasing appreciation for the animals and improving public understanding and education (Shettel-Neuber 1988; Falk and Dierking 1992). Thus, environmental enrichment efforts by zoos may improve revenue as well as conditions for inhabitants. At the same time, however, providing animals with more naturalistic habitats may also make the animals more difficult to view, resulting in disgruntled or disappointed visitors (Bitgood et al. 1988; Donahoe 1988; Falk and Dierking 1992; Morgan and Bergman, unpublished data). The challenge for zoos in this case is to achieve improvements for their animals while maintaining public support.

Another incentive for improving well-being in our captive animals is simply the desire to provide humane care. Living in captivity necessarily limits an animal's choices. Humans determine the timing and nature of food, the availability of social partners, and the opportunity to avoid others. Distress as a consequence of these limitations may bring about physical and behavioral disorders (such as hyperaggressiveness, fur-plucking, pica, or self-biting). Thus, one of the stakes at risk in our enrichment efforts is simply the day-to-day maintenance of the animals in our care, in addition to improvement of their quality of life.

Also at stake in attempts to enrich zoo environments is the successful preservation of endangered species. Increasingly, there is an urgent need for zoos to become agents of conservation. As human population growth accelerates the destruction of wildlife habitats, the natural world becomes less and less safe for many living things. For some animals, the captive population now exceeds known populations in the wild (Savage 1988). Still others exist only in captivity (e.g., Père David's deer, *Elaphurus davidianus*), having become extinct in the wild. If these species and others are to be preserved, it is imperative that zoos develop the skills, technologies, and knowledge by which to sustain healthy, breeding captive populations (Tudge 1991a). To the extent that environmental enrichment addresses the physical and psychological needs of captive animals, it is a necessary part of such an effort. The success of repatriation programs may also depend in part on environmental enrichment efforts in captivity. If the program that returns animals to the wild fails to provide them with experiences conducive to their survival on release, the time, money, and energy expended in repatriation will have been wasted (Shepherdson 1994; also see Castro et al., Chapter 8, and Miller et al., Chapter 7, in this volume).

Under these circumstances it is important that environmental enrichment is both effective and makes efficient use of limited resources. We argue for the value of seeking verification of the effectiveness of environmental enrichment efforts (see Crockett, Chapter 9, this volume). We do this here by briefly discussing some of the different epistemologies that we use daily, and how those ways of knowing may influence the decisions we make about the animals under our care, whose lives and well-being are currently at stake. We present some of the hypotheses that these ways of knowing have generated about methods used to enrich the lives of captive animals. And we argue for the need to make use of one particular way of knowing—empiricism—to test these hypotheses in situations in which the stakes are particularly high. We do this in part by reviewing some of our work attempting to document the effectiveness of several protocols suggested for enriching the lives of captive primates. Specifically, we show that some of the most intuitively obvious techniques for improving living conditions for captive primates can be among the least effective, in terms of their ability to produce any statistically significant changes in behavior (and, in some cases, in physiology).

WAYS OF KNOWING

One of the many ways human beings have for gaining knowledge about the world is simply to rely on others to inform us. In making use of authority, one accepts something that an authority figure tells one is true *because* of the authority of that person. Knowing that chimpanzees *(Pan troglodytes)* make and use tools to extract termites from termite mounds on the basis of having read Jane Goodall's books (1971, 1986, 1990) is making use of authority as a way of knowing. Certainly, reliance on the reports of others about species-typical behaviors in wild populations is a rich source from which to draw ideas about enrichment opportunities for captive animals. At least in part on the basis of Goodall's reports of chimpanzee termite-fishing, for example, David Shepherdson provided the chimpanzees at the London Zoo in England with a length of plastic tubing from which the animals could "fish" pieces of fruit by poking and prodding them through small openings in the tube until the treats reached an opening large enough to allow their extraction. Similarly, on the basis of the literature on bear behavior in the wild, Law modified an exhibit for Himalayan bears *(Ursus thibetanus)* at the Glasgow Zoo in Scotland to allow for the performance of more species-typical behavior, such as nest-building (cited in Tudge 1991b).

Authority as a way of knowing is the very basis of most of our educational systems. It is a common, respectable, and accepted basis for knowledge. However, knowledge obtained in this manner is only as good as the authority. For example, authorities on the many species of animals that we care for may unwittingly fail to provide us with complete information. One of us clearly recalls the horror of discovering several box turtles stuck fast in the litter suggested for them by several herpetologists after that litter became damp and then hardened as it dried. Because relying solely on others has its drawbacks, particularly when the stakes involved in having correct information are high, few people depend solely on others for their knowledge. Rather, we often gain knowledge through our own personal experiences or ideas about how the world works—through our *common sense.* One of the potential problems with commonsense assertions is that they are based in our own implicit theories about reality. These personal theories may generate inappropriate or incorrect ideas about a given species. For example, we may believe that confinement in a small space is unpleasant, so we assert that it only makes sense to increase cage size as a way of improving an animal's well-being. However, for some species an increase in cage size may not be very effective as an enrichment strategy.

Commonsense explanations of animal behavior often tend to be anthropomorphic. Anthropomorphism, or attributing human characteristics, feelings, and motivations to nonhuman animals, has been frowned upon in intellectual circles since before the Dark Ages, but especially following Romanes' (1882) collection of anecdotes about animal cognition. However, a little anthropomorphizing may not always be a bad thing, especially when backed by empirical data. In fact, some researchers have suggested that a little anthropomorphizing is vital in determining what constitutes unacceptable conditions for animals and what might improve those conditions (Dawkins 1980). As an extension of the epistemology of common sense, however, anthropomorphism suffers from similar problems. Just as there is no reason to believe that what might enrich the life of one species will necessarily enrich the lives of all species, there is no reason to believe that what appeals to us humans will appeal to all our primate relatives. For example, in 1990, the Animal and Plant Health Inspection Service (APHIS) proposed that monkeys that were on protocols which prevented social housing be provided instead with the opportunity to interact socially with humans. This suggestion assumes that monkeys would prefer interacting with humans to being left alone in their cages. However, the degree to which animals appear to enjoy our presence varies greatly from species to species, and even from individual to individual within a species. Even the animals that we believe we know best may be reactive to our presence, while we remain ignorant of this fact.

For example, for one of the studies we describe here, we assessed relative activity through the use of subcutaneously implanted radio transmitters. We tracked the activity levels of seven adult female rhesus macaques *(Macaca mulatta)* before, during, and after observation by a researcher. These animals had been part of an ongoing study for over a year and were presumably habituated to the presence of a human observer. However, activity levels were significantly lower ($F = 6.4$; df = 2, 6; $p < .05$) during the observation session than they were before or after the session; that is, animals moved about their cage least while their behavior was being recorded by a human observer. The meaning of these data (with respect to welfare) remains unclear. The suggestion is that even those animals with whom we interact daily may not behave normally in our presence, if what we mean by normally is behaving as they would in our absence. Even animals that we would consider to be generally unresponsive to humans, such as snakes, lizards, and tortoises, may in fact behave differently in the presence of specific individual caretakers or observers (Bowers and Burghardt 1992; Hayes et al., Chapter 13, this volume).

The changing behavior of animals in our presence suggests that they continue to be responsive to interactions with us. Thus, we appear to be faced with another commonsense paradox: Is interaction with humans desirable for nonhuman animals, or distressful? Obviously the type of interaction is a factor in this equation, as others have often pointed out (Hemsworth and Barnett 1987; Duncan 1992). Unpleasant or unwanted interactions with humans may evoke fear and aggression, with a concomitant cascade of physiological events that affect the health, reproductive status, and well-being of our charges. For the purposes of safe handling, and for healthy, breeding populations, the stress associated with unpleasant handling is to be avoided. At the same time, keeping repatriation in mind and wishing to prevent the domestication of zoo inhabitants over many generations, we may not want animals to become so habituated to human presence that they fail to demonstrate adaptive avoidance strategies when returned to the wild (Miller et al., Chapter 7, this volume).

Proposals made about the nature, frequency, and quality of human interactions with zoo animals that are based entirely on common sense are profoundly seductive because of the intimate association of our common sense with our deepest, most personal beliefs. However, as Peter Medawar (1979, 33) so aptly pointed out, "[T]he intensity of the conviction that a hypothesis is true has no bearing on whether it is true or not." To know if an hypothesis is true or not requires one to use the epistemology of *empiricism.* The power of the empiricist method is in its ability to tell us about the likelihood that the effects of an intervention are the result of chance alone. In so doing, the empirical method provides us with more information than does common sense. Specifically, empirical assessment provides us with some odds upon which to gamble. It allows us to predict with greater certainty whether our efforts to improve the well-being of our animal charges will be rewarded. Thus, when stakes are high, and maximizing effectiveness is imperative, empirical evaluation is essential.

COMBINING COMMON SENSE WITH EMPIRICAL ASSESSMENT

Despite its limitations, common sense nonetheless is helpful in generating ideas for improving the lives of animals in captivity. Particularly in the zoo setting, personal experiences are invaluable in informing any enrichment program. Keepers, curators, and researchers work closely with "their" animals every day and know them at a greater depth than can be provided by textbook accounts. They also know the constraints of the working environment and thus may be less likely to suggest improvements that are financially or physically impossible. Soliciting the ideas of animal caretakers and researchers has been fruitful in attempts to improve animal well-being. For example, using such a technique, Bayne (1988) generated a list of nearly thirty different suggestions for enriching the lives of laboratory-housed primates. Among these suggestions were several also recommended by APHIS (1990) in the first draft of regulations to enforce the 1985 amendment to the Animal Welfare Act. For several of these recommendations, the price of implementation is estimated to be in the billion-dollar range (Holden 1988). Thus, the stakes in the ability of advised changes to improve well-being include not simply enhanced survival and welfare of animals, but also the fiscal health of institutions that maintain animal colonies (Crockett, Chapter 9, this volume).

In our work at the California Primate Research Center (CPRC) in Davis, California, from 1987 to 1991, we endeavored to establish an active research program in the assessment and enhancement of psychological well-being. We began by developing relatively objective, reliable, and valid working definitions of well-being in captive primates based on behavioral and physiological indicators. The *behavioral indicators* that we used were (1) decreases in frequency of abnormal behaviors (pacing, rocking, bizarre postures, self-biting); (2) increases in frequency of species-typical behaviors (social grooming, foraging); and (3) shifts in frequencies of species-typical behaviors toward frequencies similar to those seen in free-ranging animals. The *physiological indicators* that we used were (1) decreases in baseline heart rate and plasma cortisol; and (2) decreases in degree of heart rate and cortisol reactivity to routine stressors (restraint, cage changes, etc.).

We do not suggest that our definitions are universally applicable, exhaustive, or exclusive; rather, we view them as a starting point, a base from which to generate further discussion. Our definitions are based in part on a knowledge of species-typical behavior in natural environments drawn from other authorities, in part on suggestions from this literature, and from our own common sense and that of our colleagues. Using these definitions, we assessed the effectiveness of a variety of environmental enrichment protocols, by recording their effects on a combination of behavioral and physiological variables. A sampling of our findings is described next.

INCREASING CAGE SIZE

Increasing cage size is one of the most intuitively obvious means by which to improve the lives of captive primates, particularly in laboratory settings. It is also one of the most costly. Building new cages is a substantial portion of these costs, but acquiring additional personnel to handle these larger cages will sustain the increased cost even after the larger cages are in place. Thus, the financial losses are high if increasing cage size has little or no effect on the goal—well-being. The high risks involved in increasing the cage size, in terms of both real dollars and the welfare of the animals, demand an objective evaluation of the effectiveness of this procedure before its imposition (Bowden 1988; Novak and Suomi 1988; Crockett, Chapter 9, this volume).

Because of these high costs, we investigated the effects of moving animals into larger cages (Line et al. 1989, 1990a). The subjects were ten adult female rhesus macaques, an average of five years old, all of whom had been born and raised at CPRC. Six of the animals had subcutaneous telemetry implants that allowed us to monitor heart rate and activity continuously. During this investigation, each animal was individually housed indoors in one of three sizes of squeeze-back stainless steel lab cages: a "standard" cage measuring 0.61 × 0.66 × 0.81 meter, a "larger" cage measuring 0.86 × 0.66 × 0.81 meter, and a "double" cage measuring 0.70 × 0.90 × 1.10 meters.

The behavior of each animal was recorded during two 25-minute sessions each week, for a total of four sessions per animal in each experimental or control condition. During these sessions, continuous recordings of both frequency and duration of behavior were made, using an automatic wand and bar-code system in conjunction with a laptop computer. Behavioral measures included the frequency of aggressive displays (cage-shaking or open-mouth threats to other monkeys or to the observer), submissive behavior (lipsmacking or grimacing), cage manipulation (oral or manual manipulation of any part of the cage), cooing, grunting, screaming, abnormal behavior (self-biting, self-holding, self-sucking, bizarre postures, plucking fur, head-tossing, drinking urine, and "saluting"), and durations of grooming, stereotyped locomotion (circling or pacing), rocking, sitting, and standing. Each monkey was placed in each of the two larger-sized cages twice for a one-week period each time, using a counterbalanced design to control for possible order effects.

Comparisons of behavior and baseline heart rate of subjects in each of the two larger cage sizes with behavior and heart rate in the standard cage size yielded no significant effects. The failure in this study to obtain results that might indicate improved well-being as a consequence of increased cage size may be caused by the relatively limited range of cage sizes employed. However, other workers have also found that increasing cage size fails to result in any measurable changes in behavior (Crockett et al. 1993; Crockett, Chapter 9, this volume), even if that increase is more than 600 times the standard size (Goosen 1988). The data obtained from these empirical assessments suggest that increasing cage size as a means by which to enrich and enhance an animal habitat may not be worth the cost, at least under conditions in which the size of the cage is the only aspect that is altered. It is our opinion that if significant resources are to be invested in improving living conditions for captive animals, then they should be invested in those protocols that have been shown to have demonstrably positive effects. In the case of cage size, common sense does not appear to have generated a protocol that has the desired effect.

INCREASING SOCIAL OPPORTUNITIES

One of the most frequent suggestions for improving the well-being of social primate species in captivity is to house them in species-typical groups (Reinhardt et al. 1987). At the very least, it is suggested that members of species that are not typically solitary be housed in compatible pairs. Certainly social housing provides the opportunity for expression of species-typical behaviors that are not available to singly housed animals. Also, currently, the law in the United States mandates social housing whenever possible for primates living in captivity (APHIS 1992). However, the risks of grouping animals that have been previously singly housed for some time are many (Line 1987; Novak and Suomi 1988). Death and injury as a result of fighting, and weakness and debilitation as a result of failure to compete adequately for food, are just some of the consequences of resocializing previously isolated captive primates that we have seen in our work. In our studies of attempts to resocialize singly caged laboratory and zoo primates, the effectiveness of such efforts in improving well-being is not clear (but see Reinhardt et al. 1987, 1988; Crockett et al. 1994; and Crockett, Chapter 9, this volume, for other opinions).

In one pilot investigation, for example, six juvenile rhesus macaques that had been gang-housed with peers were placed individually into cages with aged adult females. These adults had been displaying relatively high rates of abnormal behavior, and it was hoped that the companionship provided by a juvenile cage mate would reduce some of these stereotypies. In all but one case, the pairs formed in this manner were compatible. The adult partners in several of the pairs were observed cradling and grooming their younger cage mates. However, frequencies of abnormal behavior in the adults, as measured in daily fifteen-minute observation sessions, did not decrease. Rather, several of the juveniles were observed to adopt the "favorite stereotypy" of their adult cage mate. For example, one juvenile that previously showed no pacing began to pace when placed with an adult who frequently paced; the two often paced together. Thus, while we were successful in terms of forming apparently compatible pairs, we were not so fortunate in terms of reducing abnormal behavior. In fact, we unwittingly increased the number of animals showing some forms of it.

In another study, we formed an outdoor group of mixed-sex animals from thirteen aged rhesus macaques (six males and seven females; mean age, 23.3 years) (Line et al. 1990b). Before their participation in our study, these animals had been singly caged for an average of 13.1 years. Baseline data were collected on rates of abnormal and other behaviors, using the same bar-code and laptop computer protocol just described. In addition to these behavioral measures, we monitored a series of physiological variables, such as body weight, skin-fold thickness, nail growth rate, heart rate, blood pressure, and complete blood count. For the purposes of comparison, similar measures were obtained from a mixed-sex group of twelve younger animals (mean age, 10.5 years). The aged subjects had been born and spent the first several years of their lives in a large outdoor social group, but at the time of our study they had been housed indoors in single cages for an average of 5.5 years.

We prepared our aged subjects for social living by familiarizing same-sex members with one another. Familiarization was accomplished in a series of brief pair tests. A pair test consisted of taking two monkeys, each in a wire-mesh transport cage, to a quiet room and placing the two cages adjacent to one another, about 7 centimeters apart. For a half hour, several times a week, animals were "paired" in this way, close enough to see and, to a limited extent, to touch one another. Each animal was also released individually into the future group cage for several hours in the weeks before group formation to familiarize it to the new setting. Thus we did not attempt this resocialization without a considerable investment of preparation time and energy.

Nevertheless, group formation was not accomplished without some substantial losses. Fights were common on the first day, with rates of aggression as high as 8.41 fights per hour among the males and 2.1 fights per hour among the females. Although rates of fighting dropped by the second day (0.89 fight per hour among the males and 0.54 fight per hour among the females), within the first week one of the females had been killed, and three other animals (one female and two males) had to be removed for treatment of wounds or for depressed behavior.

In fact, the group had to be divided into two before stable hierarchies were achieved. On a more positive note, the animals that were able to remain in the group did show a wider range of species-typical behaviors, and abnormal behavior decreased in those animals that had frequently shown it when singly caged.

Among the physiological variables measured, only one—nail growth rate—showed a significant effect of group formation. Before group formation, the nail growth rate of the aged monkeys was significantly less than that of the younger comparison monkeys. However, after group formation, nail growth rate did not differ significantly between the aged and the younger animals. Nail growth rate is commonly less in aged macaques (Short et al. 1987) and has been correlated with some measures of immune function, such as natural killer cell activity (Coe 1989). Thus, one interpretation of this result is that the immune status of the aged animals was enhanced by exposure to the new physical and social environment. Alternatively, the minor wounds that animals sustained during fighting may have stimulated immune system activity and thus affected nail growth. In any case, resocialization of previously isolated animals was not without costs in terms of animal health, although there did appear to be benefits for at least some individuals.

Other studies by our research group on the effectiveness on animal well-being of increasing social opportunities also obtained mixed results. In one experiment, we modified ten standard stainless steel laboratory cages to allow the passage of adjacently caged animals from one cage to another when a barrier was removed. Twelve adult female rhesus monkeys were observed, using the computer-assisted technique described previously, in each of three conditions: (1) alone in their home cage with the barrier in place; (2) alone in their home cage with a perforated barrier in place (allowing animals to see and, to a limited extent, to touch one another); and (3) in pairs with the barrier between their home cage and that of their neighbors removed. A total of ninety minutes of observation were obtained for each animal in each condition. When animals were in the paired condition, a relative rank (high versus low) was determined based on the direction of aggressive and submissive behaviors observed outside of scheduled focal observations. In addition to behavioral measures, eight of the twelve animals were fitted with subcutaneous radiotelemetry implants that allowed us to monitor heart rate and activity continuously.

Although abnormal behavior decreased when the barriers between adjacent cages were removed, the reduction was not significant. A rank by cage condition interaction reached significance ($F = 5.53$; df = 1, 1; $p = .05$), with low-ranking animals showing less abnormal behavior in the paired caged condition, and much more in the single-cage condition. Autogrooming, a behavior that occurs among singly caged captive primates at much higher rates than it does in the wild, decreased in the paired-caging condition, although not significantly. Also, high-ranking animals did more self-grooming overall than did low-ranking animals ($F = 8.05$; df = 1, 1; $p = .02$). Heart rate was elevated significantly in the paired condition ($F = 12.58$; df = 2, 7; $p = .0007$), perhaps because of arousal resulting from pair formation.

A similar study conducted with twenty-two adult female cynomolgous monkeys *(Macaca fascicularis)* yielded more positive results (Line et al. 1991a). Using the same behavioral data collection method described earlier, 100 minutes of data were collected on each animal in one of three conditions: (1) alone in the home cage before removal of the barrier separating adjacent cages; (2) together with the adjacent cage mate in the first two weeks after the barrier was removed; and (3) two months after pairs were formed. Pair formation was associated with a significant decrease in frequency of abnormal behavior ($F = 4.88$; df = 2, 20; $p = .01$). Two months after pair formation, mean rates of abnormal behavior appeared to be increasing from the low levels reached in the first two weeks. There was no significant difference, however, between these two points in time. Autogrooming also decreased in frequency in the first two weeks following pair formation, and while its mean rate appeared to increase again two months later, this change was not significant. Finally, allogrooming significantly decreased two months after pair formation, compared to the rates seen in the first two weeks of social housing ($F = 16.71$; df = 1, 20; $p = .0006$). Changes in frequency of allogrooming two months after pair formation may reflect a return to equilibrium, following the exaggerated response of this variable to pair formation.

Most recently, under the guidance of one of us (Morgan), undergraduate Betsey Brewer of Wheaton College, in Norton, Massachusetts, completed a case study of the effects of pairing two zoo-maintained infant male monkeys of two different but sympatric species (a squirrel monkey, *Saimiri boliviensis,* and a capuchin monkey, *Cebus albifrons*) on the relative rate of abnormal behaviors displayed by these animals (Brewer and Morgan, manuscript in preparation). Because of maternal rejection, these infants had been hand-reared in small, indoor cages and isolated from other monkeys. Before the animals were paired, modified frequencies were obtained for the behaviors of rocking, pacing, head-tossing, self-clasping, foraging, grooming, and several social behaviors, in three weekly half-hour observation sessions. Data were collected in half-hour blocks, using a one-zero sampling technique (Altmann 1974), with thirty-second intervals. Animals were observed once a week for a total of nine weeks: for the three weeks immediately preceding their pairing, the three weeks immediately following their pairing in a larger indoor cage, and the subsequent three weeks after transfer to an even larger outdoor exhibit.

In the first three weeks after pairing, both animals showed species-appropriate prosocial behaviors, such as allogrooming or social play. The frequency of these behaviors decreased, however, in the second three weeks following the move to the outdoor exhibit. The mean number of intervals in which the capuchin infant was observed engaged in some prosocial behavior was 30.0 in the first three weeks after pairing, but only 21.3 in the second three weeks. Similarly, the mean number of intervals in which the squirrel monkey infant showed prosocial behaviors was 30.0 intervals in the first three weeks after pairing, but only 13.0 intervals in the second three weeks. While this decrease in socializing may be no more than a reflection of the distraction of the animals caused by the new cage setting, observations of increased aggression suggest another interpretation. The mean number of intervals in which the capuchin infant was observed engaging in some aggressive behavior was only 7.0 in the first three weeks after pairing, but rose to 18.0 in the second three weeks. Observations of the squirrel monkey infant in the first three weeks after pairing initially yielded no recordings of aggression, while in the second three weeks aggression was recorded in an average of 6.0 intervals. While the small number of subjects prohibited statistical analysis of the data, the trend suggested a shift in the animals' relationships with one another. And, in fact, the pair had to be separated after a few months.

Despite their incompatibilities, abnormal behaviors in these animals did decrease after social pairing. During the period that the monkeys were socially isolated, the mean number of intervals in which some abnormal behavior was recorded was 17.2 (range, 5-40). In the first three weeks of social housing, abnormal behavior initially decreased to an average of 8.3 intervals per observation. However, after the first three observation sessions it began to increase again (average number of intervals, 12.5). This was especially obvious in the squirrel monkey. When isolated, abnormal behavior in this animal was recorded in an average of 17.1 intervals per observation session. This fell to a mean of only 10.3 intervals in the first three weeks of social housing. In the last three weeks, however, the squirrel monkey's abnormal behavior increased again to an average of 15.1 intervals per observation session. This increase was concomitant with the increase in aggression directed at the squirrel monkey by the capuchin. Thus, as in the studies described earlier, pairing animals may have mixed effects on psychological well-being: decreasing abnormal behavior and increasing prospects for prosocial interactions while at the same time contributing to the stress of captivity by increasing an individual's exposure to inescapable aggression and potential injury. It may also be that demonstrable positive effects do not appear until after animals have been paired for some time and have worked out their dominance relationships.

In general, increasing social opportunities did increase the frequency of species-typical behaviors, thus normalizing behavior to a limited extent. The effects of this strategy on frequency of abnormal behavior, however, varied greatly. While for the most part compatible pairs could be formed successfully from simply housing randomly selected strangers with one another, aggression was not uncommon, and injuries sufficient to require hospitalization, as well as one death, resulted. The mixed results of these empirical assessments of the effects of social housing on the well-being of captive primates raises questions about how we are to improve the quality of life of our captive animals while at the same time protecting them from harm (Markowitz and Spinelli 1986). Objectively and carefully documenting the outcomes of attempts to resocialize isolated animals is one step toward answering these questions, and it is certainly a necessary step if we are to minimize the costs to the animals of providing them with a richer life.

PROVIDING TOYS OR OTHER ENRICHMENT APPARATUS

Simple Toys

Common sense suggests that providing a variety of simple toys designed for dogs or children is an inexpensive way to enrich the environments of captive nonhuman primates. The relatively low cost of introducing these objects in comparison to other proposed enrichment strategies makes it an attractive option and reduces the risk involved if this strategy is not successful. Among the data most commonly reported is use of the object by the animals, and such use has been well documented (Renquist and Judge 1985; Watson et al. 1989). Monitoring use of various toys may be all that is possible under some conditions, and so long as providing toys is not the exclusive enrichment protocol employed by an institution (see following), the low costs associated with this protocol may not require intensive proof of its effectiveness. However, if it could be shown that a particular object was associated with changes in variables that might be taken as indicators of improved well-being, then the argument for employing these objects as devices for environmental enrichment would be strengthened. In our work at the CPRC, we investigated the effects that access to a number of different toys had on the relative frequency of abnormal behavior displayed by singly caged nonhuman primates.

In a series of studies (Line and Morgan 1991; Line et al. 1991b), we tested the effectiveness of wooden sticks, nylon balls, and a variety of rubber dog toys in reducing frequency of abnormal behavior or increasing activity in singly caged rhesus monkeys. No statistically significant changes in any of the eighteen behavioral variables measured were attributable to the provision of a toy in any of these studies. Use of the objects also dropped radically after the first day of introduction, although captive-born animals used the objects significantly more often than did wild-caught animals. In one study (Line and Morgan 1991), sticks were used somewhat more often than nylon balls, primarily for gnawing. In addition, substantial individual differences were observed in the responses of animals to the toys and sticks. Taken together, these results suggest that, as others (Paquette and Prescott 1988) have recommended, frequent rotation of toys and novel objects might be necessary to sustain the interest of the animals. Mere access to simple toys had minimal effects on behavioral indicators of well-being. Thus, some of the inexpensive solutions to the problem of environmental enrichment may be limited in their effectiveness, indicating that additional efforts are required.

Interactive Devices

The relative lack of control that captive animals have over any aspect of their environment may be one of the experiences of captivity that they find most stressful (Markowitz 1982). In one of our enrichment projects, therefore, we strove to provide animals with some (albeit limited) control over aspects of their own immediate environs by providing each of ten adult female rhesus monkeys with a music-feeder device. This battery-operated device allowed the monkeys to turn a radio on and off, and to operate a food pellet dispenser by simply touching one of three rods that extended from the device into the animal's cage. The schedule of food delivery per effort expended by the monkey could be manipulated by caretakers, which helped to maintain the challenge of the task. Behavioral data were collected on all ten monkeys, using the bar-code and laptop computer system described previously. Baseline plasma cortisol levels and changes in blood cortisol in response to routine, stressful events were also obtained. In addition, six of the monkeys had telemetry implants that allowed us to monitor heart rate and activity. Twelve weeks of baseline data were collected and compared to data from the twelve weeks following the introduction of the device.

All monkeys began using the music-feeder apparatus rapidly after its introduction, and use increased dramatically over time. Use of the feeder rod increased from roughly 2,000 responses per day in the first week to more than 9,000 responses per day in the next week. The radio was also used throughout the twelve weeks of the experimental condition. Thus animals appeared to be very interested in interacting with the device, and this interest was sustained.

Abnormal behavior decreased significantly in the presence of this interactive apparatus (Wilcoxon test: $Z = 2.8$; $p = .005$). Nonstereotypic movement also increased, from an average of fifty-eight seconds in each observation session to a mean of eighty seconds in each session. Thus, providing animals with an interactive device that gives them control over some aspects of their environment does seem to have demonstrable effects on rates of abnormal behavior.

Physiological indicators suggest that such a device may also improve the ability to cope. Baseline cortisol was significantly lower when the apparatus was available to the animal ($F = 12.6$; df = 1, 9; $p < .005$), and heart rate reactivity to brief restraint in the squeeze mechanism of the home cage was also significantly less when the music-feeder was present ($F = 15.9$; df = 1, 5; $p < .003$). Taken together, the results of the music-feeder study suggest that providing more complex manipulable objects, particularly those that allow the animals control over some aspect of their otherwise limited environments, may be an especially effective strategy for improving captive environments.

DISCUSSION

The cumulative results from our work and the work of our students demonstrate the importance of empirical assessment in determining which of a variety of environmental strategies are likely to be most successful. We recognize that under some conditions such assessment is neither practical nor possible. We certainly do not advocate abandoning other ways of determining what might improve the lives of animals living in captivity. However, when the costs incurred in the implementation of a particular proposal will be high, or when the risks to the animals from failure are substantial, then empirical assessment is vital before the plan's implementation. Without such assessments, what may appear reasonable solutions to the difficult problem of improving the lives of captive animals and sustaining their populations may risk inadequacy. Perhaps Peter Medawar (1973, 110) said it best: "The welfare of animals depends upon an understanding of animals, and one does not come by this understanding intuitively—it must be learned."

CONCLUSIONS

There is more than one good method for gathering ideas for improving the well-being of captive animals, but only one—empiricism—allows us to draw conclusions about the relative effectiveness of these ideas. Using empirical assessment, some ideas for improving well-being in nonhuman primates have been shown to be effective while others have not. For example, observers, visitors, and caretakers may affect monkeys much more than has been previously believed or acknowledged. Increasing the size of a monkey's cage by as much as 100 percent may have little effect on levels of abnormal behavior. Simple toys or sticks may have a positive effect on the behavior of captive monkeys only for the first few days after these objects are provided. Interactive devices appear to sustain decreases in abnormal behaviors in nonhuman primates. Social pairing is associated with high health risks to monkeys and may yield mixed results in terms of improvements to well-being, and thus is a protocol that demands careful monitoring and assessment once in place. When effectiveness must be maximized, or when the costs or risks of failure are particularly high, empirical assessment is required.

ACKNOWLEDGMENTS

The research reported here was supported in part by National Institutes of Health grant RR00169, and by a grant from the California Primate Research Center. We thank the many students who assisted in data collection and analysis, especially Sharon Strong, Carmel Stanko, Mark Nakazono, Patt Stine, and Mike Riddell. Chris Tromborg, Warren Miller, Jim Wetterer, and Tommasina Gabriele made helpful comments on previous drafts of this manuscript. Thanks also to two anonymous reviewers for their suggested improvements, and to G. D. Mitchell for the use and support of his laboratory facilities in the preparation of this manuscript.

REFERENCES

Altmann, J. 1974. Observational studies of behaviour: Sampling methods. *Behaviour* 49:227–265.

APHIS (Animal and Plant Health Inspection Service). 1990. Proposed rules for animal welfare standards. *Federal Register* 55:33448–33531.

———. 1992. *Animal Welfare Regulations.* Document 311-364/60638. Washington, D.C.: U.S. Government Printing Office.

Bayne, K. 1988. Resolving issues of psychological well-being and management of laboratory nonhuman primates. In *Housing, Care, and Psychological Well-Being of Captive and Laboratory Primates,* ed. E. Segal, 183–199. Park Ridge, N.J.: Noyes Publications.

Bitgood, S., D. Patterson, and A. Benefield. 1988. Exhibit design and visitor behavior: Empirical relationships. *Environment and Behavior* 20:474–491.

Bowden, D. M. 1988. Primate research and "psychological well-being." *Science* 240:12.

Bowers, B. B., and G. M. Burghardt. 1992. The scientist and the snake: Relationships with reptiles. In *The Inevitable Bond: Examining Scientist-Animal Interactions,* ed. H. Davis and D. Balfour, 250-263. Cambridge: Cambridge University Press.

Coe, C. 1989. What immunology can tell us about primate behavior. Paper presented at the American Society of Primatologists 12th Annual Meeting, Mobile, Ala., August 27-30, 1989.

Crockett, C. M., C. L. Bowers, D. M. Bowden, and G. P. Sackett. 1994. Sex differences in compatibility of pair-housed adult longtailed macaques. *American Journal of Primatology* 32:73-94.

Crockett, C. M., C. L. Bowers, G. P. Sackett, and D. M. Bowden. 1993. Urinary cortisol responses of longtailed macaques to five cage sizes, tethering, sedation, and room change. *American Journal of Primatology* 30:55-74.

Dawkins, M. S. 1980. *Animal Suffering: The Science of Animal Welfare.* London: Chapman & Hall.

Donahoe, S. 1988. Visitor data is a three-way street. In *Visitor Studies 1988: Theory, Research, and Practice* [Proceedings of the 1st Annual Visitor Studies Conference], ed. S. Bitgood, J. T. Roper, and A. Benefield, 171-179. Jacksonville, Ala.: The Center for Social Design.

Duncan, I. J. H. 1992. The effect of the researcher on the behaviour of poultry. In *The Inevitable Bond: Examining Scientist-Animal Interactions,* ed. H. Davis and D. Balfour, 285-294. Cambridge: Cambridge University Press.

Falk, J. H., and L. D. Dierking. 1992. *The Museum Experience.* Washington, D.C.: Whalesback Books.

Goodall, J. 1971. *In the Shadow of Man.* Boston: Houghton Mifflin.

———. 1986. *The Chimpanzees of Gombe: Patterns of Behavior.* Cambridge: Belknap Press of Harvard University Press.

———. 1990. *Through a Window: My Thirty Years with the Chimpanzees of Gombe.* Boston: Houghton Mifflin.

Goosen, C. 1988. Developing housing facilities for rhesus monkeys: Prevention of abnormal behavior. In *New Developments in Biosciences: Their Implications for Laboratory Animal Science,* ed. A. C. Beyen and H. A. Solleveld, 67-70. Dordrecht: Martinus Nijhoff.

Hemsworth, P. H., and J. L. Barnett. 1987. Human-animal interactions. In *The Veterinary Clinics of North American Food Animal Practice.* Vol. 3, *Farm Animal Behavior,* ed. E. O. Price, 339-356. Philadelphia: W. B. Saunders.

Holden, C. 1988. Billion dollar price tag for new animal rules. *Science* 242:662-663.

Line, S. W. 1987. Environmental enrichment for laboratory primates. *Journal of the American Veterinary Association* 190:854-859.

Line, S. W., and K. N. Morgan. 1991. The effects of two novel objects on the behavior of singly caged adult rhesus monkeys. *Laboratory Animal Science* 41:365-369.

Line, S. W., K. N. Morgan, and H. Markowitz. 1991a. Pair formation among adult female long-tailed macaques *(Macaca fascicularis). American Journal of Primatology* 24:115-116.

———. 1991b. Simple toys do not alter the behavior of aged rhesus monkeys. *Zoo Biology* 10:473–484.

Line, S. W., K. N. Morgan, H. Markowitz, and S. Strong. 1989. Influence of cage size on heart rate and behavior in rhesus monkeys. *American Journal of Veterinary Research* 50:1523–1526.

———. 1990a. Increased cage size does not alter heart rate or behavior in female rhesus monkeys. *American Journal of Primatology* 20:107–113.

Line, S. W., K. N. Morgan, J. A. Roberts, and H. Markowitz. 1990b. Preliminary comments on resocialization of aged rhesus macaques. *Laboratory Primate Newsletter* 9:–12.

Markowitz, H. 1982. *Behavioral Enrichment in the Zoo.* New York: Van Nostrand Reinhold.

Markowitz, H., and J. Spinelli. 1986. Environmental engineering for primates. In *Primates: The Road to Self-Sustaining Populations,* ed. K. Benirschke, 489–498. New York: Springer-Verlag.

Medawar, P. B. 1973. *The Hope of Progress: A Scientist Looks at Problems in Philosophy, Literature, and Science.* New York: Anchor.

———. 1979. *Advice to a Young Scientist.* New York: Harper Colophon.

Novak, M. A., and S. J. Suomi. 1988. Psychological well-being of primates in captivity. *American Psychologist* 43:765–773.

Paquette, D., and J. Prescott. 1988. Use of novel objects to enhance environments of captive chimpanzees. *Zoo Biology* 7:15–23.

Reinhardt, V., W. D. Houser, S. G. Eisele, and M. Champoux. 1987. Social enrichment of the environment with infants for singly caged adult rhesus monkeys. *Zoo Biology* 6:365–371.

Reinhardt, V., W. D. Houser, S. G. Eisele, and D. Cowley. 1988. Behavioral responses of unrelated rhesus monkey females paired for the purpose of environmental enrichment. *American Journal of Primatology* 14:135–140.

Renquist, D. M., and F. J. Judge. 1985. Use of nylon balls as behavioral modifiers for caged primates. *Laboratory Primate Newsletter* 24:4.

Romanes, G. J. 1882. *Animal Intelligence.* London: Kegan, Paul, Trench, and Trubner.

Savage, A. 1988. Collaboration between research institutions and zoos for primate conservation. *International Zoo Yearbook* 27:140–148.

Shepherdson, D. 1994. The role of environmental enrichment in captive breeding and reintroduction of endangered species. In *Creative Conservation: Interactive Management of Wild and Captive Animals,* ed. G. Mace, P. Olney, and A. Feistner, 167–177. London: Chapman & Hall.

Shettel-Neuber, J. 1988. Second- and third-generation zoo exhibits: A comparison of visitor, staff, and animal responses. *Environment and Behavior* 20:452–473.

Short, R., D. D. Williams, and D. M. Bowden. 1987. Cross-sectional evaluation of potential biological markers of aging in pigtailed macaques: Effects of age, sex, and diet. *Journal of Gerontology* 42:644–654.

Tudge, C. 1991a. *Last Animals at the Zoo.* Washington, D.C.: Island Press.

———. 1991b. The buzz word in the best zoos is "behavioral enrichment" or how to make a captive environment more like the wild. *New Scientist* 129:26–30.

Watson, D. B. B., J. Houston, and G. E. Macallum. 1989. The use of toys for primate environmental enrichment. *Laboratory Primate Newsletter* 28:20.

11

KATHY CARLSTEAD

DETERMINING THE CAUSES OF STEREOTYPIC BEHAVIORS IN ZOO CARNIVORES

Toward Appropriate Enrichment Strategies

The technique of environmental enrichment seeks to meet the behavioral needs of zoo animals by providing them with naturalistic surroundings. Hediger (1934) long ago pointed out that we can better understand the behavioral needs of captive animals by studying their stereotyped motor reactions. The term *stereotypy* refers to behavior that is characteristically repetitive, is invariant in form, and has no obvious goal or function (Odberg 1978). Stereotypies develop in a wide variety of situations and across a broad range of species, including humans, and are heterogeneous in origin, proximate causation, and appearance (Mason 1991a). The broadest generalization one can make about the causes of stereotypy is that they are a result of an abnormal organism–environment interaction.

Some types of stereotypy seem to originate in response to pathological conditions of the organism, such as congenital defects, abnormal development, or disorders of higher-order stimulus-processing mechanisms. For example, stereotypy may develop in humans having severe mental retardation, autism, schizophrenia, and drug addiction. Isolation-reared or developmentally deprived animals may develop self-directed stereotypies such as toe sucking, rocking, and self-clasping, as often seen in laboratory primates (Berkson 1967). Many other types of stereotypy are environmentally induced, developing in situations where the animal is normal but the environment in which it lives is in some way less than optimal. These are sometimes referred to as "cage stereotypies" (Ridley and Baker 1982) and are probably the most common type observed in zoo animals. Monika Holzapfel (1938, 1939) was one of the first researchers to describe some of the behavior patterns that develop into repetitive movements in zoo animals and the situations in which this behavior is elicited.

The function of stereotypic behavior in captive animals is a widely debated issue. One strongly supported hypothesis is that the performance of stereotypic behavior physiologically or psychologically ameliorates a suboptimal environment. Stereotypies often develop in situations known, from independent behavioral and physiological evidence, to be aversive and stressful, such as low stimulus levels, physical restraint, inability to escape from fearful situations, or frustration (Mason 1991b). For this reason stereotypic behavior has long been considered an indication of poor welfare (Broom 1983; Wiepkema 1983). The "coping" hypothesis of stereotypies maintains that stereotypic behavior is a response to aversive or stressful conditions and that in some way the performance of stereotypies reduces the level of arousal the animal experiences (reviewed by Mason 1991a,b).

Research in recent years, however, has provided equivocal evidence for the coping hypothesis, and it is unlikely that all stereotypies are responses to stress. For example, research by de Pasille et al. (1991) on calves *(Bos taurus)* indicated that some types of oral stereotypy (nonnutritive sucking) affect digestive hormone secretion, suggesting that physiological systems other than those involved with stress may be important in the function of stereotypies. Furthermore, it is often difficult to demonstrate that individual animals who develop stereotypies in a given housing situation are in some way worse or better off than those individuals that, in the same situation, do not do so (reviewed by Ladewig et al. 1993). Individual behavioral styles play an important role in determining which individuals will develop stereotypy (Schouten et al. 1991). One can only conclude that, in general, the origin of stereotypy is complex and a variety of functional reasons may be responsible for the development of these behaviors.

In zoos, where animals are exhibited, stereotyped behavior is a problem because it denies the public an appreciation of how an animal's behavior is adapted to its natural environment (Hutchins et al. 1984; Shepherdson 1989). While many functional questions about stereotypy cannot yet be satisfactorily answered, we do know that appropriate environmental enrichment often will reduce stereotypy in confined animals that have been normally reared (Hediger 1950; Morris 1962). Therefore, it is worthwhile to identify why some environmental enrichment reduces stereotypy.

In most zoo animals, stereotypic activity arises from a primary behavior pattern that, over a period of time, the animal has repeatedly been motivated to perform. This occurs, however, in environments that do not allow this primary behavior to reach a normal endpoint (Holzapfel 1939; Morris 1964; Fentress 1976; Cronin 1985). Often the appropriate external stimuli that would allow the behavior to be coupled with its appropriate consequences are lacking. Even though the functional consequences of actively expressing the motivation are minimal in the absence of these stimuli, the motivation to perform these primary behaviors remains (Hughes and Duncan 1988). The primary behavior patterns are often appetitive; that is, these are the activities an animal performs in a natural environment when it is looking for some particular external stimulus. Examples include foraging for food, searching for a mate, escaping to a safe place, or seeking distance or space from conspecifics.

For stereotypies that originate as appetitive motivation, appropriate environmental enrichment would appear to be the means of eliminating these behaviors because it could provide the stimulation the animal is seeking. To provide such enrichment requires that we determine (1) what kinds of behaviors animals are motivated to perform and (2) what sorts of external stimuli can be provided to functionally satisfy this motivation. These questions can be answered by studying the diverse causes of stereotypy in zoo animals. Detailed analyses of stereotypic behaviors are needed, and they will perhaps always be needed because generalizing from situation to situation about the specific causes of these behaviors is likely always to be difficult.

For a number of years I have studied stereotypic movement patterns in several carnivore species. This chapter focuses on carnivores because stereotypy is widespread and common in this taxon. Evidence presented from bears, cat species, fennec foxes *(Vulpes zerda),* and ranch mink *(Mustela vison)* indicates that stereotypy in a given species can vary qualitatively or temporally and that these variations indicate different moods or motivation. A number of examples are also given of reduction or elimination of stereotypies by appropriate environmental enrichment.

CAUSAL FACTORS OF STEREOTYPIC BEHAVIOR

There is a large experimental literature on the causal factors of repetitive, non-goal-oriented behaviors in laboratory animals. Earlier work by experimental psychologists examined the effects of testing animals for varying periods of time in a chamber in which they were subjected to a noncontingent, intermittent schedule of food delivery (Falk 1971; Staddon 1977). At regular intervals, for example, every four minutes, a small amount of food would be delivered to a hungry rat, pigeon, or pig. The animal did not have to do anything to get the food except wait for it. With experience in this setup the animals learned to predict how long the interval was between feedings and to anticipate food arrival. In this situation two general types of "adjunctive activities" were found to be induced in the tested animals that differed both qualitatively and according to their time allocation. "Interim activities" occurred predominantly at times when food was absent and ceased before the next food delivery, and "terminal activities" occurred maximally just before food delivery. Components of feeding behavior were facilitated just before food delivery while other nonfeeding behaviors like drinking or chain-chewing emerged when the probability of food delivery was low during the interim period.

These adjunctive behaviors fit the criteria for stereotypies: They are consistent in form within an individual, they are performed repetitively, and they have no apparent function or goal. Although psychologists are interested in these behaviors for what they can learn about classically conditioned responses and their relationship to compulsive behavior, two major conclusions can be drawn from this research: (1) Parceling out food to an animal in a rigidly controlled, highly predictable, stimulus-sterile environment can cause it to develop stereotypies; and (2) at least two types of motivation, food-anticipatory and nonfeeding, are expressed as stereotypy in this kind of environment. This conclusion is relevant for confined, managed zoo animals because they are also most often on an essentially comparable schedule of noncontingent food delivery. They are out of their natural environment and living in restricted, relatively stimulus-poor surroundings in which they are fed in a highly predictable manner that requires little or no effort to obtain food. The interval between food deliveries, instead of being four minutes as in the preceding example, may be twenty-four hours.

If one looks at the twenty-four hour temporal pattern of stereotypic behaviors in relation to feeding in zoo carnivores maintained in traditional enclosures and with traditional feeding methods, a similar pattern of terminal and interim activities can be discerned in many cases. For example, such data on stereotypic pacing by each of four cat species in outdoor exhibits at the National Zoological Park in Washington, D.C.—jaguar *(Panthera onca)*, puma *(Felis concolor)*, leopard *(Panthera pardus)*, and serval *(Leptailurus serval)*—showed that the pacing occurred mainly before and during the hours that keepers placed food in the enclosure. The pacing was always at a spot in the enclosure from where the animal could see the keeper's approach. Pumas, and servals to a slight degree, showed another peak in pacing in the late afternoon. Ranch mink, an American black bear *(Ursus americanus)*, and a Geoffroy's cat *(Oncifelis geoffroyi)* showed similar patterns of stereotypy, peaking between zero and three hours before the daily feeding, with a secondary peak ten hours after feeding for the mink (de Jonge et al. 1986) and at various times during the afternoon and night for the bear and cat.

The implication is that, as in laboratory experiments, zoo animals on a regular schedule of food delivery might be exhibiting food-anticipatory stereotypies that correspond to schedule-induced terminal activities. Also, the same individual or species may express other, non-food-related stereotypies between feedings (interim activities). This does not mean that all zoo animals with a stereotypy show this pattern. However, zoo managers should be aware of the potential effects of the method and schedule of feeding. More detailed studies of stereotypy would be useful to delineate the motivational components of abnormal, repetitive behaviors and to emphasize the need to modify traditional feeding practices that are highly predictable and are not contingent on the animal's behavior. In particular, there is a need to study the following: (1) the extent to which the feeding schedule or feeding method is inducing stereotyped activity, and (2) the causes of stereotypy that occur at times other than feeding (see Mellen et al., Chapter 12, this volume).

A starting point in determining the causes of stereotypy is to ask several questions: Are feeding and nonfeeding stereotypies qualitatively different in form? What normal, natural behaviors do they resemble? What does this tell us about potentially different motivations for performing stereotypy? To answer such questions about the causes of stereotypy in zoo animals, one can examine daily and seasonal variation in the time allocated by the animals to various activities and to stereotypic behavior, examine the effects of environmental manipulation on the frequency of stereotypy, and carry out individual and species comparisons in which variation in natural behavior patterns is related to variation in stereotypic behavior. These methods are illustrated here by two studies, one with ranch mink and the other with an American black bear.

On Dutch mink farms, 149 mink were each observed for twenty-four hours via time-lapse video. About half these animals displayed a distinctly stereotyped, idiosyncratic movement pattern (de Jonge et al. 1986). The plot of the hourly frequency of stereotypy for all animals combined showed the highest occurrence from zero to three hours before feeding, as mentioned, with another, smaller peak occurring nine to ten hours after feeding. However, each individual could be classified into one of three general categories based on the movement patterns it performed: (1) running back and forth along a bottom edge of the cage, (2) figure eights or head circles along a cage side, or (3) complex movement patterns directed toward the nest box and cage top where the food was placed. Plots of the temporal pattern for each category of stereotypy also showed that all three peaked before feeding. However, animals with running and figure eight stereotypies sharply decreased their activity when food was delivered, whereas animals with complex, food-directed stereotypies remained active at a high level after food was placed on the top of the cage. As for the occurrence of stereotypies between feedings, only the running stereotypies peaked again eight to ten hours after feeding.

In this study, however, we were unfortunately never able to learn more about the different motivations for these mink stereotypies other than that older animals have the most complex forms. In a much longer term study of mink stereotypies, Mason (1993) found that different stereotypic movements appear in different contexts, and that they become less variable and more frequent with age. Mink are very active hunters, generalist carnivores taking a large range of prey in water, on land, and down burrows. Their predatory behavior is necessarily adaptable and diverse. Further investigation is needed to determine whether idiosyncratic stereotypic movement patterns are expressions of different components of the complex foraging, stalking, ambushing, prey-catching, and killing behavior patterns seen in this species, and whether other nonfeeding motivations, such as escape, are also expressed as stereotypic movements (Mason 1993).

Smokey, a singly housed male American black bear, exhibited a very strong stereotyped pacing habit along the edge of his exhibit at the National Zoological Park typified by seventeen steps in one direction, turn, seventeen steps back, turn, etc. When behavioral data over the course of a year were combined, the temporal pattern of pacing showed a prefeeding peak and another peak in the afternoon. However, when the pacing patterns were analyzed by season, these two peaks were found to predominate at different times of year. In June and July, most pacing occurred in the afternoon and evening, after feeding, but from September through November most pacing occurred before feeding. The orientation of the turns the bear made also varied between seasons; the bear may turn inward toward the area from where the keeper approaches with food (47 percent in summer, 100 percent in fall), or outward toward the public area and other bear exhibits.

The breeding season for black bears is in the late spring and early summer when males travel through their home ranges associating with females and interacting with competing males. They spend the rest of the summer and fall eating to build up fat reserves; animals in the wild normally forage for as much as eighteen hours a day during these seasons (Eagle and Pelton 1983). We speculated that pacing in June and July, characterized by an equal number of inward and outward turns and predominantly postfeeding occurrence, resulted largely from mate-seeking motivation. Pacing later in the year, characterized by 100 percent inward turns and prefeeding occurrence, resulted from foraging motivation (Carlstead and Seidensticker 1991). Both these examples, the mink and black bear, demonstrate that within a species or an individual at least two possible types of appetitive motivation for stereotypy can be discerned.

PREVENTION OF STEREOTYPIC BEHAVIOR

It is known that environmental enrichment can, in many cases, reduce stereotypy. It does so in a variety of ways: for example, by providing the stimuli an animal is motivated to seek out, by reducing the motivation to perform a behavior, or by making the environment more unpredictable and variable. In the wild, many species spend most of their waking hours looking for, pursuing, gathering, handling, or hiding food, or looking for or avoiding conspecifics or predators. In many captive situations animals are fed, by human caretakers, in one or several daily meals, expend no effort to acquire food, and consume it quickly. Their social and physical environment is mostly unchanging. Examples follow from my own research with carnivores that demonstrate the effects on stereotypy of providing more unpredictable, complex environments.

Experiments were carried out to see if providing relevant novel stimuli to the environment would reduce the stereotypic pacing of Smokey, the American black bear. Olfactory stimuli from other bears were provided during the period of mate-seeking, and foraging opportunities were provided during both the breeding season and the "foraging" season (Carlstead and Seidensticker 1991). On a number of days during May and June, commercially available hunting lures made of male and female bear urine were sprayed on objects in the enclosure. Pacing frequency was significantly reduced, by 33 percent, compared to days in the same season before scents were sprayed. Hiding most of the bear's daily food ration throughout the exhibit in substrates through which he had to forage or dig to obtain his food also reduced pacing by 25 percent during the breeding season, but was more effective in reducing pacing, by 77 percent, during late summer through fall (Carlstead and Seidensticker 1991). It should be added that *foraging opportunity,* that is, being able to manipulate objects and substrates to find food, was the critical stimulation that virtually eradicated this bear's stereotypy during the summer and fall months. In another experiment, the bear was provided the same amount of food at random intervals from an automated feeder tree; he needed only to lick up the food from one of six wells on this tree, and his pacing was not reduced (Carlstead et al. 1991).

We were able to provide appropriate additional stimuli to the environment of this black bear to reduce his pacing behavior. Despite the small sample size, one can interpret these data to suggest that increasing foraging opportunities in the bear's yard provided the stimulation necessary to carry out his intrinsic foraging behavior. Spraying bear scents in the yard during the breeding season may have likewise functioned to provide novel stimulation relevant to his sexual motivation.

Perhaps housing with another, compatible bear would also reduce or prevent pacing during those months.

Another example of how enrichment may provide environmental stimuli animals are motivated to seek out, as indicated by their stereotypy performance, is an experiment with four individually housed leopard cats *(Prioailurus bengalensis)*. When housed in a barren enclosure in a building that also held lions *(Panthera leo)*, tigers *(Panthera tigris)*, and pumas, stress levels (measured by urinary cortisol) and stereotypic pacing frequency were chronically elevated in these cats. When the bare enclosure was enriched with several hollow logs, boxes, branches, and platforms, cortisol levels and stereotypic pacing significantly declined. We hypothesized that stress and pacing were reduced because the cats previously had no places in which to hide in the presence of perceived predators (the other big cats in the building) (Carlstead et al. 1993). We speculated that hiding behavior motivated by "fear" could be consummated when the environmental feature the cats were seeking out was provided to them.

Enrichment may also serve to reduce the motivation to perform certain behaviors. Fennec foxes experienced many environmental events that precipitated stereotypic running in the males of two pairs at the National Zoological Park. Each male responded to these factors to a differing extent. Postfeeding stereotypies were common in these animals: Male 1 performed them for a mean of twenty-five minutes in the hour after feeding and male 2 for thirteen minutes. Postfeeding stereotypy in these animals was believed to be related to the motivation to cache food, as fennec foxes do in the wild with uneaten prey. In their exhibits, however, there was insufficient space or substrate for proper food-caching. To reduce the motivation to cache, food was presented experimentally in small chunks rather than in one large ball as it was usually given. The animals were then not confronted with an oversized kill that needed caching. This method significantly reduced the postfeeding stereotypy of male 1, who showed the highest levels of this behavior, to a mean of six minutes; no effect on male 2 was observed (Carlstead 1991).

The previously mentioned role of imposed, regular feeding schedules in inducing stereotypic behavior suggests that varying the schedule is another way to modulate stereotypy in captive animals. In a further experiment with the same four leopard cats, the feeding schedule was made more unpredictable. Even though their physically enriched environment reduced stereotypic pacing, they still paced very often starting after feeding time in the early afternoon and continuing all evening and night until the keepers returned in the morning. Because this pattern seemed related to the predictable single daily feeding event, feeding was increased from once to four times per day. The increase in frequency of feedings led to a significant reduction in the total amount of pacing and a significant increase in exploratory behavior (Shepherdson et al. 1993). Further, hiding their food four times per day in a pile of bushes (instead of just placing the food in a pan on the floor) led to a qualitative change in pacing. With this method of feeding, pacing bouts were significantly shorter and exploring bouts significantly longer than with nonhidden feedings. In traditional feeding practices, the animals receive food in a noncontingent manner. However, more naturalistic approaches involve hiding food and making animals work to get it. There are many examples of this in the literature, and since the early 1980s (Hancocks 1980; Hutchins et al. 1984) the practice has been gradually but increasingly encouraged in zoo animal management.

CONCLUSIONS

Although based on small sample sizes, all these data appear to indicate that the method and frequency of feeding is of primary importance in carnivore stereotypies. Food delivery can be manipulated in a variety of ways, depending on the species, to reduce stereotypy associated with food anticipation, foraging, and caching motivation. If the food has been hidden, feeding is contingent on the animal's own behavior rather than its anticipation of when the keeper will bring the food. The animal thus is released from a very unnatural way of obtaining food: an intermittent schedule of noncontingent food delivery, as discussed at the beginning of this chapter. Many environmental enrichment treatments have been concerned with directing appetitive activity, which in the wild would be devoted to foraging, to actual or simulated food acquisition. As Hediger (1966) pointed out, feeding is psychological as well as physiological; he recognized the need to provide animals with some kind of substitute for their normal foraging activity.

The data presented here also suggest that zoo professionals should look for ways of providing environments that allow the animals to remove themselves from stressful or disturbing stimuli. Much stereotypic pacing and running may arise from a motivation to escape from a threatening situation. Escape is also an appetitive behavior; if there is no appropriate hiding place to satisfy the motivation, the animal can only run repetitively back and forth.

Finally, a detailed analysis of stereotypic behavior patterns may allow animal managers to also consider the other possible types of stimuli (e.g., social, nesting, and grooming substrates) that are lacking in the environment but which an animal might be repeatedly motivated to seek out, as evidenced by its stereotypy patterns.

Individuals of carnivore or other species will vary in genetic background, rearing and social experience, coping styles, and reactivity to their physical and social environment. Such factors can be expected to affect the development and occurrence of stereotypy in a given confined environment. Stereotypic behavior is a good indicator that an environment is suboptimal for a particular individual, so that it prevents that animal from performing some highly motivated behavior. In a zoo, the public always perceive an animal with a stereotypy as being stressed, anxious, bored, or frantic, and the presence of these behaviors detracts from the educational value of the exhibited animal. In zoo animals, these undesirable behaviors can be reduced or eliminated with careful behavioral analyses and appropriate environmental enrichment.

REFERENCES

Berkson, G. 1967. Abnormal stereotyped motor acts. In *Comparative Psychopathology: Animal and Human,* ed. J. Zubin and H. F. Hunt, 76–94. New York: Grune & Stratton.

Broom, D. M. 1983. Stereotypies as animal welfare indicators. In *Indicators Relevant to Farm Animal Welfare,* ed. D. Schmidt, 8–87. The Hague: Martinus Nijhoff.

Carlstead, K. 1991. Husbandry of the Fennec fox, *Fennecus zerda:* Environmental conditions influencing stereotypic behaviour. *International Zoo Yearbook* 30:202–207.

Carlstead, K., J. L. Brown, and J. Seidensticker. 1993. Behavioral and adrenocortical responses to environmental changes in leopard cats *(Felis bengalensis). Zoo Biology* 12:321–331.

Carlstead, K., and J. Seidensticker. 1991. Seasonal variation in stereotypic pacing in an American black bear *Ursus americanus. Behavioural Processes* 25:155–161.

Carlstead, K., J. Seidensticker, and R. Baldwin. 1991. Environmental enrichment for zoo bears. *Zoo Biology* 10:3–16.

Cronin, G. M. 1985. The development and significance of abnormal stereotyped behaviours in tethered sows. Ph.D. dissertation, Agricultural University of Wageningen, The Netherlands.

de Jonge, G., K. Carlstead, and P. R. Wiepkema. 1986. *The Welfare of Ranch Mink.* COVP Issue 08. Beekbergen, The Netherlands: Het Spelderholt.

de Pasille, A. M. B., R. J. Christopherson, and J. Rushen. 1991. Sucking behaviour affects post-prandial secretion of digestive hormones in the calf. In *Applied Animal Behaviour: Past, Present, and Future,* ed. M. C. Appleby, R. I. Horrel, J. C. Petherick, and S. M. Rutter, 130–131. Potters Bar, U.K.: Universities Federation for Animal Welfare.

Eagle, T. C., and M. R. Pelton. 1983. Seasonal nutrition of black bears in the Great Smoky Mountains National Park. *International Conference on Bear Research and Management* 5:94–101.

Falk, J. L. 1971. The nature and determinants of adjunctive behavior. *Physiology and Behavior* 6:577–597.

Fentress, J. C. 1976. Dynamic boundaries of patterned behaviour: Interaction and self-organization. In *Growing Points in Ethology,* ed. P. P. G. Bateson and R. A. Hinde, 135–167. Cambridge: Cambridge University Press.

Hancocks, D. 1980. Bringing nature into the zoo: Inexpensive solutions for zoo environments. *International Journal for the Study of Animal Problems* 1:170–177.

Hediger, H. 1934. Über Bewegungsstereotypien bei gehaltenen Tieren. *Revue Suisse de Zoologie* 41:349–356.

———. 1950. *Wild Animals in Captivity: An Outline of the Biology of Zoological Gardens.* New York: Dover.

———. 1966. Diet of animals in captivity. *International Zoo Yearbook* 6:37–58.

Holzapfel, M. 1938. Über Bewegungsstereotypien bei gehaltenen Saugern. I. Mitt. Bewegungsstereotypien bei Caniden und Hyaena. *Zeitschrift für Tierpsychologie* 2:46–72.

———. 1939. Die Entstehung einiger Bewegungstereotypien bei gehaltenen Saugern und Volgeln. *Revue Suisse de Zoologie* 46:567–580.

Hughes, B. O., and I. J. H. Duncan. 1988. The notion of ethological "need," models of motivation, and animal welfare. *Animal Behaviour* 36:1696–1707.

Hutchins, M., D. Hancocks, and C. Crockett. 1984. Naturalistic solutions to the behavioral problems of captive animals. *Zoologische Garten* 54:28–42.

Ladewig, J., A. M. B. de Pasille, J. Rushen, W. Schouten, E. M. C. Terlouw, and E. von Borell. 1993. Stress and the physiological correlates of stereotypic behaviour. In *Stereotypic Animal Behaviour: Fundamentals and Applications to Welfare,* ed. A. B. Lawrence and J. Rushen, 97–118. Wallingford, U.K.: CAB International.

Mason, G. J. 1991a. Stereotypies: A critical review. *Animal Behaviour* 41:1015–1037.

———. 1991b. Stereotypies and suffering. *Behavioural Processes* 25:103–115.

———. 1993. Age and context affect the stereotypies of mink. *Behaviour* 127:191–229.

Morris, D. 1962. Occupational therapy for captive animals. In *The Environment of Laboratory Animals,* vol. 2, 7–42. Carshalton, U.K.: MRC Laboratories.

———. 1964. The response of animals to a restricted environment. *Symposia of the Zoological Society of London* 13:99–120.

Odberg, F. 1978. Abnormal behaviours (stereotypies). In *Proceedings of the First World Congress on Ethology Applied to Zootechnics,* ed. J. Garsi. Madrid: Industrias Graficas.

Ridley, R. M., and H. F. Baker. 1982. Stereotypy in monkeys and humans. *Psychological Medicine* 12:61–72.

Schouten, W., J. Rushen, and A. M. de Passille. 1991. Stereotypic behavior and heart rate in pigs. *Physiology and Behavior* 50:617–624.

Shepherdson, D. 1989. Stereotyped behaviour: What is it and how can it be eliminated? *Ratel* 16:100–105.

Shepherdson, D., K. Carlstead, J. M. Mellen, and J. Seidensticker. 1993. The influence of food presentation on the behavior of small cats in confined environments. *Zoo Biology* 12:203–216.

Staddon, J. E. R. 1977. Schedule-induced behavior. In *Handbook of Operant Behavior,* ed. W. K. Honig and J. E. R. Staddon, 125–152. Englewood Cliffs, N.J.: Prentice-Hall.

Wiepkema, P. R. 1983. On the significance of ethological criteria for the assessment of animal welfare. In *Indicators Relevant to Farm Animal Welfare,* ed. D. Schmidt, 71–79. The Hague: Martinus Nijhoff.

12

JILL D. MELLEN, MARC P. HAYES,
AND DAVID J. SHEPHERDSON

CAPTIVE ENVIRONMENTS FOR SMALL FELIDS

Enrichment plans for captive carnivores, in contrast to those for other animals, continue to be among the most difficult to develop. If a major goal of enrichment is to promote behaviors and activities characteristic of their wild counterparts (Seidensticker and Forthman 1994; see also Shepherdson, Chapter 1, this volume), then, among carnivores, providing enrichment opportunities for solitary felids is perhaps the most problematic. Wild felids acquire food using a "stalk-rush-kill" sequence (Leyhausen 1979) and dispose of their prey with either a killing bite to the back of the neck or a suffocating bite to the trachea (Ewer 1973). Welfare concerns about the prey and negative reactions of the public to providing live prey preclude feeding live birds or mammals to felids in most North American zoos (Shepherdson et al. 1993). Moreover, concern with providing a nutritionally balanced diet for captives has resulted in development of a prepared meat diet that only remotely resembles their natural prey. Because genetics and demography guide breeding programs (e.g., Species Survival Plans) and enclosure space in zoos is limited, breeding and rearing of offspring are tightly controlled. Wild felids range over large distances to find prey or potential mates and use scent marks and vocal displays to communicate with conspecifics. In captivity, enclosures are invariably smaller than natural home ranges and contact with conspecifics includes only a few individuals maintained in close proximity.

Given these significant constraints, how can exhibits be optimally designed for captive felids? What exhibit components encourage behaviors more characteristic of wild felids, and which components discourage behaviors perceived as undesirable by the public, such as pacing? An equally important question: How do we assess the adequacy of captive environments? This chapter evaluates variability in the housing of small cats (see Table 12.1 for the species list) in an attempt to begin answering these questions. Specifically, the amount of time spent in various activities was evaluated for sixteen felid taxa across a range of exhibit designs. Because time spent pacing may be a useful measure of well-being (Mason 1993), the relationship between pacing and several variables that characterize the physical and social milieu of small captive felids was investigated.

METHODS

Subjects

One of us (Mellen) observed sixty-eight individuals representing sixteen small felid taxa during a total of 465.89 hours. Table 12.1 provides details regarding the taxa observed and group composition (based on sex and group size).

Data Collection

Scan sampling at thirty-second intervals (Altmann 1974) was used to estimate the percentage of time spent in various activities. Observation periods ranged from forty-five to ninety minutes in length and were made during regular visitor hours at the respective zoos between 9 A.M. and 5 P.M. (local time). These data were collected during daylight hours because of a constraint of the original study for which they were obtained, namely, to determine whether reproductive behaviors occurred during hours when zoo personnel could monitor them (Mellen 1989). Table 12.1 also provides details regarding the dates and numbers of hours of observation as well as the institutions at which observations were made. Eleven behavioral categories were recorded (Mellen 1989), but data from only four (pacing, resting, resting but alert, and not visible) are reported here. Pacing, the only behavioral category analyzed by itself, was defined as a repetitive ambulatory movement; the minimum criterion used to identify pacing was traversing the same pathway at least twice. (See Mellen 1989 for definitions of the remaining ten behavioral categories.) Data were also obtained on seven independent variables that characterized significant features of each cat's captive environment (Table 12.2).

Data Analysis

A measure of activity was derived from the eleven behavioral categories Mellen (1989) had originally scored. Collectively, all categories summed represented the

Table 12.1
Information on Felids Observed

Enclosure number	Species	No. of animals (male.female)	Institution[a]	Observation interval (dates)	Total hours[b]	Enclosure size (m^3)
1	Pampas cat, *Oncifelis colocolo*	1.1	CIN	17.10.88-28.10.88	3.25	27
2	Pampas cat, *Oncifelis colocolo*	1.1	CIN	17.10.88-28.10.88	4.17	6
3	Jungle cat, *Felis chaus*	1.1	SAC	17.02.85-27.04.85	37.47	12
4	Geoffroy's cat, *Oncifelis geoffroyi*	1.1	WPZ	02.11.86-10.01.87	25.63	65
5	Geoffroy's cat, *Oncifelis geoffroyi*	1.1	NZP	07.07.85-05.02.85	27.69	18
6	Pallas' cat, *Octocolobus manul*	1.2	BRK	24.06.84-18.08.84	25.50	782
7	Sand cat, *Felis margarita*	1.2	BRK	24.06.84-18.08.84	7.48	229
8	Sand cat, *Felis margarita*	1.1	BRK	24.06.84-27.07.84	16.59	15
9	Sand cat, *Felis margarita*	1.1	BRK	24.06.84-27.07.84	14.37	7
10	Sand cat, *Felis margarita*	1.1	WPZ	23.02.86-10.05.86	21.71	62
11	Black-footed cat, *Felis nigripes*	1.1	CIN	17.10.88-28.10.88	4.00	37
12	Ocelot, *Leopardus pardalis*	1.1	ASM	27.01.88-05.02.88	6.00	130
13	Ocelot, *Leopardus pardalis*	1.1	CIN	17.10.88-28.10.88	3.90	41
14	Ocelot, *Leopardus pardalis*	1.1	PTL	02.04.89-15.04.89	6.50	255
15	Rusty-spotted cat, *Prionailurus rubiginosus*	1.1	CIN	17.10.88-28.10.88	3.50	6
16	Serval, *Leptailurus serval*	1.1	WPZ	18.01.87-28.03.87	13.15	501
17	Serval, *Leptailurus serval*	1.1	SDZ	30.03.88-09.04.88	6.00	78
18	Serval, *Leptailurus serval*	1.1	SAC	15.02.85-25.04.85	32.73	12
19	Serval, *Leptailurus serval*	1.1	NZP	01.07.85-08.09.85	31.03	621
20	Scottish wildcat, *Felis silvestris grampia*	1.2	SDZ	30.03.88-09.04.88	6.00	78
21	Indian desert cat, *Felis silvestris ornata*	1.1	PTL	02.04.89-15.04.89	4.00	203
22	Asian golden cat, *Catopuma temminckii*	1.1	WPZ	23.06.87-29.08.87	21.92	83
23	Asian golden cat, *Catopuma temminckii*	1.1	PTL	02.04.89-15.04.89	4.50	505
24	Fishing cat, *Prionailurus viverrinus*	1.1	WPZ	02.03.86-10.05.86	17.05	94
25	Fishing cat, *Prionailurus viverrinus*	1.1	SDZ	30.03.88-09.04.88	7.50	78
26	Fishing cat, *Prionailurus viverrinus*	1.1	PTL	02.04.89-15.04.89	4.00	236
27	Jaguarundi, *Herpailurus yaguarondi*	1.1	ASM	26.01.88-05.02.88	8.97	125
28	Jaguarundi, *Herpailurus yaguarondi*	1.1	CIN	17.10.88-28.10.88	3.25	44
29	Caracal, *Caracal caracal*	1.1	SAC	03.04.85-17.05.85	22.84	12
30	Caracal, *Caracal caracal*	1.1	SAC	24.02.85-18.05.85	35.52	12
31	Caracal, *Caracal caracal*	1.1	CIN	17.10.88-28.10.88	3.42	36
32	Caracal, *Caracal caracal*	1.1	PTL	02.04.89-15.04.89	4.50	596
33	Canadian lynx, *Lynx canadensis*	1.1	SAC	03.03.85-18.05.85	31.75	12
Overall total		33.36		24.06.84-15.04.89	465.89	

[a]ASM, Arizona-Sonora Desert Museum, Tucson; BRK, Brookfield Zoo, Chicago; CIN, Cincinnati Zoo, Cincinnati, Ohio; NZP, National Zoological Park, Washington, D.C.; PTL, Port Lympne, Kent, Great Britain; SAC, Sacramento Zoo, Sacramento, Calif.; SDZ, San Diego Zoo, San Diego, Calif.; WPZ, Metro Washington Park Zoo, Portland, Ore.

[b]Total number of hours that the enclosure group was observed.

Table 12.2
Variables Analyzed in Captive Environment of Felids

Name of variable	Description and scoring method
Enclosure climate	A binary categorical variable. Enclosures were either indoor (climate controlled) or outdoor (exposed to the local climate)
Enclosure size	A continuous variable measured in cubic meters (m^3). Enclosures at which observations were made ranged in size from 6 m^3 to 782 m^3 (see Table 12.1)
Visual barriers	An integer variable recorded as the number of barriers within an enclosure. A barrier was defined as an object within an enclosure where a cat could be hidden completely from the view of a conspecific in that enclosure (Mellen 1991). As defined here, visual barriers ranged from one to ten per enclosure, but no enclosures examined had six, eight, or nine barriers
Den sites	An integer variable recorded as the number of den sites to which cats within an enclosure had access. The number of den sites ranged from zero to four
Group size	A binary integer variable recorded as the size of the conspecific group in an enclosure. Group size was always two or three
Husbandry	A categorical variable coded as an integer ranging from one to five. This variable provided a qualitative assessment of the degree of interaction between the animal caretakers responsible for captive felids within a given enclosure. A "one" indicated little or no interaction between caretakers and cats, whereas a "five" indicated a high degree of interaction between the same. Mellen (1991) provides additional detail regarding the treatment of this variable
Diet diversity	An integer variable recorded as the number of different diet items captive felids were individually provided (Mellen 1991). This variable, which ranged in value from one to seven, provided an assessment of dietary diversity. For some conspecific groups, diet diversity differed among the individuals within their enclosure. Diet diversity was scored for each cat

entire time budget for the indicated observation. The proportion of time spent inactive was calculated by summing the proportion of time spent in three behavioral categories: resting, resting but alert, and not visible. Activity patterns based on the proportion of time spent inactive were addressed on individual-, enclosure-, and taxon-level bases.

The different types of variables, ranging from binary categorical (e.g., enclosure climate) to continuous ratio scale (e.g., enclosure size) and to the more complex variables having multimodal distributions (e.g., visual barriers) or highly skewed distributions (e.g., pacing), required using nonparametric analyses for all comparisons. All analyses were done at the level of individuals or a subset thereof. Although sixty-eight animals were observed, sixty-nine individual-level data points exist because one male sand cat was the male member of two different groups (enclosures 7 and 9 in Tables 12.1 and 12.3). A few discontinuous variables (e.g., visual barriers) were collapsed into broader categories to enable performing a Kruskal-Wallis nonparametric analysis of variance (KW-ANOVA). We employed Sidak's multiplicative inequality (Zar 1974) to adjust the alpha level where more than one statistical test was used on at least some of the same data. We recognized biases and sources of variation that might confound analyses and addressed the more important ones, those that are sex-, taxon-, and time-specific.

RESULTS

Inactivity

We found a large range in the percentages of time individual cats were inactive, from 3 percent in a male serval to 99 percent in a female Scottish wildcat (Table 12.3). Despite this variability, only thirteen of sixty-nine individuals (ten males and three females) displayed inactivity levels of less than 50 percent. Inactivity in groups of three was not significantly greater than in groups of two (Mann-Whitney U-test: $n_2 = 60$ [cats in groups of two], $n_3 = 9$ [cats in groups of three], and z [corrected for ties] $= -1.971$; $p = .049$). Excluding individuals in groups of three, inactivity of males was not significantly different than that of females (Wilcoxon signed-rank test: $n = 30$, $z = -2.056$; $p = .040$). The average percentage of time that males in groups of two were inactive was 58 percent, whereas the average percentage of time females in such groups were inactive was 67 percent (Table 12.3). Averages for males and females in groups of three also differed by 9 percent, but individuals in groups of three were too few for statistical comparison. Overall analysis among taxa revealed no significant differences in inactivity (KW-ANOVA: df = 15, $n = 69$, H [corrected for ties] $= 16.009$; $p = .382$). Critical probability for the three aforementioned tests was $p = .017$.

Pacing

Of the seven independent variables examined, two showed no significant relationship with pacing: enclosure climate (Mann-Whitney U-test: $n_i = 34$ [inside], $n_o = 35$ [outside], $z = -0.157$; $p = .876$) and den sites (KW-ANOVA: df = 4, $n = 69$, $H = 5.765$; $p = .217$). Pacing was inversely correlated with enclosure size (Kendall's tau: $n = 69$, $z = -4.494$, $t = -0.370$; $p = .0001$), and the

Table 12.3
Percentage of Time Cats Spent Inactive

Enclosure number	Species	Male[a]	Female(s)[b]
1	Pampas cat, *Oncifelis colocolo*	48	73
2	Pampas cat, *Oncifelis colocolo*	24	61
3	Jungle cat, *Felis chaus*	70	65
4	Geoffroy's cat, *Oncifelis geoffroyi*	68	62
5	Geoffroy's cat, *Oncifelis geoffroyi*	56	85
6	Pallas' cat, *Octocolobus manul*	69	73, 75
7	Sand cat, *Felis margarita*	83	84, 82
8	Sand cat, *Felis margarita*	69	73
9	Sand cat, *Felis margarita*	85	69
10	Sand cat, *Felis margarita*	69	51
11	Black-footed cat, *Felis nigripes*	89	95
12	Ocelot, *Leopardus pardalis*	63	79
13	Ocelot, *Leopardus pardalis*	43	44
14	Ocelot, *Leopardus pardalis*	74	72
15	Rusty-spotted cat, *Prionailurus rubiginosus*	39	69
16	Serval, *Leptailurus serval*	32	55
17	Serval, *Leptailurus serval*	3	61
18	Serval, *Leptailurus serval*	69	71
19	Serval, *Leptailurus serval*	72	71
20	Scottish wildcat, *Felis silvestris grampia*	55	56, 99
21	Indian desert cat, *Felis silvestris ornata*	73	73
22	Asian golden cat, *Catopuma temminckii*	41	53
23	Asian golden cat, *Catopuma temminckii*	74	73
24	Fishing cat, *Prionailurus viverrinus*	70	31
25	Fishing cat, *Prionailurus viverrinus*	38	92
26	Fishing cat, *Prionailurus viverrinus*	72	74
27	Jaguarundi, *Herpailurus yaguarondi*	21	66
28	Jaguarundi, *Herpailurus yaguarondi*	75	88
29	Caracal, *Caracal caracal*	50	57
30	Caracal, *Caracal caracal*	45	43
31	Caracal, *Caracal caracal*	74	62
32	Caracal, *Caracal caracal*	73	73
33	Canadian lynx, *Lynx canadensis*	70	80
Mean, all groupings		59	69
Mean, cats housed in pairs		58	67
Mean, cats housed in trios		69	78

[a]The male sand cat in enclosures 7 and 9 is the same individual.

[b]For the enclosures with two females, a percentage is given for each cat.

relationship has a negative logarithmic distribution (best-fit equation: percentage of time spent pacing = -8.174 log [enclosure size] + 23.806). Pacing also displayed an inverse relationship with the number of visual barriers (KW-ANOVA: df = 3, $n = 69$, $H = 16.593$; $p = .0009$); seven or more visual barriers were associated with very low levels of pacing. Significantly less pacing was seen among individuals in groups of three than among individuals in groups of two (Mann-Whitney U: $n_2 = 60$, $n_3 = 9$, $z = -3.185$; $p = .0014$). Level of pacing was inversely correlated with husbandry (Kendall's tau: $n = 69$, $z = -3.495$, $t = -0.288$; $p = .0005$).

A significant relationship was found between diet diversity and pacing (KW-ANOVA: df = 3, $n = 69$, $H = 20.113$; $p = .0002$). Post hoc analysis showed that individuals with the two lowest levels of diet diversity (1 and 2) paced significantly more than individuals at the next two levels (3 and 4). The same pattern was found in the same direction between individuals in diet diversity levels 3 and 4 and greater than 5. Critical probability for these seven tests was $p = .007$.

DISCUSSION

Inactivity

Evaluation of time spent during daylight hours revealed that small captive felids were inactive more than 57 percent of their time on average and, conversely, were active 43 percent of their time. If a major goal of enrichment is to provide environments that allow captives to mimic the activities of their wild counterparts (Shepherdson, Chapter 1, this volume), examining the activity patterns of free-ranging cats is revealing (Seidensticker and Forthman 1994). Data on day and night activities for most small felids are lacking, but where data exist, free-ranging cats spent 57 percent or more of their daylight hours inactive: Geoffroy's cat, 75 percent (estimated from Johnson and Franklin 1991), and ocelots, 57 percent (Ludlow and Sunquist 1987; Crawshaw and Quigley 1989). These results may be similar to those provided here, but our data were collected over only a portion of daylight hours, which precludes direct comparisons.

Similarities in daytime inactivity among small captive felids and their wild counterparts may indicate that the duration of daytime inactivity is largely fixed. More important, attempts to increase daytime activity among captive felids may be futile and unnatural (Hutchins et al. 1984). Thus, effort expended using enrichment to increase the period of daytime activity might be better spent on changing how felids partition their active time (e.g., to encourage less time spent pacing and more time spent investigating their environment). More specifically, we believe that species-specific enrichment opportunities should address the potential for repartitioning the manner in which felids spend their active periods.

Pacing and Enclosure Complexity

One of the most interesting results of this study was that cats spent significantly less time pacing in exhibits that were more complex. Although time spent pacing was variable (average time spent per enclosure ranged from 0 to 37 percent) and several parameters of exhibit complexity were measured, pacing varied more with one variable, namely the number of visual barriers within an exhibit, than with the other parameters measured. Where seven or more visual barriers were present, pacing was reduced or nonexistent.

This result is consistent with an experimental study of leopard cats *(Prionailurus bengalensis)* in which pacing time was reduced by 50 percent when places to hide were provided (Carlstead et al. 1993a). Adding places to hide presumably decreased pacing by increasing exhibit complexity in a manner similar to the presence of numerous visual barriers. It should be emphasized that visual barriers do not simply serve to physically prevent pacing because the addition of barriers that did not provide a visual screen did not prevent pacing (Mellen, personal observation). Our results also suggest that there is a complexity threshold above which the presence of more visual barriers may have little or no effect. This hypothesis deserves further investigation because it has crucial bearing on optimal exhibit design.

Pacing: Function and Motivation

Should the elimination of pacing in captive felids be a goal? Existing data indicate that small cats pace less when provided with some visual barriers. Some pacing may occur, however, even with barriers. Should we assume that the presence of any pacing is somehow linked to reduced well-being? Could pacing in cats have other motivations?

Felids may be motivated to perform feeding-related appetitive (i.e., hunting) behaviors in the absence of any desire to satisfy hunger (Eaton 1972; Leyhausen 1979). Deprivation of opportunities to perform these activities may result in behavior deemed abnormal (Shepherdson et al. 1993). In the wild, small cats hunt on the basis of two strategies: (1) patrolling their home range until prey is encountered (Emmons 1987), or (2) waiting to ambush prey in concealment (Kitchener 1991). Because most prey taken are small, small cats typically repeat the food acquisition sequence several times daily (Sunquist and Sunquist 1991).

In a predictive model, Hughes and Duncan (1988) suggested that the motivation to perform appetitive behavior remains high. This motivation, when removed from its functional consequences, may lock an animal into a closed feedback loop in which the behavior is performed more frequently in a stereotyped fashion. Thus, stereotypies such as pacing in felids may develop from frustrated appetitive behaviors (Shepherdson et al. 1993).

In contrast, pacing has also been construed as a stress-reducing or coping mechanism (Mason 1991). In a study of leopard cats that addressed both behavior (including pacing) and urinary hormone levels (cortisol), Carlstead et al. (1993a) monitored enclosures with and without places to hide. Both pacing and levels of cortisol were reduced when places to hide were added. A study of domestic cats *(Felis catus)* (Carlstead et al. 1993b) also supports this interpretation. Domestic cats that hid more often tended to have lower average cortisol levels, which indicates that hiding may be important for reducing stress in an unpredictable environment. Collectively, these studies suggest that a more complex environment reduces pacing in captive felids and increases their well-being. Well-being may also be linked to other aspects of the captive environment, namely diet and husbandry, both of which are addressed in the next section.

While the results of this study may appear to support the stress reduction alternative, they do not exclude the hypothesis that pacing is derived from appetitive behaviors. If pacing is exclusively a manifestation of an inability to hide (i.e., a coping mechanism) (Mason 1991), one would predict that it would not occur in a complex environment. In the present study, pacing still occurred in large complex enclosures that provided numerous opportunities to hide. The subjective perceptions of one of us (Mellen) may support the appetitive hypothesis. Pacing was often easily disrupted by external stimuli (e.g., a keeper walking by, a bird landing inside the enclosure), suggesting that pacing was not exclusively stress motivated. Pacing may represent a form of searching behavior in captive felids, but the object of that searching behavior may differ in various captive environments. Cats in a complex environment may be searching for food, patrolling their home range, or searching for mates (e.g., Freeman 1983), whereas those in a relatively barren environment may be searching for hiding places as well. Thus, exhibit design may strongly influence the motivation behind the occurrence of pacing.

From a welfare perspective (Carlstead et al. 1992), motivation is a crucial consideration. If a cat is motivated to pace because of a need to search for food, to patrol its home range, or to search for mates, we as caretakers may view the behavior as being a natural expression of a normal behavior in a spatially limited environment. Alternatively, if pacing is motivated by a need to hide or in direct response to some perceived stressor, pacing may indicate reduced well-being. Because these alternatives are unlikely to be mutually exclusive (Carlstead, Chapter 11, this volume), understanding how each varies over a variety of exhibit designs is difficult. To reveal the motivation for pacing, future research must examine experimentally a broad range of taxa for which sequential and temporal links to the occurrence of pacing are measured.

Pacing and the Captive Management Regime

We examined three variables that relate directly to how captive felids are managed: group size, husbandry, and diet diversity. The relationship between the amount of time spent pacing and each variable was addressed. For each case, the amount of time spent pacing was inversely related to the measured variable. The consequences of this relationship were addressed individually for each variable.

The finding that cats spend less time pacing in groups of three than in pairs was surprising. One might have predicted the opposite trend given the solitary felid lifestyle (Kitchener 1991). In spite of the small number of trios analyzed ($n = 3$), it might be tempting to suggest that small felids could be maintained in groups of three. However, examination of the proportion of time spent active revealed that *all* activity decreased when cats were housed in trios. One of us (Mellen 1991) examined the relationship between group size and reproductive success for these same sixty-eight cats. In that analysis, cats housed in trios were less likely to reproduce than those maintained in pairs. These data reinforce the idea that cats housed in trios show depressed activity, probably related to forced proximity and social stress. For these reasons, we suggest that no more than two small felids should be housed together.

Greater interaction between caretakers and cats was associated with less time spent pacing. The caretaker-cat interaction variable was complex and attempted to address the rapport caretakers had with the cats under their care. Interactions typically included a caretaker talking to, scratching, or playing with a cat through a mesh door or fencing. In most cases, the caretaker did not physically enter the enclosure when interacting with a cat. Caretakers did enter enclosures for cleaning and maintenance, but did not interact with the cats at those times if the cats were still present within the enclosure. Entry into an enclosure was not a prerequisite for a high level of interaction between caretakers and cats (see Mellen 1991 for additional data on this variable).

Our results are consistent with that of other recent studies that have addressed caretaker styles. With this same group of sixty-eight cats, Mellen (1991) found reproductive success to be positively correlated with the level of caretaker interaction. Carlstead et al. (1993b) found that a qualitatively poor caretaking regime profoundly stressed domestic cats. They suggested that domestic cats when confined are particularly sensitive to the manner, especially the predictability, with which humans behave toward them. Carlstead et al. (1993b) emphasized that the quality of caretaking can be critical to the welfare of caged domestic cats and, by implication, extended this concern to nondomestic cats. Alternatively, if pacing is a manifestation of a feeding-related appetitive behavior, one could argue that the reduced pacing observed in this study reflects an inhibitory (stress-related) response to the higher levels of caretaker interaction.

We believe that the greater reproductive success and the lack of obvious stress-indicating behaviors in cats who receive more caretaker attention (Mellen 1991, and personal observation) contradict this argument. However, the predictability of the caretaking routine, as Carlstead et al. (1993b) emphasized, rather than the level of caretaker interaction may be the element of caretaking responsible for the patterns observed here. Existing data cannot distinguish between these alternatives. It should be emphasized that we do not believe that caretakers should make pets of exotic felids under their care. Rather, caretaking should be provided in a manner that minimizes the potential for long-term stress. Future research must focus on systematically evaluating the impact of caretakers on the animals under their care to identify precisely what causes chronic stress (see Estep and Hetts 1992 for a review of issues related to human-animal bonds).

Greater dietary diversity (i.e., greater number of food types offered) was also associated with less time spent pacing. Dietary studies of captive felids have focused almost exclusively on meeting their physiological needs, an issue Hediger (1966) pointed out more than thirty years ago. Recent attention has been focused on the nonnutritional aspects of diet (Lindburg 1988; Bond and Lindburg 1990; Lindburg, Chapter 15, this volume). Maintaining animals exclusively on a prepared diet, even if it is nutritionally balanced, may result in tooth decay, dental pathologies, muscle atrophy, and generally poor health (Bond and Lindburg 1990). Feeding of whole or partial carcasses may help break up the monotony of captive felid existence (Hediger 1966; Hutchins et al. 1984). Bond and Lindburg (1990) found that carcass-fed cheetahs *(Acinonyx jubatus)* had larger appetites, fed for longer intervals, and displayed greater possessiveness of food than those fed a commercial diet. More time spent feeding was related to increased approach (more time spent smelling food) and handling times (more time spent chewing and using molars to slice food) in the carcass-fed animals.

Feeding carcasses to felids may also enhance psychological well-being as well as physical well-being (Hutchins et al. 1984; Bond and Lindburg 1990), perhaps by reducing the boredom of the less diverse captive environment. When rabbit, chicken, or ungulate carcasses were added to the diet of cheetahs, Lindburg (1988) found that sometimes even bouts of play were centered around the carcass. He interpreted these changes in behavior as indicators of increased well-being. If pacing is an appetitive behavior, results of this study may indicate that time spent processing food could replace some of the time spent pacing. If this reasoning is correct, cats fed a variety of unprocessed food may be less motivated to search for food (i.e., to pace).

ENVIRONMENTAL ENRICHMENT FOR FELIDS

No single form or type of enrichment is effective indefinitely, largely because felids, like many other organisms, habituate relatively rapidly to novel conditions or devices. Like the ever-changing, complex environment that wild felids face, enrichment must be dynamic and constantly modified to effectively induce the behaviors in captives that are more characteristic of their wild counterparts. To evaluate the effectiveness of any enrichment regime, some kind of monitoring is necessary (Morgan et al., Chapter 10, this volume). Bloomsmith et al. (1991) suggested using a daily check sheet to monitor use of enrichment items, but they also urge that more detailed records be kept to determine the effectiveness of enrichment on behavior and well-being. Observational or experimental evaluation of enrichment regimes is critical to improving enrichment alternatives. We next discuss some components of captive environments for small cats with suggestions on how those environments might be enhanced. We also suggest areas in which research is needed to better understand the effects of captive environments on the behavior and activity of felids.

Providing "Hunting" and Feeding Opportunities

Elsewhere in this volume, Shepherdson (Chapter 1) contends that an optimal environment in captivity should provide opportunities for animals to exhibit the same kind and degree of behaviors they might in the wild (see also Seidensticker and Forthman 1994). Because feeding live mammal or bird prey is not typically desirable or feasible, other techniques must be employed to provide appetitive or "hunting" opportunities for captive felids. An obvious technique is the feeding of humanely killed whole animals (e.g., rats, mice), gutted carcasses (e.g., chickens, rabbits), or carcass fragments (e.g., shanks of sheep or calf) (Law 1993). On receiving whole or partial carcasses, many cats exhibit all or part of the stalk-rush-kill sequence (Richardson 1982; Mellen, personal observation; Shepherdson, personal observation). In addition to the potential for improved psychological well-being (Lindburg 1988), feeding whole carcasses may enhance physical well-being (see also discussion in previous section).

Feeding of live fish was very successful in increasing the activity of fishing cats. Shepherdson et al. (1993) observed a 60 percent decline in sleeping and an increase in enclosure utilization in a female fishing cat presented with live fish in a small pool. Besides fishing cats, several other felids (tiger [*Panthera tigris*], jaguar [*P. onca*], and ocelot [*Leopardus pardalis*]) have displayed hunting behavior when presented with live fish (Mellen, personal observation). Sand cats and black-footed cats have shown interest in hunting live crickets (Mellen, personal observation). A bucket with holes placed near the bottom and set on top of (and outside) a cat's exhibit can be used as a simple cricket feeder. Crickets placed inside the bucket eventually fall into the exhibit as they escape through the holes in the bucket.

The mode of presentation of food can be as enriching as the type of food presented. A log or brush pile in a cat's exhibit can stimulate hunting behavior. Multiple feedings of food hidden in small brush piles increased the amount of exploratory behavior in leopard cats from 6 percent to 14 percent, and the diversity of behaviors increased as well (Shepherdson et al. 1993). Similar success was achieved with a pair of jaguars for whom small bits of dried fish were hidden in a large log pile (Law 1993; Menche et al. 1993). Suspending large pieces of meat near the top of an enclosure has also been observed to stimulate hunting as well as spectacular leaping behavior (Law 1993).

Results from this study suggest that the diet for small captive felids should consist of at least three or four different types of food items. Three other aspects of diet also should be considered: (1) the form or texture of diet items, (2) the presentation of those items, and (3) the schedule of feeding. Attempts should be made to provide diet items with the closest resemblance to whole prey as possible. Health and nutritional considerations may not allow the presentation of whole carcasses, but occasionally offering gutted carcasses would be preferable to feeding an exclusively commercial diet. Further, attempts should be made to select modes of presentation that allow the small felids to expend some effort hunting for or searching out their food. For example, freshly killed mice could be hidden in randomly selected spots throughout the animal's enclosure. Schedule of feeding (i.e., number of feedings per day, consistency of feeding times) is an important consideration and needs to be evaluated systematically. Many cats show increased pacing just before a regular feeding time. Research similar to that described by Carlstead et al. (1993b) is needed to assess the effects of a rigid versus a random feeding schedule.

Novel Objects and Odors

Novel objects introduced to an enclosure can also induce elements of hunting behavior, even if they are not associated with food. Stalking and pouncing behavior can be induced with Boomer Balls or other large plastic objects. Pumpkins placed in an enclosure may be stalked and attacked (Lewis 1992). To encourage exploratory behavior, Boomer Balls can also be partially cut open and large bones wedged inside. Animal skins elicited a wide range of desirable behaviors and maintained interest longer than plastic objects (B. Holst, personal communication).

Novel odors can produce intense interest. Felids use olfaction to obtain information about conspecifics (Kitchener 1991). They show great interest in both attending to and leaving "smelly signposts" in the form of urine marks, feces scrapes, and claw "sharpening" (Mellen 1993). When wild felids patrol their home range, they monitor the scent marks of others and may leave their own marks (Kitchener 1991). Numerous novel scents are available to potentially enhance the olfactory milieu of captive felids, for example, hunter's commercial mule deer musk, spices (especially mace, allspice, cumin, nutmeg), fresh catmint (=catnip), lanolin, rose petals, or feces from prey animals (e.g., zebra feces for lions). As with food, the mode of presentation of novel odors can be as important as the type of odor used, both spatially and temporally.

Enclosure Complexity

Based on our data, particular attention should be paid to visual barriers. A visual barrier should allow a small felid to completely hide from view. Numerous visual barriers can conflict with public exhibition needs, but creative exhibit design can result in exhibits that both meet the needs of the small felids and provide the public an opportunity to view them. Our results suggest that small felid enclosures should have a minimum of seven visual barriers, but experimental evaluation of that suggestion is needed.

Enclosure size is another aspect of enclosure complexity that should be addressed. Although the enclosure-size needs of small captive felids may vary, our results suggest that enclosures of at least 200 square meters may be sufficient. This size criterion should be viewed as tentative because our data are based on relatively few large enclosures, and we addressed neither enclosure shape nor taxon-specific variation in enclosure size. In general, complexity may be more important than size alone. Enclosures for felids should provide opportunities for patrolling and hunting, objects that can be stalked and pounced upon, olfactory novelties, and numerous vertical and horizontal pathways (branches, logs) that encourage the cats to climb or leap and allow the animals to use the vertical dimension of their enclosure.

Enrichment Value of Conspecifics

Although most species of felids are solitary, zoos routinely house them in pairs or trios. Housing a cat with a conspecific might be construed either as providing company—that is, as enrichment—or as a potential source of chronic stress. This concept is further confounded by species and individual differences in behavior; some cats appear to be quite compatible when housed together while others appear fearful in the constant presence of a conspecific. Perceptions of keeping staff and systematic observations can provide useful information in decisions about the potential enrichment value of conspecifics. See Mellen (1989, 1991) for further discussions of effects of management regimes on reproductive success.

At the Metro Washington Park Zoo in Portland, Oregon, an incompatible pair of Siberian tigers *(Panthera tigris altaica)* is alternated on exhibit each day. The female is given access to the exhibit for half the day; she is then brought into a holding area, and the male is given access to the exhibit until the following morning. Each tiger investigates the urine, feces, and cheek-rub markings of the other and subsequently adds its own scents. The public has an opportunity to see more active tigers exhibiting scent-marking behaviors. Similar rotation techniques could be used to stimulate natural social conditions for small felids.

Educating the Public about Enrichment

Public education is the primary goal of zoos. Graphics that describe and explain the purpose of environmental enrichment can enhance the public's understanding of this management technique and its benefits to captive animals. Signs could also be developed to educate the public about how wild and captive cats spend their time, the function of pacing, the importance of enrichment procedures (e.g., feeding whole food items), and that development of enrichment is an ongoing, scientifically based endeavor.

FUTURE AREAS OF RESEARCH

Determining the function of pacing is critical to differentiating the negative from the positive aspects of a variety of husbandry and enrichment regimes and alternatives. Motivation for pacing will only be revealed through experimental and observational examination of the sequential and temporal relationship between its occurrence and components of the captive environment.

Results from the present study suggest that an increased level of caretaker–cat interaction may be desirable, but precisely what this means is not at all clear. Our husbandry variable subsumed a variety of undoubtedly interacting factors, one of which was the predictability of the caregiver's interaction with the cat as described by Carlstead et al. (1993b). These factors need to be experimentally partitioned before we can be confident about what level of caregiver interaction should be provided and under precisely what conditions.

We have speculated that the amount of time cats spend sleeping or generally being inactive may be genetically fixed and that attempts to increase activity during regular zoo hours may be difficult, perhaps unnatural. However, at least some field workers have indicated that significant differences exist in foraging times among some felids (Emmons 1987). Unfortunately, current time-budget data are not available for most felids. Undoubtedly, the activity levels of wild felids are influenced by season, prey availability, climatic conditions, presence of humans, reproductive status, etc. Nevertheless, such information could generate a range for activity patterns and provide guidelines for optimal activity levels in captive felids.

A suite of novel (e.g., olfactory) enrichment items have been offered to captive animals (including small felids), but most of these have been offered without systematic analyses to evaluate their physiological or behavioral effects. This information is critical to our understanding of when and how to use these enrichment alternatives properly and in a taxon-specific way. We need to evaluate the effects of our enrichment techniques by studying measurable changes in behavior and activity rather than simply assuming that the enrichment is effective. In short, we need to test our assumptions.

ACKNOWLEDGMENTS

We gratefully acknowledge the eight institutions that supported this research by allowing J.D.M. access to their facilities: Brookfield Zoo (Chicago Zoological Society), Brookfield, Illinois; National Zoological Park, Washington, D.C.; Metro Washington Park Zoo, Portland, Oregon; San Diego Zoo, San Diego, California; Cincinnati Zoological Gardens, Cincinnati, Ohio; Sacramento Zoo, Sacramento, California; Arizona-Sonora Desert Museum, Tucson, Arizona; and Port Lympne Zoo Park, Lympne, Kent, Great Britain. The Chicago Zoological Society and the Friends of the National Zoo each provided summer internships for which J.D.M. is extremely grateful. The Institute of Museum Services is gratefully acknowledged for providing a conservation grant that allowed J.D.M. to visit and collect data at four of these institutions: San Diego Zoo, Cincinnati Zoological Gardens, Arizona-Sonora Desert Museum, and Port Lympne Zoo Park. We are especially indebted to the animal keepers at all these institutions; each provided great insight into the cats under their care. We thank Bengt Holst for his insights in the use of animal skins for enrichment. Finally, we thank John Seidensticker, Michael Hutchins, an anonymous reviewer, and especially Kathy Carlstead for their comments and suggestions on this paper.

REFERENCES

Altmann, J. 1974. Observational study of behaviour: Sampling methods. *Behaviour* 49:227–267.

Bloomsmith, M., L. Brent, and S. Schapiro. 1991. Guidelines for developing and managing an environmental enrichment program for nonhuman primates. *Laboratory Animal Science* 41:372–377.

Bond, J., and D. Lindburg. 1990. Carcass feeding of captive cheetahs *(Acinonyx jubatus):* The effects of a naturalistic feeding program on oral health and psychological well-being. *Applied Animal Behaviour Science* 26:373–382.

Carlstead, K., J. Brown, S. Monfort, R. Killens, and D. Wildt. 1992. Urinary monitoring of adrenal responses to psychological stressors in domestic and nondomestic felids. *Zoo Biology* 11:165–176.

Carlstead, K., J. Brown, and S. Seidensticker. 1993a. Behavioral and adrenocortical responses to environmental changes in leopard cats *(Felis bengalensis). Zoo Biology* 12:321–332.

Carlstead, K., J. Brown, and W. Strawn. 1993b. Behavioural and physiological correlates of stress in laboratory cats. *Applied Animal Behaviour Science* 38:143–158.

Crawshaw, P., and H. Quigley. 1989. Notes on ocelot movement and activity in the Pantanal region, Brazil. *Biotropica* 21:377–379.

Eaton, R. L. 1972. An experimental study of predatory behavior and feeding behavior in the cheetah *(Acinonyx jubatus). Zeitschrift für Tierpsychology* 31:270–280.

Emmons, L. H. 1987. Comparative feeding ecology of felids in neotropical rainforest. *Behavioral Ecology and Sociobiology* 20:271–283.

Estep, D. Q., and S. Hetts. 1992. Interactions, relationships, and bonds: The conceptual basis for scientist–animal relations. In *The Inevitable Bond: Examining Scientist–Animal Interactions,* ed. H. Davis and D. Balfour, 6–26. Cambridge: Cambridge University Press.

Ewer, R. F. 1973. *The Carnivores.* Ithaca, N.Y.: Cornell University Press.

Freeman, H. 1983. Behavior in adult pairs of captive snow leopards *(Panthera uncia). Zoo Biology* 2:1–22.

Hediger, H. 1966. Diet of animals in captivity. *International Zoo Yearbook* 6:37–58.

Hughes, B., and I. Duncan. 1988. The notion of ethological need, models of motivation, and animal welfare. *Animal Behaviour* 36:1696–1707.

Hutchins, M., D. Hancocks, and C. Crockett. 1984. Naturalistic solutions to the behavioural problems of captive animals. *Zoologische Garten* 54:28–42.

Johnson, W., and W. Franklin. 1991. Feeding and spatial ecology of *Felis geoffroyi* in southern Patagonia. *Journal of Mammalogy* 72:815–820.

Kitchener, A. 1991. *The Natural History of the Wild Cats.* New York: Comstock.

Law, G. 1993. Cats: Enrichment in every sense. *Shape of Enrichment* 2:3–4.

Lewis, C. 1992. Cat nips. *Shape of Enrichment* 1:1–2.

Leyhausen, P. 1979. *Cat Behavior: The Predatory and Social Behavior of Domestic and Wild Cats.* New York: Garland STPM Press.

Lindburg, D. 1988. Improving the feeding of captive felines through the application of field data.

Ludlow, M., and M. Sunquist. 1987. Ecology and behavior of ocelots in Venezuela. *National Geographic Research* 3:447–461.

Mason, G. 1991. Stereotypies: A critical review. *Animal Behaviour* 41:1015–1037.

———. 1993. Forms of stereotypic behaviour. In *Stereotypic Animal Behaviour: Fundamentals and Application to Welfare,* ed. A. Lawrence and J. Rushen, 7–40. Tucson: University of Arizona Press.

Mellen, J. 1989. Reproductive behavior of small captive exotic cats (*Felis* spp.). Ph.D. dissertation, University of California, Davis.

———. 1991. Factors influencing reproductive success in small captive exotic felids (*Felis* spp.): A multiple regression analysis. *Zoo Biology* 10:95–110.

———. 1993. A comparative analysis of scent-marking, social and reproductive behavior in 20 species of small cats *(Felis). American Zoologist* 33:151–166.

Menche, E., D. Shepherdson, C. Lewis, P. Prewett, and J. Rorman. 1993. Large cat enrichment: Providing foraging opportunities for captive jaguars. Poster paper presented at the First Conference on Environmental Enrichment, Portland, Ore., July 1993.

Richardson, D. 1982. Wild felid management at Howletts Zoo Park. *Animal Keepers' Forum* 9:362–365.

Seidensticker, J., and D. Forthman. 1994. Planning for the species: Incorporating behavioral and ecological data. In *Proceedings of the American Zoo and Aquarium Association Annual Conference,* 39–45. Wheeling, W.Va.: AZA.

Shepherdson, D., K. Carlstead, J. Mellen, and J. Seidensticker. 1993. The influence of food presentation on the behavior of small cats in confined environments. *Zoo Biology* 12:203–216.

Sunquist, F., and M. Sunquist. 1991. Ocelots and servals. In *The Great Cats,* ed. J. Seidensticker and S. Lumpkin, 156–161. Emmaus, Pa.: Rodale Press.

Zar, J. H. 1974. *Biostatistical Analysis.* Englewood Cliffs, N.J.: Prentice-Hall.

Part Three

ENVIRONMENTAL ENRICHMENT IN CAPTIVE MANAGEMENT, HUSBANDRY, AND TRAINING

13

MARC P. HAYES, MARK R. JENNINGS, AND JILL D. MELLEN

BEYOND MAMMALS

Environmental Enrichment for Amphibians and Reptiles

During the past decade, increased concern for the psychological as well as the physical well-being of animals has allowed environmental enrichment to mature as a focal concept in their captive management (Shepherdson 1989, 1991a,b, 1992). As the concept has developed, its scope has expanded to include almost any variable linked to the perceptual universe, or *umwelt* (von Uexküll 1909), of the captive animal (Shepherdson 1992). Despite this rapid conceptual growth, environmental enrichment remains an approach applied largely to mammals (Warwick 1990a; Shepherdson 1992; King 1993); other groups, such as amphibians and reptiles, are rarely addressed. For example, a recent extensive catalogue of enrichment ideas (Copenhagen Zoo 1990) devoted only a single page to these taxa. In other reviews, amphibians and reptiles either are not mentioned (duBois 1991; Griede 1992; Shepherdson 1991a; Tudge 1991) or appear only briefly in passing discussion (Markowitz 1982). Nascent efforts have been made to address the psychophysiological problems of reptiles in captivity (Bels 1989; Warwick 1990a,b; Burghardt and Layne 1995), but these are the salient exceptions.

The purpose here is to provide a foundation for designing environmental enrichment programs for amphibians and reptiles. Because amphibians and reptiles are species-rich groups (more than 4,000 species of amphibians and 6,000 species of reptiles are currently recognized; Zug 1993) and the literature on their behavior is vast, this review is not intended to be comprehensive. Rather, we address selected topics that provide a basis for exploring enrichment opportunities. In particular, we review the requirements of captive amphibians and reptiles in three areas: (1) contact with conspecifics, (2) interaction with other species, and (3) physical characteristics of the captive environment. Further, we interpret experimental and observational data that address these areas in an enrichment context and identify opportunities to improve the well-being of captive amphibians and reptiles. Finally, because the maturation and socialization of captive-reared amphibians and reptiles may require some form of enrichment (see Miller et al., Chapter 7, this volume), we emphasize the pivotal role of enrichment in captive-breeding programs that have repatriation as a goal (i.e., repopulation of extirpated wild populations as per Reinert 1991; see also Nielsen 1988).

CONTACT WITH CONSPECIFICS

Of the three critical areas of captive management we discuss, opportunities for contact with conspecifics may be the best understood. Traditionally, housing single-species rather than multispecies groups has been favored because risk of significant injury or death from aggression or predation was thought to be minimized (Pawley 1967; McKeown 1985). Unfortunately, this approach often entirely ignored the social environment of captive amphibians and reptiles, probably because amphibians and reptiles exhibit few of the observable indications of socially induced stress found among mammals. Moreover, full appreciation of how stress affects the physiology of nonmammalian vertebrates is a very recent development (Bels 1989; Lance 1990, 1992).

Captive environments often differ from those in the wild in that spatial constraints (often linked to the needs of the viewing public) frequently increase population densities. Responses to elevated densities, which often result in more frequent contact among conspecifics, are varied. In the wild, the densities at which amphibians and reptiles occur reflect a range of spatial organizations that is often a direct consequence of their social system. Social systems range from species possessing well-defined territories (e.g., many lizards; Stamps 1977), in which obvious spatial segregation exists, to those that lack resource-defense behavior (e.g., some frogs; Wells 1977a), among which spatial organization can vary but individuals are not segregated. Differences in spatial organization can reflect sex-specific, other intraspecific, or even interspecific differences in social organization and behavior. At one extreme, the response of amphibians and reptiles to elevated densities in the captive environment can jeopardize survival of conspecifics.

Two different conditions of elevated density put captive amphibians and reptiles at survival risk: (1) housing aggressively territorial conspecifics or (2) housing significantly different size classes together in the same enclosure. Without an opportunity to escape, subordinate individuals in both these categories risk immediate or protracted death. For example, close-proximity housing (direct auditory and visual contact) of male Goliath frogs *(Conraua goliath),* a species in which males in the wild are highly territorial (A. Koffman, personal communication), ultimately resulted in the death of all males that assumed a subordinate role (M. Hayes, personal observation). Indeed, the only male goliath frog that survived more than five years in captivity was housed alone (R. Pawley, personal communication). Similarly, cannibalism of larger conspecifics on smaller ones is well documented among amphibian and reptile populations (Auffenberg 1981; Simon 1984; Polis and Myers 1985).

Although cannibalism has been historically viewed as an aberrant behavior, growing evidence indicates that it is a common density-dependent regulatory mechanism in many free-ranging populations (Simon 1984; Rootes and Chabreck 1993). Frog "farmers" were among the first to recognize the phenomenon in the captive environment because the high captive densities needed to turn a profit also produced very high rates of cannibalism among size classes (Schorsch 1933). Nevertheless, documentation of cannibalism or other extreme responses among group-housed captive amphibians and reptiles remains sparse. Rare documentation of such responses may reflect, in part, the intelligent choice made by many amphibian and reptile caretakers: to avoid group-housing conspecific amphibians and reptiles that differ significantly in size or which are known to exhibit intraspecific agonistic behavior.

Elevated densities in the captive environment can frequently result in less extreme responses. Captive amphibians and reptiles often establish dominance hierarchies (frogs [Boice and Witter 1969; Boice and Williams 1971]; toads [Boice and Boice 1970]; salamanders [Keen and Reed 1985]; lizards [Colnaghi 1971; Stamps 1977; Alberts 1994]; snakes [Barker et al. 1979]; and turtles [Evans and Quaranta 1951; Boice 1970; Harless and Lambiotte 1971]), a pattern that is less evident among free-ranging animals (Alberts 1994). A hierarchy per se may not indicate density-associated stress, but subordinates may have less access to food (Boice and Witter 1969; Boice 1970; Boice and Boice 1970; Boice and Williams 1971; Colnaghi 1971; Keen and Reed 1985) or may be more restricted in their use of space than dominant individuals (Alberts 1994). This may be the result of direct antagonism from dominant individuals (e.g., aquatic turtles; Warwick 1990a) or the indirect result of physiological stress (e.g., lizards; Alberts 1994). Subordinates also may fail to reproduce (Evans and Quaranta 1951); however, if this is not physiologically harmful, it could actually be a desirable method to control reproduction in captive populations. Alternatively, manipulations that encourage reproduction among subordinates may have the advantage of increasing genetic variability in captive populations.

Identifying physiological stress linked to higher densities in amphibians and reptiles frequently requires that some independent variable be evaluated against an adequate control. Among more promising measures of stress are growth rates and plasma corticosteroid levels (Greenberg and Wingfield 1987). Growth rates of hatchling snapping turtles *(Chelydra serpentina)* raised in isolation were higher than growth rates of hatchlings raised in groups (McKnight and Gutzke 1993), presumably because dominance hierarchies were established within the groups as a result of competition for food (Froese and Burghardt 1974). Growth rates were depressed and plasma corticosteroid levels were elevated in juvenile American alligators *(Alligator mississippiensis)* raised at high densities (Elsey et al. 1990). Identification of stress using hormones or other methods (see Lance 1990 for alternatives) is often problematic because most taxa lack baseline data from wild populations; also, when examined, these data can differ markedly from mammalian patterns. For example, plasma corticosteroid levels reflecting presumably low stress conditions in a free-ranging ranid frog have been observed to be much higher than the levels that mammals display (Reinking et al. 1980). Moreover, the array of environmental insults that many wild populations now endure increases the likelihood that other factors will confound any data that are gathered.

Limited understanding of most variables one might measure on captive amphibians and reptiles requires that work in this area be viewed as experimental. Difficulties notwithstanding, identification of density-linked stresses among captive amphibians and reptiles is badly needed, particularly when comparing captive with wild populations. When such stresses are linked to increased densities, reductions in density that limit the behavioral or physiological manifestations of stress may be desirable. We caution, however, that provision of such enrichment should be context specific because stress-free captive environments may be as undesirable as extremely stressful ones. For example, the short-term stress of temporary housing at high density may actually benefit an animal that is to be repatriated to a high-density congregation (see Kreger et al., Chapter 5, this volume).

Amphibians and reptiles housed in isolation may be faced with the opposite problem: the captive environment may provide insufficient contact with conspecifics. The latter concern may not appear to present a problem for many taxa, but experimental confirmation of this hypothesis is needed. Such contact is important to those taxa that may require conspecific contact during their ontogeny for adequate socialization or to permit recognition of conspecifics or kin. For example, kin recognition is thought to occur among some larval amphibians (Blaustein and O'Hara 1986), but it is not known whether rearing in isolation interferes with this recognition ability nor whether inability to recognize kin (but not conspecifics) is a detriment in a captive situation. We cannot overemphasize that adequate socialization is crucial in captive breeding programs in which the goal is repatriation (see Miller et al., Chapter 7, this volume). Introducing improperly socialized individuals to the wild may result in the total failure of carefully planned repatriation programs. For example, inadequate socialization may have been the cause of poor survivorship among repatriated desert tortoises *(Gopherus agassizii)*; captive-reared tortoises displayed behaviors indicating too great a dependence on humans to survive in the wild (Cook et al. 1978; G. Stewart, personal communication).

Careful evaluation of the socialization requirements of juvenile tortoises and marine turtles in so-called headstarting and repatriation programs is needed. Such evaluations are absolutely necessary because current approaches in many such programs may prevent individuals from repatriating through various potential mechanisms (Dodd and Seigel 1991; Reinert 1991). We emphasize that such assessments be taxon specific; experimental evidence suggests that isolated neonates of some reptiles feed better and grow faster than conspecific counterparts without evidence of long-lasting social consequences (Burghardt and Layne 1995).

For potential reproductive partners, the increased contact resulting from captive confinement can habituate individuals to one another, inducing reproductive lethargy. This pattern has been postulated to occur in captive snakes, a group that depends heavily on olfaction to identify conspecifics (Gillingham 1987). The basis of this pattern is poorly understood, but the near-space of captives may be continuously saturated with the odors of conspecific partners. As a consequence, olfactory saturation may inhibit them from following conspecific odor trails, which are probably more precise in the wild and through which male snakes can locate females over long distances (Duvall et al. 1990a,b). Notably, novel olfactory stimuli may induce reproduction in snakes and turtles that have been housed together for some time. Manipulations that seem to have induced reproduction include (1) removal and replacement of one sex, often the female, for varying intervals (in snakes, replacement of the removed individual during or immediately after a shedding cycle has been observed to elicit a greater response) (Copenhagen Zoo 1990; Radcliffe and Murphy 1983); (2) the addition of one or more conspecifics, especially males and especially where ritualized male combat can be induced; (3) changing environmental parameters (e.g., simulating rainshowers or altering enclosure features); and (4) moving a pair of individuals to a novel enclosure (Murphy and Campbell 1987). All these manipulations could be altering the olfactory environment, but whether odors or other features of environment are really altered and what combination of features induces reproduction, if any, needs to be addressed. Until the relationship between reproduction and the aforementioned manipulations is experimentally addressed, precisely how and when to manipulate captives or the captive environment to facilitate reproduction will remain elusive.

Olfaction can also be an important mode of communication among conspecifics of different ages. Juvenile green iguanas *(Iguana iguana)* avoid the femoral secretions of adult conspecifics (Morgan and Nee 1993), a response that should be considered when housing mixed-age groups of green iguanas or other species that display such behavior. Based on this response, femoral secretion extracts or their analogs might also be used to manipulate the behavior of juveniles.

A related but completely unexplored area involves those species in which the sense of hearing is important to reproduction (e.g., many frogs; Wells 1977a,b). Acoustical signals used by frogs are thought to provide sex, age, quality, and perhaps other information about conspecifics (Wells 1988; Wagner 1992). Consequently, enormous potential exists to manipulate the auditory environment of captive frogs using the taped calls of conspecifics in a manner that may facilitate or inhibit reproduction or modify other aspects of frog social behavior. To our knowledge, the only attempt to use calls in a captive situation has been in the captive breeding program for the Puerto Rican crested toad, *Peltophryne lemur* (Johnson 1991); recorded calls were used successfully to stimulate breeding activity. Crocodilians could also be enriched using conspecific sounds because significant crocodilian communication is vocal (Herzog and Burghardt 1977; Lang 1989).

INTERACTION WITH OTHER SPECIES

One of the ways in which the environment of captive amphibians and reptiles likely differs the most from that of their wild counterparts is in the level of interaction with other species. In recent years, some zoological institutions have expended considerable effort in developing elaborate multispecies displays that include at least some amphibians and reptiles (e.g., tropical rain forest exhibits at the Bronx Zoo/Wildlife Conservation Park in New York City and the Woodland Park Zoological Gardens in Seattle) (see also Stockton 1992), but most institutions still use single-species displays for captive amphibians and reptiles (Pawley 1967; McKeown 1985). As noted previously, housing groups of conspecifics has not been favored because of risk of injury from interspecific aggression or predation. Regardless of whether that reasoning was well founded, it created captive environments with little opportunity for contact with other species.

When active, wild amphibians and reptiles are continuously exposed to diverse auditory, gustatory, olfactory, visual, and perhaps other unrecognized stimuli from an array of sympatric organisms. The composition and nature of these stimuli influence the degree to which free-ranging individuals are alert and responsive to their environment. While we can conceive of many ways in which such stimuli might contribute to the well-being of amphibians and reptiles, it is beyond the scope of this chapter to address the bewildering array of alternatives. Instead, we focus on two key classes of interaction with other organisms: (1) situations in which other organisms are the prey for captives, and (2) situations in which other organisms are predators of captives.

Prey

Captive environments are poor in potential prey organisms for captive amphibians and reptiles for several reasons. Commercial availability of potential prey organisms for use in a captive environment is simply not adequate to continuously provide the array of prey that might be encountered in the wild. For example, in North American institutions, most prey available commercially for insectivorous amphibians and reptiles consists of four insect taxa: crickets *(Gryllus domesticus),* fruit flies *(Drosophila melanogaster),* mealworms *(Tenebrio molitor),* and waxworms *(Galleria mellonella)*. Moreover, not all these groups are suitable for all insectivorous species; larger (e.g., basilisk lizards, *Basiliscus* spp.) or slower (e.g., true chameleons, *Chameleo* spp.) species may ignore or be unable to capture tiny flying (volant) insects (e.g., fruit flies). Prey that are more difficult to capture can also be enriching (Copenhagen Zoo 1990) if the cost or frequency of capture does not create an energy deficit or other stresses. Where used, living prey are often provided on a regular schedule, which may differ from the sometimes opportunistic appearance of prey in the wild.

Living prey can pose a risk to captive amphibians or reptiles if the prey organism is not immediately eaten. Giving small live mammals to unattended captive reptiles under the assumption that they will be eaten later is an error that caretakers of mammal-eating reptiles usually avoid; anecdotes of captive reptiles (especially snakes) injured, maimed, or even killed when small live mammals were left unattended in their enclosures are legion (Mattison 1982; Frye 1991). Less well known is that nonmammalian prey can do the same thing. In different enclosures, one of us (Hayes) observed recently molted mealworms that were provided in excess of what could be eaten at one feeding kill a captive ring-necked snake *(Diadophis punctatus)* and injure a southern alligator lizard *(Elgaria multicarinata);* in the former case, mealworm larvae chewed away neck scales and soft tissue to the point of exposing vertebrae, and in the latter, a hole was opened through the side into the body cavity. Amphibians and reptiles may be particularly vulnerable to uneaten live organisms provided to them during intervals when they are inactive. Public distaste for the feeding of live bird or mammal prey and concern for the well-being of prey organisms can conflict with the provisioning needs of captives (see Kreger et al., Chapter 5, this volume).

Given these restrictions, enriching the environment of captive amphibians and reptiles with prey organisms or their equivalent can take several approaches, some of which depend on the major sensory modality of the species being enriched. First, for insectivorous taxa, insect-attracting devices (e.g., opened ripe fruit or scents such as clove or peppermint oil placed on stationary objects) inside the enclosure may provide opportunities for captives to stalk or take insects that are not part of their regular diet. The effectiveness of this approach depends on several requirements: (1) enclosures should not be insect proof; (2) enclosures should already be located in areas that are relatively insect rich; (3) devices should not attract insects or other organisms that place the captives at risk; and (4) devices should not create other conditions that are undesirable in the captive environment (e.g., potential health hazard).

Second, scheduling of feeding could be modified to increase its variability. Because different species are known to respond differently to variation in the timing of feeding, modifications should be approached experimentally on a taxon-specific basis. Third, when feeding prey organisms that pose a potential risk to the captive, several alternatives exist. Because escape routes are typically not available in captivity, provisions should be made to allow escape of such live organisms if the predator is disinterested, gives prolonged chase, or does not mortally wound the prey. A caretaker should monitor the feeding, and the amount fed should be completely consumed in a short time. Well-being of the prey as well as that of the predator should be addressed, and public education of this approach should be emphasized.

For amphibians and reptiles slated for repatriation, some experience with live prey may be necessary. For snakes that depend primarily on olfaction, feeding with live organisms may be undesirable for other reasons than the risk to captive snakes or public antipathy against feeding bird or mammal prey. In the wild, prey frequently leave scent trails that can be relatively precisely tracked. In a captive environment, a prey organism can rapidly distribute its scent throughout an enclosure, which may make locating prey more difficult. In such instances, feeding dead prey to bird- or mammal-eating snakes may be advantageous. To create a scent trail, it may be desirable to rub the dead prey over a precise, predetermined path. Rattlesnakes (*Crotalus* spp.) will not follow a scent trail unless they have had an opportunity to strike at the prey (strike-induced chemosensory searching; Chiszar et al. 1977). Allowing such species to strike at dead prey before the prey item is placed in a remote part of the enclosure or after the dead prey has been rubbed over a predetermined path are approaches that have been routinely used in some reptile collections (S. McKeown, personal communication).

Novel enrichment that addresses the prey aspect of the captive environment of amphibians and reptiles requires attention to the sensory modality of the species being enriched. Provision of a prey-simulating stimulus in that modality may be as enriching as providing the prey itself. For example, in species in which olfaction is the primary modality, periodically providing the scent of prey may be as effective in behavior induction as using the prey itself (Chiszar et al. 1990; Schell et al. 1990). Frogs may cue on the calls of frogs that they prey upon (Jaeger 1976; M. Hayes, personal observation), so using the calls of prey species to enrich acoustically focused predatory frogs or even crocodilians may be a possibility. Determining the most desirable form of enrichment requires experimental examination of the frequency and sequence of presentation.

An enrichment approach to the diversification of diet and diet presentation is particularly important for those captives that are intended for repatriation (Miller et al., Chapter 7, this volume). Marmie et al. (1990) demonstrated that juvenile wild-caught Baja California rattlesnakes *(Crotalus enyo)* were more vigorous predators than captive-raised juveniles. They postulated the difference might reflect the experience of wild-caught snakes having killed and trailed natural prey some distance, whereas captive-raised juveniles never searched more than a few centimeters for their prey (see also Scudder et al. 1992). This idea has not been experimentally addressed. Burger (1991) showed that hatchling pine snakes *(Pituophis melanoleucus)* with experience in eating mice displayed a scent-tracking ability for mice; inexperienced hatchlings lack this ability, which reinforces the importance of early experience.

For captives not intended for repatriation, enrichment approaches that provide experience may not be desirable. For example, captive-hatched horned lizards (*Phrynosoma* spp.) can be reared successfully on a non-ant diet, while attempts to provision wild-caught horned lizards with a non-ant diet, if they have become accustomed to an ant diet, may fail (Montanucci 1989). In this case, a period during posthatching development may determine dietary flexibility. Sherbrooke (1987) also reported rearing several captive-hatched regal horned lizards *(Phrynosoma solare)* on a non-ant diet without hibernation to an age of two years, but the value of his approach over the long term is difficult to assess because no individuals survived to reproduce. Which dietary enrichment approaches to use, if any, require assessment of conditions related to the captives in question as well as the taxon involved.

Predators

With rare exceptions involving some of the aforementioned mixed-species exhibits, captive environments lack predators. The restricted space available in most captive environments simply would not permit the survival of amphibians or reptiles with one of their predators, nor do we imply that housing amphibians and reptiles with their predators should be a fundamental part of environmental enrichment. However, an emerging idea is that total isolation from selected stimuli, such as predator-associated stimuli, may also stress captives (Moodie and Chamove 1990; Beck 1991; Shepherdson 1992). The concept proposes that lack of predator-associated stimuli assists the general pattern of lethargy common among individuals, and that a certain level of such stimuli contributes rather than detracts from their general well-being. The difficulty in applying this concept is determining the level of stimulation appropriate, if any, on a taxon-specific basis. We simply indicate here that numerous, sensory modality-specific ways exist to provide stimuli that captive amphibians and reptiles might recognize as predator associated. Some include (1) adding the feces or the shed skin of a predator (e.g., a snake) or rubbing the extract of a predator's scent that an olfaction-oriented captive could recognize; (2) using mobile shadow models that simulate visually the presence of a predator; and (3) placing the predator in an adjacent enclosure with either visual or only olfactory access to the captive (Stanley and Aspey 1984). Experimental assessment of how much and how often the appropriate stimulus or stimuli should be applied needs taxon-specific determination.

Linked to the predator simulation approach is the idea that visually oriented amphibians and reptiles may identify predators or the risk of predation based on some aspect of the potential predator. Because a size (predators are recognized as larger than themselves, particularly in the vertical plane) or feature threshold, such as the eye size of the predator (Burger et al. 1991) or the manner in which a potential predator approaches them (Burger and Gochfeld 1991) may be key, captive amphibians and reptiles may view caretakers or members of the viewing public (if the latter can be seen) as predators. Lizards, among other visually oriented organisms, reduce their activity in the presence of a predator (Sugerman and Hacker 1980; Sugerman 1990), presumably to minimize the likelihood of detection. At the Metro Washington Park Zoo, a concealed researcher noted that when the building housing the exhibit of the slender-snouted crocodile *(Crocodylus cataphractus)* was empty, the adult male displayed a wide range of activities. When zoo visitors appeared, this animal ceased moving; activity resumed when the visitors vacated the building (B. Houck, personal communication).

Reduction of activity conflicts with the fundamental approach of enrichment and zoological institution exhibitry. Because captives can habituate to predator-simulation stimuli if actual predatory episodes do not follow, how much reduction of activity observed among amphibians and reptiles in zoological institutions can be attributed to caretaking regimes or the viewing public is not known. However, addressing it is worthwhile because significant implications exist for both potentially adjusting caretaking regimes and the design of exhibits.

Providing stimulation to predator-naive captives should be a crucial part of amphibian and reptile breeding programs the ultimate goal of which is repatriation (Miller et al., Chapter 7, this volume). One of the greatest shams ever foisted on the taxpaying public is the multi-million-dollar captive-rearing programs for commercially important food or game fishes that were intended to increase productivity of exploited populations. The failure to recognize the predator-naive nature of rearing conditions has resulted in an expensive food subsidy for wild predators (Meffe 1992), a situation that has had early parallels among captive breeding programs for charismatic endangered species (Snyder et al. 1989; Moodie and Chamove 1990). Although amphibian or reptile antipredator behavior has been experimentally examined in only one snake (Herzog 1990), those data indicate that a less vigorous antipredator response can result from a more benign captive environment.

This consideration has particular importance to a variety of "head-starting" or similar programs that have been implemented for green iguanas *(Iguana iguana)*, a variety of tortoises and aquatic and marine turtles, the Houston toad *(Bufo houstonensis)* (Dodd and Seigel 1991), and the Puerto Rican crested toad (Johnson 1991), but also applies to any captive-breeding program with a repatriation goal. Many such programs have not satisfactorily addressed the predator-depauperate aspect of rearing conditions, an approach that may be reflected in the failure to identify significant numbers of reproducing individuals from the wild that were captive reared (Dodd and Seigel 1991). We emphasize that predator simulation under captive-rearing conditions may not be a simple task because captives also habituate to simulation stimuli if not periodically provided with less "benign" stimuli (Suboski and Templeton 1989; Burghardt 1991). Finally, we believe that instilling adequate predator awareness or escape skills in captive-reared amphibians and reptiles (as well as other organisms) is so important to repatriation programs that implementation of adequate predator-awareness training should be among the primary evaluating criteria used by those organizations that control program funding.

PHYSICAL CHARACTERISTICS OF THE CAPTIVE ENVIRONMENT

Of the three areas of captive requirements we address, the physical structure of the captive environment is perhaps the least neglected. This area is enormous in scope, so we restrict our discussion to four major aspects of the physical environment: space, temperature, light, and water. Some understanding of how to approach enrichment for those aspects of the physical environment that we do not address can be gained from understanding the approaches we indicate here for these four areas.

Space

Of the several ways in which one might consider the importance of space to captive amphibians and reptiles, we address three here: enclosure size, spatial refuges, and spatial familiarity. Potential constraints imposed by enclosure size have been frequently ignored because of a general belief that amphibians and reptiles are insensitive to changes in the size of the captive enclosure. Discussing reptiles, Warwick (1990a) suggested that a threshold minimum enclosure dimension that prevents either basic locomotion or rapid movements (for fast-moving taxa) can result in physiological imbalances, loss of conditioning, overgrown claws, or accidental collisions. Such thresholds may exist. Lawler (1992) indicated that good reproductive success in captive San Esteban Island chuckwallas *(Sauromalus varius)* was achieved, in part, because of the size of the enclosure, but his results need experimental verification.

Moreover, the critically restrictive environments (CREs; basic locomotion inhibiting) and overly restrictive environments (OREs; rapid-movement inhibiting) that Warwick (1990a) postulates are probably taxon specific and may be individually specific (on the basis of age or other parameters) for certain taxa. To our knowledge, only one experimental test of enclosure size has addressed Warwick's postulates. Marmie et al. (1990) found no significant differences between young Baja California rattlesnakes that had been raised in large cages compared to those raised in small plastic boxes or between either group and wild-caught young rattlesnakes, based on a measure of their ability to explore a novel environment. We emphasize that one should not assume that the lack of difference observed in this analysis automatically translates to all other measures. Space-constrained individuals might perform more poorly than less-constrained individuals if other measures of competence were used.

A feature of the spatial environment that is more frequently addressed is refuges. Many amphibians and reptiles, particularly snakes and fossorial species, are highly contact oriented; they function poorly unless they can spend most of their time in contact with a substrate on at least two, and often more, sides. Placing contact-oriented species in enclosures with refuges that lack these characteristics can be stressful (Warwick 1990a). Some snakes with fossorial tendencies (e.g., coral snakes, *Micrurus* sp.) could not be historically maintained on exhibit unless they were provided with refuges having spaces constrained to particular dimensions. Some caretakers have solved this problem by placing a piece of plexiglas parallel to the floor and close enough to the substrate to allow snakes to maintain substrate contact with more than just the surface upon which they rest (S. McKeown, personal communication). Because snakes may not display the light sensitivity of many lizards, using clear plexiglas can permit viewing them in their refuges (Chiszar et al. 1987).

The optimal number, orientation, and size of such refuges appear to be taxon specific. A refuge does not have to represent a structure in or under which the captive can find concealment; it may simply be a vertically oriented or aerial structure or a more cryptic background color. Juvenile green tree pythons *(Chrondropython viridis),* which undergo an ontogenetic change from brown or yellow (as juveniles) to green (as adults), were found to prefer an elevated location with a dark background over one with a light background (Garrett and Smith 1994). Although captive husbandry literature is replete with examples of taxon-specific refuge requirements (Kaplan 1993; see also many examples in Mattison 1982, 1993), experimentation with the variation that different taxa will tolerate in refuge characteristics is largely lacking.

Perhaps the least recognized aspect of how captive amphibians and reptiles utilize space is spatial familiarity, the familiarity amphibians or reptiles have with their immediate surroundings, their "near-space" environment. The captive environment differs from the wild in that most amphibian and reptile enclosures probably consist mostly of near-space (from the amphibian and reptile perceptual point of view). Chiszar et al. (1993) describe a near-space manipulation in which the contents of the enclosures of one-half the snakes in a study were rearranged (rocks and water bowls were moved around). In the control series, enclosure contents were touched but not rearranged. Following this manipulation, spitting cobras (species unspecified) tongue-flicked more than four times as frequently as did conspecifics in the control series. As Chiszar et al. (1993) have elegantly pointed out, this result implies cognitive abilities that parallel some mammals and begs further investigation of what other responses amphibians and reptiles may display to enclosure modification.

Temperature

Temperature is a fundamental feature of the captive environment of amphibians and reptiles because, as ectothermic organisms, the temperature of their environment determines when and in part how well they function. In the wild, the variation in temperatures available to amphibians and reptiles in their local habitats and the temperature changes individuals undergo are typically relatively diverse (Huey 1982; Rome et al. 1992); in the absence of artificially induced temperature modulation, however, the less variable temperatures found in their enclosures often limit captive amphibians and reptiles. As a consequence, captives generally have less opportunity to select among temperatures or gradients in their environment.

The ability to provide some sort of temperature gradient has constraints linked to general thermodynamics; it becomes increasing difficult (ultimately impossible) to create some sort of variation in temperature or gradient as the size of an enclosure decreases (Ross and Marzec 1990). As a result, maximizing the size of the enclosure is essential if creation of temperature variation or gradients is intended. Manipulation of temperature variation or gradients has been suggested as an enrichment approach to induce activity for reptiles (Copenhagen Zoo 1990), or to mimic a temperate regime not found at the latitude where the captives are housed (Lawler 1992).

Seasonal manipulation of temperature to induce artificial hibernation seems to increase the probability of mating success in some temperate-latitude snakes (Radcliffe and Murphy 1983), and a similar pattern is postulated for temperate and subtropical amphibians (Laszlo 1979, 1983). However, the precise manipulations needed to trigger these responses have not been experimentally examined. Substrate heating sources (i.e., heating pads or elements) are frequently used for reptiles, but their enriching potential has not been experimentally evaluated. Substrate heating sources used alone have been criticized as potentially promoting unnatural behavior because continuous bottom heat will keep a location warm constantly, whereas sun heat disappears when the animal has been lying on a heated location for some time (Copenhagen Zoo 1990). Multiple substrate heating sources on timers that switch the sources of heat on and off at intervals have been used to temporally vary the pattern of intraenclosure temperatures (D. Pate, personal communication) but have not been experimentally evaluated.

Opportunities exist to design elegant enclosures that could take advantage of the varying sun angle seasonally and allow captive reptiles to bask in locations that are simultaneously advantageous to their well-being and exhibitry, but implementation of this approach is rare.

Disease-stressed captive reptiles are claimed to alternatively seek out higher, then lower, preferred body temperatures depending on whether captives are in early or later stages of disease, respectively (Warwick 1991). Data needed to adequately evaluate this observation are not available. Because this pattern has been postulated to be linked to immune system function (Warwick 1991), it should be adequately documented before an experimental study of the relationship is attempted.

Incubation temperatures can influence the quality of posthatchling individuals. Snapping turtles hatched from eggs incubated at 27°C grew faster than those incubated at cooler or warmer temperatures (McKnight and Gutzke 1993). Hatchling pine snakes incubated at 33°C were able to detect potential prey scent trails faster than those incubated at 28°C (Burger 1991). As a consequence, incubation temperatures have the potential to indirectly influence survivorship, so their taxon-specific experimental analysis is important to captive-rearing programs, especially those designed for repatriation. Despite its importance, temperature is a variable poorly understood from the point of view of environmental enrichment for captive amphibians and reptiles.

Light

Light is another key environmental variable for captive amphibians and reptiles that can interact with temperature in inducing selected behaviors. In the wild, amphibians and reptiles are exposed to the daily and seasonal variation in ambient light, but captive amphibians and reptiles are exposed to light regimes that are enclosure dependent. In open-air or outdoor enclosures, captives experience light regimes relatively similar to individuals in the wild except that microhabitat variation in ambient light quality may be more limited. For indoor enclosures or enclosures that provide only some ambient light, captives are exposed to a light regime that may include some ambient light, often with selected wavelengths attenuated, as well as artificial light of selected wavelengths.

Unless particular attention is paid to the quality of ambient light, artificial light environments can be poor in longer-range ultraviolet (UV) wavelengths (Blatchford 1986; Frye 1991). The latter environments may also differ in photoperiod if the artificial light regime does not match the ambient light regime. Quality of light, especially selected UV wavelengths, is critical for activation of intrinsic DNA photorepair mechanisms and, for some amphibians and reptiles (especially certain lizards), for synthesis of selected cofactors and vitamins, especially D complex vitamins (Moyle 1989). Photoperiod and perhaps light quality can be critical to a variety of seasonal rhythms (Licht 1972; Michaels 1987), including annual reproductive cycles, a pattern found in most species examined. Continuous light, light deprivation, and inappropriate photoperiods can cause varying symptomatic manifestations ranging from lethargy to sterility and even death.

Attempts at enrichment should focus on taxon-specific light quality and photoperiod needs. Recent upgrades of the Honolulu Zoo reptile exhibits included customized skylighting to provide selective access to sunlight (S. McKeown, personal communication). In the absence of experimental data (which applies to most species), light quality and photoperiod requirements should be based on existing conditions within the known natural latitudinal range of the taxon in question (Copenhagen Zoo 1990). Jones (1978) provides a table for easy determination of photoperiod given latitude, month, and day. Until tolerance to variation in light quality and photoperiod are experimentally addressed, amphibians and reptiles kept at latitudes outside their natural geographic range should be provided with artificial lighting regimes that attempt to match those found in their natural range.

Water

Water is a critical aspect of the habitat for most amphibians and a fundamental need for all amphibians and reptiles. Water requirements vary greatly, ranging from taxa that obtain nearly all their water from their diet (e.g., some reptiles) (Bradshaw 1986) to those that live their entire lives in water (e.g., some amphibians) (Boutilier et al. 1992). Many amphibians require a water source they can use periodically (Shoemaker et al. 1992), but some taxa must have access to a water source at least every few hours. Water available in the captive environment can differ substantially from that in the wild; the captive environment constrains individuals to one or (rarely) a few sources of water, whereas sources of water are often diverse in the wild. As a consequence of having only one or a few sources of water, captive amphibians and reptiles can exercise little or no choice as to where they obtain their water. For those taxa for which obtaining free water is a survival necessity, the quality of water provided can be crucial. Indeed, water quality may be the parameter that has most often discouraged keeping amphibians in captivity and, ironically, is often overlooked in captive enclosures (Odum 1985).

To detail the water quality requirements that are currently known for amphibians and reptiles is beyond the scope of this chapter, and we refer the reader to reviews that address this topic (Boutilier et al. 1992; Shoemaker et al. 1992). Here, we focus on aspects of water quality that appear especially problematic for amphibians and reptiles. Three aspects of the life history of some amphibians and reptiles complicate maintaining high water quality: (1) aquatic feeding, (2) aquatic defecation and, in some cases, urination, and (3) shedding of epidermal skin. Concerning aquatic feeding, many amphibians, such as ranid frogs (*Rana* spp.), and some reptiles, for example, sea snakes (Hydrophiidae), and aquatic or marine turtles cannot or will not feed unless in an aqueous environment. Most of these taxa eat nitrogen-rich animal food and (especially the turtles) often consume it in a coarse-grained manner (i.e., significant pieces of food are often uneaten). Rehydration induces defecation in some amphibians. As a consequence, such amphibians often leave nitrogen-rich feces in the water where they rehydrate.

Many amphibians and reptiles also use water, if it is available, to assist shedding their epidermal skin. For the amphibians, shedding frequency is relatively high (some cycle every few days). Amphibians often eat their shed skin, but their coarse-grained mode of feeding can result in another nitrogen input to their water source. These patterns make many amphibians and some reptiles vulnerable to nitrogen-induced toxemia and associated septic conditions from fouling their own water if no provision exists for cleaning the water continuously or at frequent intervals (Mattison 1982, 1993; Odum 1985; M. Jennings, personal observation). Less often realized is that nonaquatic postmetamorphic amphibians appear more sensitive to ammonia poisoning than their larval counterparts, which may because the larvae cannot leave the water. Nevertheless, among taxa in which blood levels of ammonia tolerance have been examined, even larval amphibians can tolerate no more than slightly elevated blood ammonia levels (Dole 1967). In captivity, the smaller the water source available for rehydration, the greater the potential for a problem.

The manner in which water is presented can be as important as its quality. Lapping water droplets from vertically oriented substrates is the way that a variety of subtropical and tropical reptiles (e.g., some anoles, *Anolis* sp.) obtain their water. Such taxa will rarely take water presented from a large source, such as a water dish, especially if the source is provided on the ground. Rather, water must typically be misted or sprayed on surfaces (Mattison 1993), which can be done by condensation from a commercial nebulizer or mister. A number of tropical terrestrial amphibians, such as dart poison frogs (Dendrobatidae), and reptiles, such as some varanids (*Varanus* sp.), green iguanas, and anacondas (*Eunectes* sp.), require high temperatures and corresponding humidity to obtain sufficient amounts of water from the environment (Ross and Marzec 1990;

Mattison 1993). Insufficient humidity and suboptimal temperatures can produce stress sufficient to cause death. Timed nebulizers have been used to modulate humidity relatively precisely for goliath frogs, poison dart frogs, and other tropical amphibians (R. Pawley, personal communication). Some aquatic breeding amphibians (e.g., the spadefoot toads *Scaphiopus* spp. and *Spea* spp.) appear unable to reproduce successfully unless a sound stimulus mimicking heavy rainfall along with the application of water breaks a dormant period.

Additional considerations are important to amphibian and reptile life stages that live mostly or completely in water. First, many larval amphibians, some snakes, some turtles, and some crocodilians can recognize conspecifics or kin (Blaustein and O'Hara 1986), prey, and predators in an aquatic environment (Wilson and Lefcort 1993; C. Hawkins, personal communication). If chemical cues detected through olfaction or taste without contact with the target organism are known or suspected, it is important to consider how the mode of maintenance of the aquatic environment may alter these cues. The use of chemical cues also presents an opportunity for enrichment that to our knowledge has not been addressed experimentally; chemical cues have the potential to alter behavior in aquatic captives that ranges from foraging activities to social interactions through the use of selectively inoculated water sources.

Second, aqueous environments carry sound more effectively than terrestrial environments, so the acoustic milieu of aquatic life stages of amphibians and reptiles needs consideration. Acoustic communication in air is reasonably well understood among crocodilians (Lang 1989), but its counterpart in water is virtually unexplored among amphibians or reptiles, particularly in context of the social milieu. Males of some frogs (e.g., northern red-legged frogs, *Rana aurora aurora*) are known to vocalize primarily underwater, so opportunities may exist to manipulate the social environment of captives that call underwater.

Last, amphibians may require an aquatic overwintering site (e.g., northern leopard frogs, *Rana pipiens*). Temperate taxa that overwinter in aquatic environments are vulnerable to death from water quality-related toxemia, so sites with high water quality are typically selected on the basis of high levels of oxygen and low levels of dissolved solutes. Therefore, if captives are temperate taxa that are allowed to overwinter, special attention needs to be paid to the indicated water quality parameters. Some caretakers place amphibians and reptiles in dishes of water for short periods of time after emergence from an inactive or overwintering period (Frye 1991). Such actions may assist organisms in hydration and removal of toxins accumulated during inactivity. Experimental verification of the enrichment value of these procedures or behaviors is needed.

DISCUSSION

The reason that attempts to enrich captive amphibians and reptiles have been infrequent has not been formally addressed, but several factors have probably contributed. First, the fundamental differences that exist between amphibians and reptiles and the more familiar mammals probably have led to interpretational biases regarding the needs of the former (see Warwick 1987, 1990a for reptiles). The metabolic requirements and movement-mediated activity that characterize the physiology of most amphibians and reptiles are considerably less than those of mammals (Pough 1983; Gatten et al. 1992). For example, stereotypies, repetitive movements with a long and controversial association with suboptimal captive conditions in mammals (Mason 1991), are infrequently observed or at least are rarely recognized among amphibians and reptiles (but see Warwick 1990a for postulates of environmentally encouraged modified behaviors [EEMBs] and environmentally induced traumas [EITs] in reptiles). Furthermore, amphibians and reptiles are generally characterized as performing behaviors that are less plastic than those which mammals display (Burghardt 1977; Carpenter and Ferguson 1977; Ferguson 1977; see also Wilson 1975). The concept that the repertoire of amphibians and reptiles is less plastic likely encouraged the perception that opportunities for enriching the environment of amphibians and reptiles are limited or, in the extreme, the interpretation that amphibians and reptiles do not need enrichment (Bels 1989; Warwick 1987, 1990a,b,c for reptiles).

Second, amphibians and reptiles are far less familiar to humans than are mammals, a function of their evolutionarily more distant relationship (Gauthier et al. 1988; Hoff and Maple 1982) and the cryptic behavior of many species (Fitch 1987; Gillingham 1987; Pough et al. 1992). This lesser familiarity may have made amphibians and reptiles a less desirable choice if enrichment was considered.

Third, amphibians and reptiles have rarely been the focus of living zoological collections, so less opportunity has existed to enrich them. Of the more than 200,000 living vertebrates cataloged in the 465 zoological collections that are members of the International Species Information System (ISIS) in the most recent summary available (ISIS 1993a,b), only 17 percent were amphibians and reptiles (2 percent being amphibians and 15 percent reptiles; Table 13.1). Our parallel analysis of the 167 American Zoo and Aquarium Association-accredited institutions in North America listed by Boyd (1994) indicates that reptiles represent only 16 percent of total collections on the basis of species and 5 percent on the basis of individuals; amphibians represent 3 percent on the basis of species

Table 13.1
Demographic Data on 200,000 Living Vertebrates Held in Member Institutions of the International Species Information System (ISIS)

Taxon	Total no. in captivity (*N*)[a]	New births in captivity		Deaths of recently captive-born neonates[b]	
		No. (*n*)[a]	% of *N*	No.	% of *n*
Mammals	76,822	13,547	17.6	3,470	25.6
Birds	90,637	14,426	15.9	3,536	24.5
Reptiles	29,595	436	1.5	380	87.2
Amphibians	4,632	600	13.0	65	10.8

Sources: ISIS 1993a,b.

Note: All data are for the 12-month period from January 1 to December 31, 1992.

[a] Cumulative total, not accounting for losses from death.

[b] Neonates that died within 30 days of birth.

and 2 percent on the basis of individuals. Whatever the precise basis of the limited focus of environmental enrichment of amphibians and reptiles, experimental and observational data indicate that these taxa not only may benefit significantly from much of the enrichment that has been attempted for mammals but may benefit from novel approaches as well.

Environmental enrichment for amphibians and reptiles is clearly a nascent field. Among the more revealing data indicating this fact are basic demographic features of captive vertebrate populations (see Table 13.1). The percentage of captive "births" for amphibians and reptiles is lower than that for birds and mammals. We recognize that a variety of population management programs (e.g., Species Survival Plans, Regional Collection Plans) may confound this analysis. However, the fact that the average clutch or litter size among amphibians or reptiles is considerably larger than that found in birds or mammals bodes especially poorly for the latter two groups. Moreover, although the amphibian percentage seems closer to that of birds and mammals, the total number of amphibian captive births is small enough to potentially represent less than one clutch for many amphibian taxa.

Examination of the proportion of neonates that die within the first 30 days is even more revealing. Reptiles, the lower vertebrate group for which the better captive management data exist, sustain losses that constitute nearly 90 percent of captive births. This figure is astounding considering it indicates mortality in the presumed absence of predators. The amphibian rate appears better, but in amphibians, much environmental or physiological death not attributable to predators occurs after the first thirty days. We do not wish to belabor these figures, but simply indicate that despite the considerable efforts of caretakers among ISIS-member institutions, we are still novices in the captive management of amphibians and reptiles.

The value of enrichment extends far beyond the survival of captive amphibians and reptiles suggested by the aforementioned facts. Enhanced welfare of individual animals resulting from enrichment not only translates into reduced veterinary care, but it can improve exhibitry, facilitate captive husbandry, enhance reproduction among captive animals, and improve public education and perception (see Kreger at al., Chapter 5, this volume). Enrichment itself provides a basis for enhancing education about taxa such as amphibians and reptiles that are less familiar to the public, because enrichment approaches are founded on the perceptual environment of the captive, which helps reduce anthropocentric biases. Most important, such educational approaches have a significant conservation value because they inform the public about the value of animals. In particular, this helps humans understand the ecological value of these animals and their contribution to biodiversity.

There are too many kinds of amphibians and reptiles for us to be able to provide species-specific recommendations on environmental enrichment, but we can provide recommendations that address enrichment on other bases. First, nothing is more fundamental than an adequate knowledge of the natural history of the taxon addressed. Numerous authors have made this point (Johnson 1991; Lawler 1992; see especially Chiszar et al. 1993 and Dodd and Seigel 1991). We emphasize that "adequate" is the key word, which means having an understanding of the biological constraints (e.g., demographic, habitat; see Dodd and Seigel 1991) that limit a taxon and, most importantly, how those constraints may cause differences between the behavior of an animal in the captive environment and that in the wild.

Second, the possibility should always be entertained that variation in behavior may have intraspecific geographic-, population-, and perhaps even individual-level bases requiring adjustment of what was traditionally considered a taxon-specific management technique. For example, the intraspecific geographic variation identified in food-cue preferences of neonatal garter snakes *(Thamnophis sirtalis)* (Burghardt 1970) would mean that snakes born of parents from different geographic areas would feed more readily on different prey. Thus, providing them with different diets might ensure greater survivorship. One should keep in mind that the systematics of many amphibian and reptile groups are far from stable, so identification of intraspecific variation in behavior may have phylogenetic significance. Collaborative efforts with herpetological systematists have the potential to contribute significantly to both captive management and systematics.

Third, approaches to environmental enrichment should always be founded in the sensory modalities (e.g., olfactory, tactile) a taxon uses to assess its environment and the organisms around it. We realize that an understanding of the primary sensory modalities has not been adequately established for most amphibian and reptile taxa, so approaches should be viewed as experimental. Known taxon group or subgroup responses can serve as a guide (e.g., the primary modality of snakes is olfactory), and caretakers should be encouraged to experiment within the scope of existing knowledge. Significant opportunity exists for greater collaboration between caretakers or zoo research personnel and outside investigators to address experimental verification of sensory modality-specific enrichment and the causal bases of these approaches (Chizar et al. 1993).

Fourth, although primary sensory modalities should be a focus, one should be open to possibilities for enrichment using stimuli that may not address the focal sensory modality for a taxon. For example, western toads *(Bufo boreas)* can learn to identify prey insects by smell (Dole et al. 1981). Thus, enriching western toads using odors may be a possibility. In a similar vein, we recommend exploring enrichment with approaches that have worked for closely related species. Nevertheless, different taxa, however closely related, may require modified or even entirely different regimes.

Fifth, we have pointed out many areas that could benefit from research, but a few of these deserve special mention either because they have been ignored or because we believe them to be important. They are discussed next.

The Captive's Perception of the Near-Space Environment

Whether amphibians and reptiles actually perceive the space immediately surrounding them is a difficult question that will require collaborative efforts and years of research. However, we do not have to wait until many of the answers from this research are available. Experiments can be designed to indirectly assess perception based on the subtle responses of captives, as Chizar et al. (1993) have pointed out. There are enormous opportunities for novel experimental analyses in this area that can be directly incorporated in caretaking regimes and the next neglected area we address: exhibit design.

Exhibit Design

Many strides have been made in the design of exhibits, and opportunity exists to advance into a new age of exhibit design for amphibians and reptiles. While many taxa (e.g., snakes) may not be sensitive to light in their refuges, others may be. Exhibits could be designed with "dark-room" refuges lighted with wavelengths that are not perceived by the animals but allow enough visibility for humans. Moreover, such exhibits could be constructed so humans can observe animals in the dark-room refuge from above. Education as to a priority of exhibit design should promote the priorities Hediger (1964) originally suggested: (1) needs of captive animals, (2) needs of caretakers, and (3) needs of the viewing public, rather than the reverse, which as he so accurately pointed out has been the historical pattern.

Ignored Sensory Modalities

Enormous opportunities also exist for enriching the acoustic milieu of crocodilians and frogs and the gustatory-olfactory environment of snakes and turtles. We did not address the tactile milieu, but it is likely significant opportunities exist in this area, especially for snakes and other fossorial forms.

Adequate Socialization

We agree with Chiszar et al. (1993) and Dodd and Seigel (1991) that socialization is the aspect most important and perhaps most ignored. We will even go so far as to predict that most of the major impediments to keeping a variety of amphibian and reptile taxa in captivity will ultimately be traced to their socialization needs. The physical parameters of success in the captive organism can be impressive (e.g., growth rate; Swingle et al. 1993) but may mean little outside the context of adequate socialization. Recognition of and the ability to interact with conspecifics are fundamental to rearing nonisolated individuals in captivity, particularly if reproduction of captives is desired. This requirement becomes an absolute necessity where the goal is repatriation. Indeed, we believe that repatriation programs should address four areas of adequate socialization: (1) recognition of, and interaction with, conspecifics; (2) recognition of the food or prey base that may be encountered in the wild (failure of early repatriation attempts with desert tortoises has been attributed in part to a failure to recognize or to use wild plants as forage because of entrainment on an iceberg lettuce diet; G. Stewart, personal communication); (3) recognition of predators and adequate escape skills to avoid them; and (4) avoidance or minimization of bonding to human caretakers or learning patterns that will interfere with survival skills on release. Such an approach will minimize the repatriation problems seen elsewhere (Snyder et al. 1989; Dodd and Seigel 1991; Meffe 1992).

CONCLUSION

We emphasize that the scope of this chapter represents a fraction of the possibilities for environmental enrichment. Much elegant work on amphibians and reptiles could not be included here, but this does not imply that it lacks importance. We encourage those who are attempting environment enrichment to search the amphibian and reptile literature in a hierarchical manner. Information extending from the specific taxon of interest to the taxon group as well as ideas from unrelated groups may be useful when developing an enrichment protocol for a specific amphibian or reptile.

ACKNOWLEDGMENTS

David J. Shepherdson patiently clarified basic aspects of environmental enrichment and provided key references. Chuck Hawkins provided some unpublished observations, Janice L. Hixson provided several important references, Becky Houck provided the slender-snouted crocodile anecdote, and Catherine E. King kindly provided a draft of her now-published manuscript that addresses enrichment in birds. Andy Koffman and Ray Pawley generously shared their information on goliath frogs. Sean McKeown gave information on fossorial snake husbandry, provided a key reference, and directed us to key resource people. Kathleen N. Morgan allowed us to cite unpublished information on green iguanas and detailed her ongoing research on this taxon. Dennis Pate provided key information on substrate heating sources and information about selected zoo exhibits. Glenn R. Stewart clarified data on early desert tortoise repatriations. Michael Hutchins, Sean McKeown, Ray Pawley, and an anonymous reviewer gave suggestions that significantly improved the manuscript.

REFERENCES

Alberts, A. C. 1994. Dominance hierarchies in male lizards: Implications for zoo management programs. *Zoo Biology* 13:479–490.

Auffenberg, W. 1981. *The Behavioral Ecology of the Komodo Monitor.* Gainesville: University Presses of Florida.

Barker, D. G., J. B. Murphy, and K. W. Smith. 1979. Social behavior in a captive group of Indian pythons, *Python molurus* (Serpentes, Boidae) with formation of a linear social hierarchy. *Copeia* 1979:466–477.

Beck, B. B. 1991. Managing zoo environments for reintroduction. In *Proceedings of the American Association of Zoological Parks and Aquariums Annual Conference,* 260–264. Wheeling, W.Va.: AAZPA.

Bels, V. 1989. Analysis of the psychophysiological problems of reptiles in captivity. *Herpetopathologia* 1:11-18.

Blatchford, D. 1986. Environmental lighting. In *Proceedings of the United Kingdom Herpetological Societies Symposium on Captive Breeding,* 87-97. London: United Kingdom Herpetological Societies.

Blaustein, A. R., and R. K. O'Hara. 1986. Kin recognition in tadpoles. *Scientific American* 254:108-116.

Boice, R. 1970. Competitive feeding behaviours in captive *Terrapene c. carolina. Animal Behaviour* 18:703-710.

Boice, R., and C. Boice. 1970. Interspecific competition in captive *Bufo marinus* and *Bufo americanus* toads. *Journal of Biological Psychology* 12:32-36.

Boice, R., and R. C. Williams. 1971. Competitive feeding behaviour of *Rana pipiens* and *Rana clamitans. Animal Behaviour* 19:544-547.

Boice, R., and D. W. Witter. 1969. Hierarchical feeding behaviour in the leopard frog *(Rana pipiens). Animal Behaviour* 17:474-479.

Boutilier, R. G., D. F. Stiffler, and D. P. Toews. 1992. Exchange of respiratory gases, ions, and water in amphibious and aquatic amphibians. In *Environmental Physiology of the Amphibians,* ed. M. E . Feder and W. W. Burggren, 81-124. Chicago: University of Chicago Press.

Boyd, L. J., ed. 1994. *Zoological Parks and Aquariums in the Americas, 1994-95.* Wheeling, W.Va.: AZA.

Bradshaw, S. D. 1986. *Ecophysiology of Desert Reptiles.* North Ryde, Australia: Academic Press Australia.

Burger, J. 1991. Response to prey chemical cues by hatchling pine snakes *(Pituophis melanoleucus):* Effects of incubation temperature and experience. *Journal of Chemical Ecology* 17:1069-1078.

Burger, J., and M. Gochfeld. 1991. Risk discrimination of direct versus tangential approach by basking black iguanas *(Ctenosaura similis):* Variation as a function of human exposure. *Journal of Comparative Psychology* 104:388-394.

Burger, J., M. Gochfeld, and B. G. Murray, Jr. 1991. Role of a predator's eye size in risk perception by basking black iguanas, *Ctenosaura similis. Animal Behaviour* 42:471-476.

Burghardt, G. M. 1970. Intraspecific geographical variation in chemical food cue preferences of newborn garter snakes *(Thamnophis sirtalis). Behaviour* 36:246-257.

———. 1977. Learning processes in reptiles. In *Biology of the Reptilia.* Vol. 7, *Ecology and Behavior,* ed. C. Gans and D. W. Tinkle, 555-681. New York: Academic Press.

———. 1991. Cognitive ethology and critical anthropomorphism: A snake with two heads and hognose snakes that play dead. In *Cognitive Ethology: The Minds of Other Animals,* ed. C. A. Ristau, 53-90. Hillsdale, N.J.: Lawrence Erlbaum.

Burghardt, G. M., and D. Layne. 1995. Effects of ontogenetic processes and rearing conditions. In *Health and Welfare of Captive Reptiles,* ed. C. Warick, F. L. Frye, and J. B. Murphy, 165-185. London: Chapman & Hall.

Carpenter, C. C., and G. W. Ferguson. 1977. Variation and evolution of stereotyped behavior in reptiles. I. A survey of stereotyped reptilian behavioral patterns. In *Biology of the Reptilia.* Vol. 7, *Ecology and Behavior,* ed. C. Gans and D. W. Tinkle, 335-403. New York: Academic Press.

Chiszar, D., T. Melcer, R. Lee, C. W. Radcliffe, and D. Duvall. 1990. Chemical cues used by prairie rattlesnakes *(Crotalus viridis)* to follow trails of rodent prey. *Journal of Chemical Ecology* 16:79-86.

Chiszar, D., J. B. Murphy, and H. M. Smith. 1993. In search of zoo-academic collaborations: A research agenda for the 1990s. *Herpetologica* 49:488-500.

Chiszar, D., C. W. Radcliffe, T. Boyer, and J. L. Behler. 1987. Cover-seeking behavior in red spitting cobras *(Naja mossambica pallida):* Effect of tactile cues and darkness. *Zoo Biology* 6:161-167.

Chiszar, D., C. W. Radcliffe, and K. M. Scudder. 1977. Analysis of the behavioral sequence emitted by rattlesnakes during feeding episodes. I. Striking and chemosensory searching. *Behavioral Biology* 21:418-425.

Colnaghi, G. 1971. Partitioning of a restricted food source in a territorial iguanid *(Anolis carolinensis). Psychoneural Science* 23:59-60.

Cook, J. C., A. E. Weber, and G. R. Stewart. 1978. Survival of captive tortoises released in California. In *Proceedings of the Desert Tortoise Council Symposium,* 130-135. San Diego, Calif.: Desert Tortoise Council.

Copenhagen Zoo, ed. 1990. *Behavioural Enrichment: A Catalogue of Ideas.* Copenhagen, Denmark: Copenhagen Zoo.

Dodd, C. K., and R. A. Seigel. 1991. Relocation, repatriation, and translocation of amphibians and reptiles: Are they conservation strategies that work? *Herpetologica* 47:336-350.

Dole, J. W. 1967. The role of substrate moisture and dew in the water economy of leopard frogs, *Rana pipiens. Copeia* 1967:141-149.

Dole, J. W., B. B. Rose, and K. H. Tachiki. 1981. Western toads *(Bufo boreas)* learn odor of prey insects. *Herpetologica* 37:63-68.

duBois, T. 1991. Behavioral enrichment: Labors of love. *Zoo View* 25:8-11.

Duvall, D., D. Chiszar, W. K. Hayes, J. K. Leonhardt, and M. J. Goode. 1990a. Chemical and behavioral ecology of foraging in prairie rattlesnakes *(Crotalus viridis viridis). Journal of Chemical Ecology* 16:87-101.

Duvall, D., M. Goode, W. K. Hayes, J. K. Leonhardt, and D. G. Brown. 1990b. Prairie rattlesnake vernal migration: Field experimental analyses and survival value. *Journal of Chemical Ecology* 16:102-118.

Elsey, R. M., T. Joanen, L. McNease, and V. Lance. 1990. Growth rate and plasma corticosterone levels in juvenile alligators maintained at different stocking densities. *Journal of Experimental Zoology* 255:30-36.

Evans, L. T., and J. V. Quaranta. 1951. A study of the social behavior of a captive herd of giant tortoises. *Zoologica* (New York) 36:171-181.

Ferguson, G. W. 1977. Variation and evolution of stereotyped behavior in reptiles. II.

Social displays of reptiles: Communications value, ultimate causes of variation, taxonomic significance, and heritability of population differences. In *Biology of the Reptilia*. Vol. 7, *Ecology and Behavior*, ed. C. Gans and D. W. Tinkle, 405–554. New York: Academic Press.

Fitch, H. S. 1987. Collecting and life-history techniques. In *Snakes: Ecology and Evolutionary Biology*, ed. R. A. Siegel, J. T. Collins, and S. S. Novak, 143–164. New York: Macmillan.

Froese, A. D., and G. M. Burghardt. 1974. Food competition in captive juvenile snapping turtles, *Chelydra serpentina*. *Animal Behavior* 22:735–740.

Frye, F. L. 1991. *Biomedical and Surgical Aspects of Captive Reptile Husbandry*, Vol. 1. Malabar, Fla.: Krieger.

Garrett, C. M., and B. E. Smith. 1994. Perch color preference in juvenile green tree pythons, *Chondropython viridis*. *Zoo Biology* 13:45–50.

Gatten, R. E., Jr., K. Miller, and R. J. Full. 1992. Energetics at rest and during locomotion. In *Environmental Physiology of the Amphibians*, ed. M. E. Feder and W. W. Burggren, 314–377. Chicago: University of Chicago Press.

Gauthier, J., A. Kluge, and T. Rowe. 1988. Amniote phylogeny and the importance of fossils. *Cladistics* 4:105–209.

Gillingham, J. C. 1987. Social behavior. In *Snakes: Ecology and Evolutionary Biology*, ed. R. A. Siegel, J. T. Collins, and S. S. Novak, 184–209. New York: Macmillan.

Greenberg, N., and J. C. Wingfield. 1987. Stress and reproduction: Reciprocal relationships. In *Hormones and Reproduction in Fishes, Amphibians, and Reptiles*, ed. D. O. Norris and R. E. Jones, 461–503. New York: Plenum Press.

Griede, T. 1992. *Two Hundred Examples of Environmental Enrichment for Zoo Animals*. Amsterdam: National Foundation for Research in Zoological Gardens.

Harless, M. D., and C. W. Lambiotte. 1971. Behavior of captive ornate box turtles. *Journal of Biological Psychology* 13:17–23.

Hediger, H. 1964. *Wild Animals in Captivity: An Outline of the Biology of Zoological Gardens*. London: Butterworths.

Herzog, H. A., Jr. 1990. Experiential modification of defensive behaviors in garter snakes *(Thamnophis sirtalis)*. *Journal of Comparative Psychology* 104:334–339.

Herzog, H. A., Jr., and G. M. Burghardt. 1977. Vocal communication signals in juvenile crocodilians. *Zeitschrift für Tierpsychologie* 44:294–304.

Hoff, M. P., and T. L. Maple. 1982. Sex and age differences in the avoidance of reptile exhibits by zoo visitors. *Zoo Biology* 1:263–269.

Huey, R. B. 1982. Temperature, physiology, and the ecology of reptiles. In *Biology of the Reptilia*, Vol. 12, ed. C. Gans and F. H. Pough, 25–91. New York: Academic Press.

ISIS (International Species Information System). 1993a. *Amphibian Abstract*. Apple Valley, Minn.: ISIS.

———. 1993b. *Reptile Abstract*. Apple Valley, Minn.: ISIS.

Jaeger, R. G. 1976. A possible prey-call window in anuran auditory perception. *Copeia* 1976:833–834.

Johnson, B. R. 1991. Conservation of threatened amphibians: The integration of captive breeding and field research. In *Proceedings of the Conference on Captive Propagation and Husbandry of Reptiles and Amphibians,* ed. R. E Staub, 33–38. Special Publication 6. Davis, Calif.: Northern California Herpetological Society.

Jones, J. P. 1978. Photoperiod and reptile reproduction. *Herpetological Review* 9:95–100.

Kaplan, M. L. 1993. An enriched environment for the African clawed frog *(Xenopus laevis). Lab Animal* 22:25–28.

Keen, W. H., and R. W. Reed. 1985. Territorial defense of space and feeding sites by a plethodontid salamander. *Animal Behaviour* 33:1119–1123.

King, C. E. 1993. Environmental enrichment: Is it for the birds? *Zoo Biology* 12:509–512.

Lance, V. A. 1990. Stress in reptiles. In *Progress in Comparative Endocrinology,* ed. A. Epple, C. G. Scranes, and M. H. Stetson, 461–466. New York: Wiley-Liss.

———. 1992. Evaluating pain and stress in reptiles. In *The Care and Use of Amphibians, Reptiles, and Fish in Research,* ed. D. Shaeffer, K. Kleinow, and L. Krulisch, 101–106. Greenbelt, Md.: Scientists Center for Animal Welfare.

Lang, J. 1989. Social behavior. In *Crocodiles and Alligators,* ed. C. A. Ross, 102–117. New York: Facts on File.

Laszlo, J. 1979. Notes on reproductive patterns of reptiles in relation to captive breeding. *International Zoo Yearbook* 19:22–27.

———. 1983. Further notes on reproductive patterns of amphibians and reptiles in relation to captive breeding. *International Zoo Yearbook* 23:166–174.

Lawler, H. E. 1992. Advanced protocols for the management and propagation of endangered and threatened reptiles. In *Proceedings of the Fifteenth International Herpetological Symposium on Captive Propagation and Husbandry,* 57–66. Seattle, Wash.: International Herpetological Symposium, Inc.

Licht, P. 1972. Photoperiodic and thermal influences on reproductive cycles in reptiles. In *Proceedings of the Fourth International Congress on Endocrinology,* 185–190. International Congress Series 273. Amsterdam: Elsevier.

Markowitz, H. 1982. *Behavioral Enrichment in the Zoo.* New York: Van Nostrand Reinhold.

Marmie, W., S. Kuhn, and D. Chiszar. 1990. Behavior of captive-raised rattlesnakes *(Crotalus enyo)* as a function of rearing conditions. *Zoo Biology* 9:241–246.

Mason, G. J. 1991. Stereotypies: A critical review. *Animal Behaviour* 41:1015–1037.

Mattison, C. 1982. *The Care of Reptiles and Amphibians in Captivity.* Poole, U.K.: Blandford Press.

———. 1993. *Keeping and Breeding Amphibians.* London: Blandford Press.

McKeown, S. 1985. The ecosystem approach: New survival strategies for managing and displaying reptiles and amphibians in zoos. In *Proceedings of the Eighth Symposium on Captive Propagation and Husbandry of Reptiles and Amphibians,* 1–5. Davis, Calif.: Northern California Herpetological Society.

McKnight, C. M., and W. H. N. Gutzke. 1993. Effects of the embryonic environment and of hatchling housing conditions on growth of young snapping turtles *(Chelydra serpentina)*. *Copeia* 1993:475-482.

Meffe, G. K. 1992. Techno-arrogance and halfway technologies: Salmon hatcheries on the Pacific Coast in North America. *Conservation Biology* 6:350-354.

Michaels, S. J. 1987. Artificial lighting for herps: A preliminary report. *Bulletin of the Chicago Herpetological Society* 22:79-83.

Montanucci, R. R. 1989. Maintenance and propagation of horned lizards (*Phrynosoma* sp.) in captivity. *Bulletin of the Chicago Herpetological Society* 24:229-238.

Moodie, E. M., and A. S. Chamove. 1990. Brief threatening events beneficial for captive tamarins? *Zoo Biology* 9:275-286.

Morgan, K. N., and S. Nee. 1993. Avoidance of adult femoral secretions by juvenile green iguana *(Iguana iguana)*. Poster paper presented at the Founder's Award Session, Animal Behavior Society Meeting, Davis, Calif.

Moyle, M. 1989. Vitamin D and UV radiation: Guidelines for the herpetoculturalist. In *Proceedings of the Thirteenth International Herpetological Symposium on Captive Propagation and Husbandry,* 61-70. Phoenix, Ariz.: International Herpetological Symposium, Inc.

Murphy, J. B., and J. A. Campbell. 1987. Captive maintenance. In *Snakes: Ecology and Evolutionary Biology,* ed. R. A. Siegel, J. T. Collins, and S. S. Novak, 165-181. New York: Macmillan.

Nielsen, L. 1988. Definitions, considerations, and guidelines for translocation of wild animals. In *Translocation of Wild Animals,* ed. L. Nielsen and R. D. Brown, 12-51. Milwaukee: Wisconsin Humane Society.

Odum, A. 1985. Water quality: An often overlooked parameter for the amphibian enclosure. In *Proceedings of the Eighth Annual International Symposium on Captive Propagation and Husbandry,* ed. R. A. Hahn, 33-58. Thurmont, Md.: Zoological Consortium, Inc.

Pawley, R. L. 1967. Mixing it up in Brookfield's reptile house. *Animal Kingdom* 70:90-95.

Polis, G. A., and C. A. Myers. 1985. A survey of intraspecific predation among reptiles and amphibians. *Journal of Herpetology* 19:99-107.

Pough, F. H. 1983. Amphibians and reptiles as low-energy systems. In *Behavioral Energetics: The Cost of Survival in Vertebrates,* ed. W. P. Aspey and S. I. Lustick, 141-188. Columbus: Ohio State University Press.

Pough, F. H., W. E. Magnusson, M. J. Ryan, K. D. Wells, and T. L. Taigen. 1992. Behavioral energetics. In *Environmental Physiology of the Amphibians,* ed. M. E. Feder and W. W. Burggren, 395-436. Chicago: University of Chicago Press.

Radcliffe, C. W., and J. B. Murphy. 1983. Precopulatory and related behaviors in captive crotalids and other reptiles: Suggestions for future investigation. *International Zoo Yearbook* 23:163-166.

Reinert, H. K. 1991. Translocation as a conservation strategy for amphibians and reptiles: Some comments, concerns, and observations. *Herpetologica* 47:357-363.

Reinking, L. N., C. H. Daugherty, and L. B. Daugherty. 1980. Plasma aldosterone concentrations in wild and captive western spotted frogs *(Rana pretiosa). Comparative Biochemistry and Physiology* 65A:517–518.

Rome, L. C., E. D. Stevens, and H. B. John-Alder. 1992. The influence of temperature and thermal acclimation on physiological function. In *Environmental Physiology of the Amphibians,* ed. M. E. Feder and W. W. Burggren, 183–205. Chicago: University of Chicago Press.

Rootes, W. L., and R. H. Chabreck. 1993. Cannibalism in the American alligator. *Herpetologica* 49:99–107.

Ross, R. A., and G. Marzec. 1990. *The Reproductive Husbandry of Pythons and Boas.* Stanford, Calif.: Institute for Herpetological Research.

Schell, F. M., G. M. Burghardt, A. Johnston, and C. Coholich. 1990. Analysis of chemicals from earthworms and fish that elicit prey attack by ingestively naive garter snakes *(Thamnophis). Journal of Chemical Ecology* 16:67–77.

Schorsch, I. G. 1933. *Ranaculture.* Philadelphia: George H. Buchanan.

Scudder, K. M., D. Chiszar, and H. M. Smith. 1992. Strike-induced chemosensory searching and trailing behaviour in neonatal rattlesnakes. *Animal Behaviour* 44:574–576.

Shepherdson, D. J. 1989. Environmental enrichment. *Ratel* 16:4–9.

———. 1991a. A wild time at the zoo: Practical enrichment for zoo animals. In *Proceedings of the American Association of Zoological Parks and Aquariums Annual Conference,* 413–420. Wheeling, W.Va.: AAZPA.

———. 1991b. Behavioural enrichment. *Lifewatch* (London Zoo) (Spring): 8–9.

———. 1992. Environmental enrichment: An overview. In *Proceedings of the American Association of Zoological Parks and Aquariums Annual Conference,* 100–103. Wheeling, W.Va.: AAZPA.

Sherbrooke, W. C. 1987. Captive *Phrynosoma solare* raised without ants or hibernation. *Herpetological Review* 18:11–13.

Shoemaker, V. H., S. S. Hillman, S. D. Hillyard, D. C. Jackson, L. L. McClanahan, P. C. Withers, and M. L. Wygoda. 1992. Exchange of water, ions, and respiratory gases in terrestrial amphibians. In *Environmental Physiology of the Amphibians,* ed. M. E. Feder and W. W. Burggren, 125–150. Chicago: University of Chicago Press.

Simon, M. P. 1984. The influence of conspecifics on egg and larval mortality in amphibians. In *Infanticide: Comparative and Evolutionary Perspectives,* ed. G. Hausfater and S. B. Hrdy, 65–86. New York: Aldine.

Snyder, N. F. R., H. A. Snyder, and T. B. Johnson. 1989. Parrots return to the Arizona skies. *Birds International* 1:40–52.

Stamps, J. A. 1977. Social behavior and spacing patterns in lizards. In *Biology of the Reptilia,* Vol. 12, ed. C. Gans and F. H. Pough, 265–334. New York: Academic Press.

Stanley, M. E., and W. P. Aspey. 1984. An ethometric analysis in a zoological garden: Modifications of ungulate behavior by the visual presence of a predator. *Zoo Biology* 3:89–109.

Stockton, K. 1992. Design and management of a naturalistic mixed-species exhibit for

Sonoran birds and reptiles. In *Proceedings of the American Association of Zoological Parks and Aquariums Regional Conference,* 260–264. Wheeling, W.Va.: AAZPA.

Suboski, M. D., and J. J. Templeton. 1989. Life skills training for hatchery fish: Social learning and survival. *Fisheries Research* (Amsterdam) 7:343–352.

Sugerman, R. A. 1990. Observer effects in *Anolis sagrei. Journal of Herpetology* 24:316–317.

Sugerman, R. A., and R. A. Hacker. 1980. Observer effects on collared lizards. *Journal of Herpetology* 14:188–190.

Swingle, W. M., D. I. Warmolts, J. A. Keinath, and J. A. Musick. 1993. Exceptional growth rates of captive loggerhead sea turtles, *Caretta caretta. Zoo Biology* 12:491–497.

Tudge, C. 1991. A wild time at the zoo. *New Scientist* 1750:26–30.

von Uexküll, J. J. 1909. *Umwelt und Innenwelt der Tiere.* Berlin: Springer.

Wagner, W. E. 1992. Deceptive or honest signaling of fighting ability? A test of alternative hypotheses for the function of changes in call dominant frequency by male cricket frogs. *Animal Behaviour* 44:449–462.

Warwick, C. 1987. Effects of captivity on the ethology and psychology of reptiles. *Herpetoculturist* 1:10–12.

———. 1990a. Reptilian ethology in captivity: Observations of some problems and an evaluation of their aetiology. *Applied Animal Behaviour Science* 26:1–3.

———. 1990b. Important ethological and other considerations of the study and maintenance of reptiles in captivity. *Applied Animal Behaviour Science* 27:363–366.

———. 1990c. *Reptiles: Misunderstood, Mistreated, and Mass Marketed.* Worchester, U.K.: Trust for the Protection of Reptiles.

———. 1991. Observations on disease-associated preferred body temperatures. *Applied Animal Behaviour Science* 28:375–380.

Wells, K. D. 1977a. The social behaviour of anuran amphibians. *Animal Behaviour* 25:666–693.

———. 1977b. The courtship of frogs. In *The Reproductive Biology of Amphibians,* ed. D. H. Taylor and S. I. Guttman, 233–262. New York: Plenum Press.

———. 1988. The effect of social interactions on anuran vocal behavior. In *The Evolution of the Amphibian Auditory System,* ed. B. Fritzsch, M. J. Ryan, W. Wilczynski, T. E. Hetherington, and W. Walkowiak, 433–454. New York: Wiley.

Wilson, D. J., and H. Lefcort. 1993. The effect of predator diet on the alarm response of red legged frog, *Rana aurora,* tadpoles. *Animal Behaviour* 46:1017–1019.

Wilson, E. O. 1975. *Sociobiology: The New Synthesis.* Cambridge: Harvard University Press.

Zug, G. R. 1993. *Herpetology: An Introductory Biology of Amphibians and Reptiles.* San Diego, Calif.: Academic Press.

14

DEBRA L. FORTHMAN

TOWARD OPTIMAL CARE FOR CONFINED UNGULATES

The purpose of this chapter is twofold. The first objective is to propose that it is time for us, as zoo professionals, to move beyond the concept of "enrichment," a term applied to describe exhibition and management practices that rightly dominated the first years of the zoo revolution of the 1970s (Murphy 1976; Markowitz 1982). As we approach a new century, however, it may be useful to recognize that this most recent iteration in the history of the exhibition of wildlife lies more than two decades behind us. Many of us now work in new or renovated facilities, and a word like enrichment, which specifically describes actions taken to "enhance" or "improve" a deprived environment (Rosenzweig and Bennett 1976), should no longer dominate our thinking as we manage or design new facilities.

We should now think neither separately of "enrichment" as an adjunctive set of actions nor of how "enrichment" influences decisions about design, veterinary care, feeding, or species selection, for example. Instead, we must plan to provide optimal conditions in every aspect of our care of confined animals. The second objective acknowledges that not all zoos have been rebuilt or redesigned with behavior in mind and thus some still do require the ameliorative actions that constitute "environmental enrichment." Accordingly, the following discussion adheres largely to the first objective, while the summary describes actions that may be useful in older facilities and new enclosures that have been traditionally designed.

More and more, zoo studies have addressed enclosure design and the management and husbandry of confined ungulates (Byers 1977; Vestal and Vander Stoep 1978; Read 1982; Popp 1984; Stanley and Aspey 1984; Walther 1984; Forthman-Quick and Pappas 1986; Hutchins et al. 1987). With respect to natural history, ecology, behavior, reproduction, and ontogeny, Eisenberg (1981, 206–207) listed a number of key references for each of the ungulate taxa. Leuthold (1977) surveyed the African ungulates, but perhaps one of the best references still is the two-volume compendium of reviews and research papers by Geist and Walther (1974) on the behavior of ungulates and its implications for management. The work presented therein offers a wealth of relevant information, whenever zoo professionals begin to succumb to the syndrome eloquently described only twenty years ago by Brambell (1973, 44): "Often the approach to animal management in captivity has been to perpetuate, without searching for improvement, methods which allow the animal to survive long enough for its death not to be an embarrassment to the manager. When conditions permit the animal to survive, there is a tendency to assume that such conditions are not only satisfactory, but are the norm of good management. This approach to management automatically focuses attention on the animals' performance under the conditions of captivity, and away from a consideration of the conditions that the species has evolved to tolerate in the wild environment."

Brambell (1973, 45) also articulated, as a management goal, to "provide an environment in which the animal, if it *could* know, *would not* know that it was in captivity." To achieve this goal, and to create functionally appropriate transpositions of wild habitats (Hediger 1964), requires a broad knowledge of the evolutionary adaptations of a species to its environment (von Uexkull and Kriszat 1934; Eisenberg 1981). This issue of understanding the confined animal in the context of its evolutionary history is discussed further elsewhere in this volume by Seidensticker and Forthman (Chapter 2).

Appendix 14.1 lists the majority of ungulate orders, with elements recommended for their optimal care organized into five relevant behavioral or ecological classes of variables: (1) physical, (2) sensory, (3) occupational, (4) feeding, and (5) social (Brambell 1973; D. Shepherdson and J. Mellen, personal communication). The categorical organization is used only to facilitate this presentation and is not meant to imply that rigid functional distinctions exist among categories. Selected examples are used here to illustrate how consideration of each variable influences design, management, and husbandry of confined ungulates.

PHYSICAL VARIABLES

Animals have evolved in response to selection pressures imposed by numerous abiotic variables; some of the factors critical to the management of ungulates are described in this section.

Geographic Range

Geographic range is discussed first, both because it is important and because its implications are frequently ignored. When the ability to provide for captive animals is already constrained, it seems unnecessary to make the task more difficult by trying to provide an appropriate environment for a Dall sheep *(Ovis dalli)* or a scimitar-horned oryx *(Oryx dammah),* for example, in a zoo located in a semitropical climate. Such pronounced mismatches between the ancestral climate and captive conditions of a species subject the animals to environmental stressors that may have either profound or subtle effects. For example, because so many of the behavioral patterns in temperate-zone species are regulated by endocrine systems entrained by circadian and circannual variations in light quality, intensity, and period, daily exposure to such light cycles may be essential for successful reproduction, to name just one function (Gwinner 1977; Jander 1977; Goss 1983). For small species that may be accommodated indoors without undue expense, artificial lighting regimes may be used to approximate any seasonal changes in light cycle in the species' range. Further, an animal's ability to thermoregulate easily within a zone of comfort should be a fundamental aspect of care. When thermoregulation begins to constrain other species-typical activities, the animal is both deprived and distressed (see following).

Body Size

The factor of body size is closely related to that of geographic range. Principles of allometry dictate that very large animals will be sensitive to heat stress and very small ones to both heat and cold stress (Western 1979; Calder 1984). Thermal assessments, particularly the evaluation of microclimate availability and variety, are important in designing any exhibit but are critical for species at each end of the size spectrum; however, with the exception of holding facilities design, this factor often receives only cursory attention (J. Fraser, personal communication). Although specifications on the thermal performance of building materials and designs exist (Watson and Labs 1983; ASHRE 1989), the communication gap between biophysicists and designers is more difficult to bridge. Progress is being made, however. For example, Thompson (1991) has reviewed applications of telemetric research on thermoregulation in confined animals, and more recently Langman and colleagues have applied their basic research techniques (Langman 1985) to elephants *(Loxodonta africana)* (Langman et al., manuscript in preparation) and sea lions *(Zalophus californianus)* (Langman et al. 1996). They demonstrated that exhibits constructed primarily of gunite become hotter (or colder) and stay that way longer than horizontal or vertical surfaces made of other manufactured materials or of earth. Not surprisingly, the gunite's color also affected its thermal characteristics.

The implications for optimal care are that (1) designers should select exhibit materials that approximate natural climatic conditions, using gunite to a limited extent in locales which experience daily temperature extremes and using moated or open-fenced exhibits with earth substrates otherwise; (2) *species-appropriate* water or mud features should be employed; (3) shelter should be provided at multiple sites in exhibits, as well as by using perimeter plantings; and (4) designers must employ methods beyond simple shelter from the sun in those climates in which the heat index (a measure of "apparent temperature" that takes into account relative humidity) may be very high. With respect to the second through fourth items, until simple methods of biophysical assessment are available, a useful rule-of-thumb is to provide multiple areas in each exhibit that provide these options: sun and wind, shade and wind, sun with no wind, and shade with no wind. Finally, body size of course influences exhibit size, which will in turn influence human-animal interactions. If a large animal is exhibited in a relatively small enclosure, caretakers will almost always be within the animal's flight distance and may provoke dangerous defensive behavior (Hediger 1968).

Life History Strategy

Body size also relates to life history strategy, the third physical factor. Many although not all ungulates are relatively large, and longevity is positively correlated with body size. Further, ungulates are herbivores and generally live longer than species of the same size at other trophic levels (Eisenberg 1981), implying that providing optimal captive environments and care for many ungulates is a long-term proposition. Additionally, as most large ungulates have small "litter" sizes, the death of even one offspring is significant to their reproductive fitness (Eisenberg 1981), a factor relevant to the propagation and management of underrepresented lineages of endangered species.

For enrichment of ungulate species, then, (1) facilities must be designed to accommodate individuals throughout their lives. (2) To approximate the temporal environmental complexity with which long-lived animals may be adapted to cope, managers should strive to provide temporal variation of exhibit objects while maintaining fundamental features that contribute to their sense of security in a familiar "home range" (Stevenson 1983). (3) Relatively stable social groups should be maintained over time to best approximate field conditions for most social species (Franke Stevens 1990), including continuity of caretaking staff. Finally, (4) managers should be required to modify diets and to provide refuges to protect immature animals from aggression (Felton 1982) and environmental extremes (Langman 1977). Techniques of early pregnancy detection will ensure that prenatal environmental modifications are completed in a timely manner.

SENSORY ECOLOGY VARIABLES

With respect to the sensory ecology of ungulates, the only safe generalization about such a diverse group is that many ungulates have acute abilities in the three primary modalities: vision, audition, and olfaction. The tactile sense must also not be overlooked, as it is extremely important in social communication (Walther 1984) and is often useful in reinforcing the human-animal bond (see following). Partly because ungulate perception may not always be the same as that of the humans who design their facilities, sensory overstimulation, or the masking of species-relevant cues by irrelevant or distressful ones, must be considered as a potential and often subtle source of distress for captive ungulates (Stoskopf and Gibbons 1993). Forthman et al. (1995) present a more general review of potential influences of abiotic factors on confined animals.

Vision

Because their eyes are laterally placed, ungulates have relatively limited binocular vision. In the horse *(Equus caballus),* at least, the ratio of rod to cone cells also suggests that color vision, while probably present, may not be crucial in ungulate visual discrimination. From this and observations of behavioral responsiveness, it appears that motion and visual contrast are of primary importance to ungulates (Waring 1983), both in predator detection and in safely negotiating rough ground or unfamiliar areas (Lingle 1993; Caro 1994). Perhaps it is important, therefore, to minimize nonessential visual contrast in exhibitry, such as dramatic discontinuities in background paint or elaborate geometric decorative schemes, which would in any case be consistent with efforts to naturalize it. Such considerations may be equally important in holding areas.

Audition

Like vision, the sensory modality of audition serves multiple important functions in ungulate life. Certain conditions associated with confinement may subject ungulates to auditory distress, either chronic or acute, predictable or random (Hanson et al. 1976; Peterson 1980; Ising 1981; Gamble 1982; de Boer et al.

1988, 1989; Thomas et al. 1990; Gold and Ogden 1991). Species whose range of auditory sensitivity differs significantly from humans, as is the case with most ungulates, particularly may suffer auditory distress. For example, many ungulate holding areas are constructed with smooth-surfaced interiors because such surfaces are easier to clean. The disadvantage of such surfaces is that these are also acoustically highly reflective (C. Piper, personal communication). If animals are exhibited in climates for which they are adapted, indoor ungulate holding areas may be unnecessary.

Olfaction

Routine cleaning of exhibits and disinfection of holding areas can deprive animals of olfactory stimuli, including urine, feces, and sebaceous gland secretions, that function in spatial orientation and social communication (Doty 1976; Muller-Schwarze and Mozell 1977; Walther 1984). Some species, such as the eland *(Taurotragus oryx),* use aromatic plants for social displays (Hillman 1979). In an extension of techniques applied to large felids (Powell 1995), Hodgden (1993) used commercial extracts, such as garlic, to introduce novel scents into a black rhinoceros *(Diceros bicornis)* exhibit. Preliminary analysis suggests that the male and female were differentially responsive. More experimental work is needed regarding the management implications of auditory and chemical enrichment in ungulates.

OCCUPATIONAL VARIABLES

Activity Cycle

Consideration of a species activity cycle also has important implications for optimal care. As the result of the interaction of their feeding, thermoregulatory, and antipredator strategies, many ungulates are crepuscular or polycyclic in their activity patterns (Jarman and Sinclair 1979). Work shifts and exhibit security should be such that managers can bring animals in twice daily for individual monitoring and to supplement their diet, then turn them out again so that most of the twenty-four hours of the day are spent in the exhibit. With appropriate feeding, this routine would permit more normal activity patterns and might also reduce the incidence of behavioral stereotypies associated with confinement, social restriction, and scheduled feeding times (Cronin et al. 1986; Dodman et al. 1987, 1988). There is also evidence from other species suggesting that exposure to changes in light intensity is important for entrainment of circadian rhythms

(Helfman 1981). Therefore, if the animals cannot be outside during dawn and dusk, holding areas should have adequate skylights or appropriate artificial lighting regimes.

Habitat Use

Habitat use is inextricably related to feeding strategy. Grazers tend to exploit relatively open habitat, while browsers are usually found in more heavily wooded environments, where, depending on body size, they exploit resources between the herbaceous layer and tree tops. Browsers may also have specialized types of locomotion to facilitate quick escape over uneven terrain or through thick vegetation (Lingle 1993). When confined, however, the large size and herbivorous diet of most ungulates often result in monotonous exhibits in which the animals are managed on compacted dirt, perhaps enhanced with a mud wallow and small patches of hardy or heavily protected vegetation.

In designing exhibits for large ungulates, therefore, it is imperative to provide access for trucks, cranes and other heavy equipment necessary to install and replace large exhibit furnishings, including entire trees, as well as to till compacted substrates, adding gravel and replanting or reseeding. Plans for adequate irrigation and careful selection of the hardiest grasses and herbs are also extremely important in the maintenance of vegetation. Surface area, terrain, and substrate are additional considerations in designing and maintaining habitats that will elicit species-typical behaviors. For example, Byers (1977) found that ibex kids *(Capra ibex sibirica)* played more frequently on the most steeply sloped areas of their exhibit. Provision of appropriate and varied substrates can eliminate the need for regular hoof care. Where the maintenance of thick vegetation in exhibits is problematical, overhanging perimeter vegetation may help, as will other forms of cover such as rocks, logs, earth berms, and ditches. Again, chute systems and open-sided shelters can be incorporated into the exhibit design so that the need for indoor holding facilities is minimized (Doherty and Gibbons 1993).

FEEDING VARIABLES

Foraging Strategy

Foraging ecology influences virtually every aspect of the animal's life history, not just its diet: ranging patterns, activity budgets, and social organization, for example (Jarman 1974). Grazers, which are adapted to exploit lower-quality forage than browsers of the same body size, accordingly spend more time feeding

(Eisenberg 1981). They are often found in seminomadic herds (sometimes very large ones), while browsers tend to be territorial and live in smaller groups (Jarman and Sinclair 1979), varying from the semisolitary life of black rhinoceros (Schenkel and Schenkel-Hulliger 1969) to the family groups of dik-dik (*Madoqua* spp.), bushbuck *(Tragelaphus scriptus),* and others, to varied types of herds, such as those found among impala *(Aepyceros melampus)* and eland (e.g., nursery groups, mixed female and juvenile groups, bachelor herds; Jarman and Jarman 1979).

The implication for optimal management is that animals that range over large areas in the wild are prone to suffer from the dissociation of appetitive and consummatory behaviors which can occur with arbitrary feeding schedules (Breland and Breland 1961; Carder and Berkowitz 1970; Garcia et al. 1973). The result is often "superstitious" and adjunctive appetitive behaviors (Jenkins and Moore 1973; Moore 1973; Brett and Levine 1979; cf. Domjan and Burkhard 1986 for discussion of the distinctions between superstitious and adjunctive behavior) in the form of stereotypic patterns of ingestion (e.g., polydipsia, cribbing, pica, fence-licking) or locomotion (e.g., pacing, weaving). Adequate distribution and continuous availability of feed can prevent these problems from developing; once these habits are acquired, the success of environmental modifications in reducing or eliminating them may vary. Although it seems clear that there is no single etiology or function of stereotypical behaviors (Mason 1991; Cooper and Nicol 1993; Rushen 1993; see also Carlstead, Chapter 11, this volume), they have been induced by administration of stimulants (Robbins et al. 1989) and reduced by administration of narcotic antagonists (Dodman et al. 1987, 1988).

It has been demonstrated that foraging and diet selection in omnivores and granivores require considerable learning (Harrison 1985; Johnson et al. 1993; Valone and Giraldeau 1993), but the level of cognitive processing involved in ungulate foraging and feeding may often be underestimated (Westoby 1974; Belovsky 1981; Owen-Smith and Novellie 1982). Consider the task: from an enormous variety of plant material, ungulates must not only select the most appropriate species to eat, but must also select those *parts* that are most nutritious while avoiding those which contain harmful concentrations of toxins—a factor that covaries with phenology (Glander 1978; Hladik 1978; Janzen 1978). Vision, odor, and taste all play roles in the process of learning to select an appropriate diet, and then in applying that knowledge over time (Garcia et al. 1977, 1985; Garcia and Garcia y Robertson 1985).

Observations of captive ungulates fed only processed food, pristine hay, and fresh fruits and vegetables may not document the complete range of their abilities. Even a superficial look at feeding under field conditions is misleading, but close observation of a foraging ungulate quickly reveals that a far more sophisticated process of inspection and selection or rejection is involved (Odberg and Francis-Smith 1977; cf. Jarman and Sinclair 1979 for a review). Therefore, zoo feeding practices may limit the proportion of the species' behavioral repertoire that will be exhibited under conditions of confinement. Intense efforts by many zoos to prevent exposure to toxic plants, even those with which a species has coevolved, may be overly conservative (Rozin and Kalat 1971; Janzen 1978); it is certainly possible to challenge captive ungulates without posing risks to their health. Admittedly, doing so would require additional labor, as well as acceptance of a certain level of food waste. Appendix 14.1 offers some suggestions for increasing the cognitive challenges and the time required for feeding in captive ungulates.

Diet

With respect to diet, the fact that ungulates are with few exceptions strictly herbivorous is both an advantage and a disadvantage. It is advantageous that many species can be maintained on a small variety of vegetables, fruits, and grasses; the disadvantage is that managers have often overlooked species-specific variations in diet quality, form, and accessibility. First, the feeds given to wild ungulates may be too high in protein and too low in fiber (Dierenfeld et al. 1995). Second, the tendency to use concentrated feeds (with guaranteed nutritional composition) necessitates reduced frequency of feeding; together, these factors can contribute to health problems such as colic as well as psychological problems such as those previously mentioned (Fiennes 1966; Hediger 1966; Dittrich 1971).

SOCIAL VARIABLES

Social Structure

The propagation of confined primates improved dramatically when species were managed more often in appropriate social groups (Benirschke 1986). However, one may still find on exhibit "herds" of two or three zebra *(Equus burchelli)*, wildebeest *(Connochaetes taurinus)*, or eland. Clearly, one major problem is providing sufficient space for appropriate social groups of large, gregarious animals. The second problem is even more difficult: placement of surplus animals (Lindburg 1991).

There are at least two solutions to the first problem. One is obvious: If insufficient space is available for an appropriate herd, the species should be excluded from the collection. With the range of ungulates from which to choose, small zoos can easily concentrate on the smaller species, on appropriate groups of fewer species, or on mixed-species aggregations (Crotty 1981; Popp 1984; Partridge 1990). Another possibility for those rebuilding old zoos or expanding relatively new ones is what I refer to as "interpolated exhibitry," in which each "exhibit" is a set of linked convex and linear spaces (spheres and pathways) that may wind through a large area overall. Smaller exhibits can be embedded within larger ones, and the largest can be situated with their convex spaces adjacent to the paths of others, achieving a more integrated and efficient use of space while permitting the animal inhabitants considerable flexibility in location and movement (Forthman et al. 1995).

Reproductive Behavior

Consideration of social structure leads readily to reproductive behavior. Many ungulate species are polygynous, and, through the process of sexual selection, have evolved a variety of forms of male combat, some highly ritualized, some less so (Geist 1971; Walther 1984). Among such species, specialized management of breeding males is necessary. Because courtship and mating (Walther 1984) and parturition and mother-infant behavior (Lent 1974) are some of the most fascinating aspects of ungulate life, it is incumbent upon zoo professionals to develop methods of permitting these activities to occur while still managing the problem of surplus animals, especially males. If possible, seasonal changes in group composition to prevent male-female contact and contraception is preferable to maintaining social ungulates in single-sex groups. Formation of bachelor herds, either single- or mixed species (with care), can be appropriate in facilities with sufficient space. Another strategy is to breed each year only some proportion of the species in which breeding is desirable.

Vulnerability of Young

Vulnerability of young in confined settings usually (except in the case of mixed-species exhibits) results from disease and inter- and intraspecific aggression rather than predation and relates to management of space and group composition (Schwede et al. 1993). In addition to routine deworming, and rotation of pastures, disease caused by parasites may be reduced by adding large terrestrial birds to the exhibits (Tong 1973). With respect to aggression, in some species, adult males and other females are tolerant of immatures, while among others, multiple infant refuges, particularly for hider species, may be a fundamental requirement (Hutchins et al. 1987). This is certainly the case in exhibits that mix species of different sizes

(Felton 1982). Age at weaning and emigration patterns must be considered and planned in advance as well. It may be optimal to maintain a stable female group in an exhibit over time and to exchange breeding males, but in most species, adolescent males must be assisted to "emigrate" before they become the target of male competitive aggression (Forthman-Quick and Pappas 1986).

Human-Animal Interactions

Finally, what is appropriate in the area of human-ungulate interactions? Traditionally, the line between hands-on and hands-off has been drawn between exotics and domesticated or semidomesticated species, although managers have sometimes employed hand-rearing of neonates to reduce adult mortalities among exceptionally flight-prone species such as gerenuk *(Littocranius walleri)* and pronghorn *(Antilocapra americana)* (Muller-Schwarze and Muller-Schwarze 1973). However, given sufficient time and a positive approach, even species reputed to be extremely high-strung can become quiet and amenable to limited contact. A word of caution, however: It is generally agreed that it is ill advised to hand-raise many of the larger ungulates, because such animals become imprinted to humans and thus as adults may direct dangerous or fatal sexual or aggressive behavior to caretakers (see Appendix 14.1).

The use of training for confinement in specially designed restraint devices is particularly useful for the large and more dangerous species (Wienker 1986). This may be accomplished by means of a dedicated daily effort, patience, and the use of a proven reinforcer, such as a highly preferred food item or tactile contact, during shaping (see Kuczaj et al., Chapter 18, and Laule and Desmond, Chapter 17, this volume). When habituated to at least a short period of confinement on a regular basis, shifting, veterinary care, research, and transport are greatly facilitated, and the mortalities associated with both are reduced (Mellen and Ellis-Joseph 1991; Priest 1991; Forthman and Ogden 1992). Ultimately, training programs involving pens, crates, or chutes constructed in exhibits may reduce or eliminate the need for holding areas in many zoos.

SUMMARY AND CONCLUSIONS

I hope that this overview of suggestions will inspire creative approaches to ungulate management, encourage more quantitative evaluations of methods now in use, and promote comparison with techniques that may be tried in future. Let me conclude by summarizing each point with practical suggestions.

With respect to geographic range, managers should consider eliminating from their collections species that are poorly suited to the zoo's local climate; with the additional space available, several adjacent pens might be gated and combined to create rotating "pastures" for the largest species retained. This would allow establishing appropriate social groups and facilitate management of rutting males or bachelor herds. Reducing the number of species would also permit the flexibility to form separate nursery herds for postparturient females, while still offering visual, olfactory, and limited tactile contact between adjacent enclosures.

If the activity cycles of each remaining species are evaluated, it may be possible to alter the turnout schedule or to institute shift-training for crepuscular or polycyclic species, providing the animals with more hours in the exhibit and increased exposure to light cycles. Levels of novel olfactory stimulation can be increased periodically, by application of aromatic extracts to exhibit perimeter fencing or interior furnishings, or by leaving a small amount of fecal material in dunghills, rather than removing everything daily. It may be difficult to reduce excessive auditory stimulation in older facilities, but replacement of some sound-reflecting concrete or metal features with pressure-treated and water-sealed lumber may be possible. More appropriate visual contrast might be achieved in holding areas by the use of darker paint on floors extending in an irregular line midway up walls, with lighter, neutral shades above.

With regard to feeding strategy, bulk feed should be available at all hours; this should be distributed in small patches throughout the enclosures. For browsers, a concerted effort should be made to provide fresh browse daily, at feeding stations at varying heights; if pens can be rotated, hydroseeding with hardy grasses may be practical. Aspects of habitat use can be addressed in old enclosures with the provision of more refuges for small species that live in closed habitats; in some cases, these could even be large potted plants (Hutchins et al. 1984).

Increasing the complexity of larger exhibits may hinge on access by heavy machinery needed to dig mud wallows or pools and to install deadfalls, rocks, or other visual barriers. Installation of metered feeders, increased variety in the diet, and the replacement of some concentrated feeds with more natural forms can be attempted even in old facilities. If executed with care, mixed-species exhibits are also a practical way to achieve more interesting and complex exhibits.

Finally, minimal training procedures can be initiated in most institutions, using habituation and desensitization procedures in such facilities as stalls, small off-exhibit paddocks, transport crates, or chute systems. Routine training will make many husbandry practices possible and will permit others to be accomplished more efficiently and safely (Forthman and Ogden 1992).

APPENDIX 14.1: ENVIRONMENTAL ELEMENTS RECOMMENDED FOR OPTIMAL CARE OF UNGULATES

The recommendations presented here were developed through discussions in the ungulate workshop held at the First Environmental Enrichment Conference (Portland, Ore., 1993), and they reflect the joint ideas and experiences of the workshop participants. Most ungulate taxa are covered, with the exception of tragulids and antilocaprids; see Eisenberg (1981) and Geist and Walther (1974) for discussions of those taxa. We have organized the recommendations for each taxon into five behavioral or ecological categories (physical, sensory, occupational, feeding, and social) based on Brambell (1973) and personal communications with D. Shepherdson and J. Mellen. These categories are not meant to be rigid functional distinctions.

PERISSODACTYLS

Equids

Physical:

Moderately xeric
Minimum 0.5 hectare
Moderate vegetational complexity
Multiple substrates
Vertical furniture
Shade
Moderately shallow stream or pond
Pasture rotation
Exterior loading/handling pens
Ground scales

Sensory:

Auditory: Radio in holding, if applicable
Olfactory: Scents applied to exhibit furniture
Visual: Views of predators and competitors
Tactile: Wet and dry areas for bathing, rolling, and rubbing

Occupational:

Work (exercise) for food
Choice between solitary and social settings
Daily husbandry training sessions

Feeding:

Ad libitum pasture or well-distributed grass hay
Processing time maximized with metered devices when feeding mineral powders, mixed grain, or other supplements
Wide variety and spatiotemporal distribution of feeds

Social:

Species-typical herds; no singles or pairs
Mixed-species exhibits, but with care
If exhibit is large, multiple pastures for bachelors and harems
If exhibit is small, single sex only
Noncontact or low-contact training (desirable)

Tapirs

Physical:

Wet
Moderate size
Dense, heavily shaded
Highly complex vegetation
Varied slope
Logs, turf, sand, mulch, and mud substrates
Both deep, moving water and shallow, quiet pools

Sensory:

Olfactory: Scents applied to exhibit furniture

Tactile: Deep litter for rooting; logs for rubbing
Auditory: Any stimulation from public kept to a minimum

Occupational:
Work (exercise) for food
In exhibit 24 hours
Provisions for diving, swimming, and sliding

Feeding:
Ad libitum browse hung at varied heights
Varied diet, fed several times per day
Food hidden in substrate and logs
Floating food plants scattered on water

Social:
Females with young for up to 2 years
Males kept separate if calves present (otherwise, female regulates access to adjacent males)

Rhinos

Physical:
Species-appropriate climate and vegetational density
Large exhibits with varied substrates and soil types
Varied forms of scratching- and scent-posts
Wet and dry wallows
Integrated program of training and restraint devices

Sensory:
Olfactory: Scents applied to exhibit furniture; some dung piles
Tactile: Varied rubbing surfaces at varied heights; both wet and dry wallows
Auditory: Any "noise" kept to a minimum both indoors (if applicable) and outdoors
Visual: Long views (to minimize incidence of startling at sudden appearance of people or objects)

Occupational:
Work (exercise) for food
Choice between solitary and social settings
Daily training sessions

Feeding:
Ad libitum bulk feed, spatially well distributed
Unpredictable temporal and spatial distribution of concentrates and mineral supplements, if/when used, or administering these through devices that require animal to work
For black rhinos, browse
For white rhinos, grasses
For Sumatran rhinos, fruits
For Indian rhinos, water plants

Social:
For black rhinos in small area, adjacent paddocks; those in large area, cow-bull *or* cow-calf pairs
White rhinos in one-male, multi-female herds, mixed with other ungulates
All species with geographically appropriate variety of birds
Noncontact or low-contact training (highly desirable)

ARTIODACTYLS

Suids and Tayassuids

Physical:
Moderately large, soft exhibits
Varied substrates suitable for burrowing
Water and mud features (essential)
Logs, stumps, and sacrificial plants

Rotation of yards (desirable)
Visual barriers and refuges (essential)

Sensory:

Olfactory: Vertical objects for trail-marking

Tactile: Varied objects for rubbing and scratching

Occupational:

Movable objects, on ground and hanging
Scent application to stimulate territorial patrol
Training with tactile and food reinforcement
Self-control over social access
Work for food
Opportunities for burrow construction and swimming
Varied objects for chewing

Feeding:

Omnivorous diet ideal for varied presentation
Snout-operated dispensers, buried food, and vertical browse
Sod, hydroponics, snails, worms, "herps," rodents, and nestling birds

Social:

Sows with young, boars separate; sows control access to boars
For all species, socialization with humans at an early age to facilitate care and training

Hippopotamus

Physical:

Large exhibits for appropriate water-to-land ratio
Mixture of water and pasture or of marsh and bush (optional)
Multiple pools, and large pond with high-volume filtration
Calving area
Mud wallows
Vertical furniture
Sloping pool sides

Sensory:

Olfactory: Vertical objects for trail-marking
Auditory: Conspecific vocalizations (important)
Tactile: No rough vertical surfaces

Occupational:

Holding kept to a minimum
Nocturnal, terrestrial grazing
Wallows or marshy areas for daytime
Manipulable objects such as "boomer balls"

Feeding:

Pasture, or well-distributed mixed hays and minerals
"Fruit-cicles" or floating fruit and vegetables during day

Social:

Conspecific contact and vocalizations (very important)
Group size ≥ 1.4 animals
Mixed-species exhibits with fish, turtles, birds
Body care by keepers
Training for shifting and dental inspection

Camelids

Physical:

Xeric-arid, often at altitude
Moderately large to large exhibits
Complex terrain and varied substrates
Large pond or recirculating stream
Much shade from protected trees; herbaceous plants
Visual barriers, escape areas, and rubbing posts
Barn or covered area with water and feed stations
For South American species, rotation of yards to maintain grass

For other species, overhanging perimeter shade and browse

Sensory:

All senses are acute

Olfactory: Use of scents to distract breeding males during separation from females

Tactile: Wet and dry bathing areas

Occupational:

Work for food

Exercise and training (essential, especially if males held alone when not breeding)

Artificial sparring partners

Restraint chutes for intractable Old World individuals; halter- and target-training for domestic species

Feeding:

Grass hays, herbs, browse, and hydroponics in multiple, small feedings

Multiple feeder locations at multiple heights

Overhanging browse outside exhibits

Social:

Mixed-species exhibits with cathartids, small canids, and felids

Group size ≥ 1.5 animals

Removal of yearling males

No hand-raising of males from birth (to avoid "berserk llama syndrome")

Training for loading

Use of restraint devices for veterinary care

Placement of breeding group in exhibit next to a bachelor herd (if possible)

Measures taken to *avoid* having caretakers perceived as members of the social group (important)

Cervids

Physical:

Geographic range and body size are highly variable

Moderately to highly complex exhibits

Herbaceous layer for neonatal cover

Varied substrates

Varied vertical furniture for antler maintenance and display

Shade, pools, and overhanging vegetation and browse

Rotation of pastures

Picket-type fencing to replicate forest view and to reduce risk of self-injury

Sensory:

Auditory and olfactory senses are most important

Tactile: Use of wallows (very important for many species' social, reproductive, and thermoregulatory behavior)

Auditory and olfactory: Stimulation kept to a minimum, especially for solitary species, which are very stress-prone

Solid verticals and overhanging components

Occupational:

Large exhibits and placement of varied resources in multiple sites to promote exercise

Feeders requiring appropriate foraging behavior, such as pawing and tongue use

Feeding:

Harvest browse in winter

Varied amounts of concentrates fed seasonally

Multiple feed stations at varied heights

Scatter feeders set to distribute feed on natural temporal schedule: morning, evening, and night

Social:

Mixed-species exhibits with other ungulates and birds

Freeze-branding for permanent identification in large herds

No hand-rearing (to avoid harm when animal reaches adulthood)

"Arms-length" policy for human interactions with animal

Quick return of animal to herd after any handling, to minimize stress

"Seeding" of a flighty herd with one to three calm animals

Restraint chutes (very useful)

Use of "creeps" to let females control male access

Giraffids

Physical:

Moderately large to large exhibits

Species-appropriate habitat (wide variation, from dense rain forest with varied understory to short- and tall-grass savanna, bush, and woodland)

For giraffes, varied substrates (mud, hardpack, and sand)

Much shade, and pools and streams

Rotation of pastures or use of exclusion areas

Overhanging browse and bamboo

Multiple shift, quarantine, calving, and introduction areas

Skylights, cross-ventilation, and ample room in holding

For giraffes, integrated program of training and use of restraint devices (essential)

Ample straight-line space for explosive flight response

Sensory:

For giraffes, visual and auditory senses are the most important; for okapi, auditory and olfactory. Infrasound may be quite important to okapi

Background noise kept to a minimum

Occupational:

Rubbing posts, other than trees, at varied heights

Food ad libitum and well distributed in connecting yards

Water at ground level

Restraint-, shift-, target-, and trailer-training (all highly desirable)

Feeding:

24-hour availability of browse

Multiple feed areas, and vertical and hanging browse feeders, filled at different times of day

Varied mixes of grass and legume hays

Multiple types of browse

Fresh produce given as treats and for training

Multiple mineral sources

Social:

Mixed-species exhibits with ungulates, birds, small canids, and felids (very effective)

Designation of individual's purpose—either for breeding or for exhibiting, not both

Hands-on training by humans mostly for habituation to shifting, loading, and targeting to floor scales; veterinary care in restraint chute

Bovids: Antelopinae

Physical:

Species-appropriate ranges (wide variation, from small territories to huge grasslands)

Exhibits longer than they are wide, to serve as a buffer from humans (essential to some species)

Multiple, varied refugia for small species

Pasture rotation for large species

Varied substrates

Multiple sources of shade and shallow water
Visual barriers with terrain variations
Refugia for calves and fawns

Sensory:

All senses are acute

Olfactory: Various furniture and vegetation for scent-marking
Tactile: Varied substrates for middens and for mud and dust baths; fiber brushes mounted on walls and fences for grooming
Visual and auditory: Long corridors away from public to accommodate explosive flight, and visible fencing to reduce risk of injury during such episodes

Occupational:

Ad libitum foraging
Hanging branches, logs, and obstacles to jump
Sufficient space for top-speed locomotion
Periodic rearrangement of some internal furniture (e.g., logs, rocks, heated shelters)
Work requirement for food supplements and for fresh produce "treats"

Feeding:

24-hour availability of feed
Multiple feeders at varied heights
Daily variation of feeder contents; periodic variation of feeder locations
Spatial dispersal of feeders to reduce competition
Varied diet of hays, herbs, flowers, bark, fruits, hydroponics, and various mineral powders

Social:

Mixed-species exhibits, if areas are large
Placement of a variety of social organizations (bachelor herds, leks, creches, and solitary males) in adjacent paddocks
Mutiple refugia for young or small species
Limited, species-appropriate contact with humans
No hand-rearing of males of aggressive species
Training to shift, load, and target to scales, and for other husbandry procedures

Bovids: Bovinae

Physical:

Large exhibits of moderate complexity
Species-appropriate and habitat-appropriate vegetational complexity
Restraint devices and electric gates for larger species
Furniture such as deadfalls and mature, protected trees
Rotation of pastures to maintain grass
Varied substrates; mud and dust wallows
Multiple vertical and overhanging components
Streams and pools
Multiple, large sources of shade

Sensory:

All senses are acute

Tactile: Varied and plentiful opportunities for bathing, horn care, and body care
Auditory: Quiet, secluded, shaded areas for rest and rumination

Occupational:

24-hour grazing
Distribution of concentrates among adjacent pastures; daily variation of configuration and contents
Concentrates and minerals, if required, administered by dispensers requiring work
Artificial sparring partners

Visual, olfactory, and auditory access to predators and competitors

Feeding:

24-hour access to mixed hays, minerals, and produce

Limited use of concentrate feeds

Social:

Wallows and shade areas large enough to accommodate several animals in physical contact

Herds kept as large as possible; bulls kept with herd if calm, or bulls held in adjacent pens

Mixed-species exhibits possible with other bovids, especially bachelor groups, and with bird symbionts and other large terrestrial birds

Bovids: Hippotraginae

Physical:

Species-appropriate exhibit size and complexity (wide variation within subfamily, from medium to very large exhibits of low to moderate complexity)

Species-appropriate and habitat-appropriate vegetational complexity (from xeric to marshy)

Chute systems for routine husbandry procedures (optional)

Furniture such as deadfalls and mature, protected trees

Varied substrates and well-spaced mounds for territorial display

For savanna species, rotation of pastures to maintain grass

For species from more wooded habitats, provision of vertical and overhanging components

Streams or pools

Multiple, large sources of shade

Sensory:

All senses are acute

Visual: Unobstructed views (beneficial to most species)

Tactile: Varied substrates for horn and body care

Occupational:

24-hour grazing for savanna species

Distribution of concentrates among adjacent pastures; daily variation of configuration and contents

Concentrates and minerals, if required, administered by dispensers requiring work

Artificial sparring partners for species with ritualized combat

Visual, olfactory, and auditory access to predators and competitors

Feeding:

24-hour access to mixed hays, minerals, and produce

Limited use of concentrate feeds

Social:

For most species, herds kept as large as possible; if space is sufficient, possibility of exhibiting multiple bulls, each with territory or lek space, as appropriate

Mixed-species exhibits possible with other bovids, especially bachelor groups, and with bird symbionts and other large terrestrial birds

Bovids: Caprinae

Physical:

Exhibiting of alpine species limited to climates with hard freezes

For forest species, more vegetation than rock

Moderate to large exhibits; complex terrain with strong vertical component for most species

Visitor area located at bottom of slope (ideal)

Brush, ponds, and various furniture for butting and rubbing

Areas of heavy shade (essential because of strong potential for heat stress)

Varied substrates from rock to pasture; terraces for bedding

Restraint chutes with floor scales

Large area for lambing

Sensory:

All senses are acute

Visual: Multiple areas for panoramic scans

Tactile: Furniture of different textures at different heights for coat maintenance and shedding

Olfactory: Only limited cleaning of dung heaps and marked furniture

Occupational:

For ovid and caprid males, opportunities to engage in ritual combat

For rupicaprids, artificial sparring partners of log, brush, or adjacent males

Varied, plentiful vegetation for scent-marking

Interconnecting yards and placement of essential resources at tops and bottoms of slopes to encourage regular exercise

Barriers and promontories to encourage species-typical play patterns

Noncontact training for husbandry procedures; use of restraints

Feeding:

24-hour availability of mixed hays and browse

Multiple, well-spaced feeders and varied distribution of feeds

Upright browse containers

Placement of grain and produce feeders or dispensers in outcrops

Multiple mineral sources

Social:

Outside rut, duplicate bachelor herds and female groups; during rut, bachelor herds and groups of females with one ram

Solid barrier between groups if needed to prevent self-inflicted injuries

Use of contraception or euthanizing of surplus males

Minimal contact with keepers

Minimal territorial intrusions by keepers or other "outsiders" during rut, as rams may redirect aggression toward group members

REFERENCES

ASHRE (American Society for Heating, Refrigerating, and Air Conditioning Engineers). 1989. *ASHRE Handbook and Product Directory. Fundamentals Volume.* Atlanta: ASHRE.

Belovsky, G. E. 1981. Food plant selection by a generalist herbivore: The moose. *Ecology* 62:1020–1030.

Benirschke, K., ed. 1986. *Primates: The Road to Self-Sustaining Populations.* New York: Springer-Verlag.

Brambell, M. R. 1973. The requirements of carnivores and ungulates in captivity. In *The Welfare and Management of Wild Animals in Captivity,* 44–49. Potters Bar, U.K.: Universities Federation for Animal Welfare.

Breland, K., and M. Breland. 1961. The misbehavior of organisms. *American Psychologist* 16:661-664.

Brett, L. P., and S. Levine. 1979. Schedule-induced polydipsia suppresses pituitary-adrenal activity in rats. *Journal of Comparative and Physiological Psychology* 93:946-956.

Byers, J. A. 1977. Terrain preferences in the play behavior of Siberian ibex kids *(Capra ibex sibirica). Zeitschrift für Tierpsychologie* 45:199-209.

Calder, W. A. III. 1984. *Size, Function, and Life History.* Cambridge: Harvard University Press.

Carder, B., and K. Berkowitz. 1970. Rats preference for earned in comparison with free food. *Science* 167:1273-1274.

Caro, T. 1994. Ungulate antipredator behaviour: Preliminary and comparative data from African bovids. *Behaviour* 128:189-228.

Cooper, J. J., and C. J. Nicol. 1993. The "coping" hypothesis of stereotypic behaviour: A reply to Rushen. *Animal Behaviour* 45:616-618.

Cronin, G. M., P. R. Weipkema, and J. M. van Ree. 1986. Endorphins implicated in stereotypies in tethered sows. *Experientia* (Basel) 42:198-199.

Crotty, M. J. 1981. Mixed exhibits or "What's that funny looking animal in with the monkeys?" *International Zoo Yearbook* 21:203-206.

De Boer, S. F., J. L. Slangen, and J. van der Gugten. 1988. Adaptation of plasma catecholamine and corticosterone responses to short-term repeated noise stress in rats. *Physiology and Behavior* 44:273-280.

De Boer, S. F., J. van der Gugten, and J. L. Slangen. 1989. Plasma catecholamine and corticosterone responses to predictable and unpredictable noise stress in rats. *Physiology and Behavior* 45:795-798.

Dierenfeld, E. S., R. du Toit, and W. E. Braselton. 1995. Nutrient composition of selected browses consumed by black rhinoceros *(Diceros bicornis)* in the Zambezi Valley, Zimbabwe. *Journal of Zoo and Wildlife Medicine* 26:220-230.

Dittrich, L. 1971. Food presentation in relation to behaviour in ungulates. *International Zoo Yearbook* 16:48-54.

Dodman, N. H., L. Shuster, M. H. Court, and R. Dixon. 1987. Investigation into the use of narcotic antagonists in the treatment of a stereotypic behavior pattern (crib-biting) in the horse. *American Journal of Veterinary Research* 48:311-319.

Dodman, N. H., L. Shuster, M. H. Court, and J. Patel. 1988. Use of a narcotic antagonist (nalmefene) to suppress self-mutilative behavior in a stallion. *Journal of the American Veterinary Medical Association* 192:1585-1586.

Doherty, J. G., and E. F. Gibbons, Jr. 1993. Managing naturalistic environments in captivity. In *Naturalistic Environments in Captivity for Animal Behavior Research,* ed. E. F. Gibbons, Jr., E. J. Wyers, E. Waters, and E. W. Menzel, Jr., 125-141. Albany: State University of New York Press.

Domjan, M., and B. Burkhard. 1986. *The Principles of Learning and Behavior,* 2nd ed. Monterey, Calif.: Brooks/Cole.

Doty, R. L. 1976. *Mammalian Olfaction, Reproductive Processes, and Behavior.* New York: Academic Press.

Eisenberg, J. F. 1981. *The Mammalian Radiations: An Analysis of Trends in Evolution, Adaptation, and Behavior.* Chicago: University of Chicago Press.

Felton, G., Jr. 1982. Aspects of mixed hoofstock species exhibits. In *Proceedings of the Annual Conference of the American Association of Zoological Parks and Aquariums,* 235-238. Wheeling, W.Va.: AAZPA.

Fiennes, R. 1966. Feeding animals in captivity. *International Zoo Yearbook* 6:58-67.

Forthman, D. L., R. McManamon, U. A. Levi, and G. Y. Bruner. 1995. Interdisciplinary issues in the design of mammal exhibits (excluding marine mammals and primates). In *Captive Conservation of Endangered Species,* ed. E. F. Gibbons, Jr., J. Demarest, and B. Durrant, 377-399. Albany: State University of New York Press.

Forthman, D. L., and J. J. Ogden. 1992. The role of applied behavior analysis in zoo management: Today and tomorrow. *Journal of Applied Behavior Analysis* 25:647-652.

Forthman-Quick, D. L., and T. C. Pappas. 1986. Enclosure utilization, activity budgets, and social behavior of captive chamois *(Rupicapra rupicapra)* during the rut. *Zoo Biology* 5:281-292.

Franke Stevens, E. 1990. Instability of harems of feral horses in relation to season and presence of subordinate stallions. *Behaviour* 112:149-161.

Gamble, M. R. 1982. Sound and its significance for laboratory animals. *Biological Review* 57:395-421.

Garcia, J., J. C. Clarke, and W. G. Hankins. 1973. Natural responses to scheduled rewards. In *Perspectives in Ethology,* ed. P. P. G. Bateson and P. Klopfer, 1-41. New York: Plenum Press.

Garcia, J., and R. Garcia y Robertson. 1985. The evolution of learning mechanisms. In *Psychology and Learning: The Master Lecture Series,* vol. 4, ed. B. L. Hammonds, 191-243. Washington, D.C.: American Psychological Association.

Garcia, J., W. G. Hankins, and J. D. Coil. 1977. Koalas, men, and other conditioned gastronomes. In *Food Aversion Learning,* ed. N. W. Milgram, L. Krames, and T. M. Alloway, 195-218. New York: Plenum Press.

Garcia, J., P. S. Lasiter, F. Bermudez-Rattoni, and D. A. Deems. 1985. A general theory of aversion learning. In *Experimental Assessments and Clinical Applications of Conditioned Food Aversions,* ed. N. Braveman and P. Bronstein, 8-21. Annals of the New York Academy of Sciences, vol. 443. New York: New York Academy of Sciences.

Geist, V. 1971. *Mountain Sheep.* Chicago: University of Chicago Press.

Geist, V., and F. R. Walther, eds. 1974. *The Behaviour of Ungulates and Its Relation to Management,* Vols. 1 and 2. IUCN publication n.s. no. 24. Morges, Switzerland: International Union for the Conservation of Nature.

Glander, K. E. 1978. Howling monkey feeding behavior and plant secondary compounds: A study of strategies. In *The Ecology of Arboreal Folivores,* ed. G. G. Montgomery, 561-574. Washington, D.C.: Smithsonian Institution Press.

Gold, K. C., and J. J. Ogden. 1991. Effects of construction noise on captive lowland gorillas *(Gorilla gorilla gorilla)* [abstract]. *American Journal of Primatology* 24:104.

Goss, R. J. 1983. Control of deer antler cycles by the photoperiod. In *Antler Development in Cervidae,* ed. R. D. Brown, 1-14. Kingsville, Tex.: Caesar Kleberg Wildlife Research Institute.

Gwinner, E. 1977. Biological clocks. In *Grzimek's Encyclopedia of Ethology,* ed. B. Grzimek, 187-198. New York: Van Nostrand Reinhold.

Hanson, J. P., M. E. Larson, and C. T. Snowdon. 1976. The effects of control over high intensity noise on plasma cortisol levels in rhesus monkeys. *Behavioral Biology* 16:333-340.

Harrison, M. J. S. 1985. Time budget of the green monkey, *Cercopithecus sabaeus:* Some optimal strategies. *International Journal of Primatology* 6:351-376.

Hediger, H. 1964. *Wild Animals in Captivity.* New York: Dover.

———. 1966. Diet of animals in captivity. *International Zoo Yearbook* 6:37-57.

———. 1968. *The Psychology and Behaviour of Animals in Zoos and Circuses.* New York: Dover.

Helfman, G. S. 1981. Twilight activities and temporal structure in a freshwater fish community. *Journal of Fisheries and Aquatic Sciences* 38:1405-1420.

Hillman, J. C. 1979. The biology of the eland (*Taurotragus oryx* Pallas) in the wild. Ph.D. dissertation, University of Nairobi, Nairobi, Kenya.

Hladik, A. 1978. Phenology of leaf production in rain forest of Gabon: Distribution and composition of food for folivores. In *The Ecology of Arboreal Folivores,* ed. G. G. Montgomery, 51-72. Washington, D.C.: Smithsonian Institution Press.

Hodgden, R. 1993. Rhino enrichment at Zoo Atlanta. Unpublished report, Zoo Atlanta, Atlanta.

Hutchins, M., D. Hancocks, and C. Crockett. 1984. Naturalistic solutions to the behavioural problems of captive animals. *Zoologische Garten* 54:28-42.

Hutchins, M., G. Thompson, B. Sleeper, and J. Foster. 1987. Management and breeding of the Rocky Mountain goat *(Oreamnos americanus)* at Woodland Park Zoo. *International Zoo Yearbook* 26:297-308.

Ising, H. 1981. Interaction of noise-induced stress and Mg decrease. *Artery* 9:205-211.

Jander, R. 1977. Orientation ecology. In *Grzimek's Encyclopedia of Ethology,* ed. B. Grzimek, 145-163. New York: Van Nostrand Reinhold.

Janzen, D. H. 1978. Complications in interpreting the chemical defenses of trees against tropical arboreal plant-eating vertebrates. In *The Ecology of Arboreal Folivores,* ed. G. G. Montgomery, 73-84. Washington, D.C.: Smithsonian Institution Press.

Jarman, P. J. 1974. The social organization of antelopes in relation to their ecology. *Behaviour* 58:215-267.

Jarman, P. J., and M. V. Jarman. 1979. The dynamics of ungulate social organization. In *Serengeti: Dynamics of an Ecosystem,* ed. A. R. E. Sinclair and M. Norton-Griffiths, 185-220. Chicago: University of Chicago Press.

Jarman, P. J., and A. R. E. Sinclair. 1979. Feeding strategy and the pattern of resource-partitioning in ungulates. In *Serengeti: Dynamics of an Ecosystem,* ed. A. R. E. Sinclair and M. Norton-Griffiths, 130-163. Chicago: University of Chicago Press.

Jenkins, H., and B. Moore. 1973. The form of the auto-shaped response with food or water reinforcers. *Journal of the Experimental Analysis of Behavior* 20:163-181.

Johnson, D. F., J. Triblehorn, and G. Collier. 1993. The effect of patch depletion on meal patterns in rats. *Animal Behaviour* 46:55-62.

Langman, V. A. 1977. Cow-calf relationships in giraffe. *Zeitschrift für Tierpsychologie* 43:264-286.

———. 1985. Heat balance in the black rhinoceros *(Diceros bicornis). National Geographic Research Reports* 21:251-254.

Langman, V. A., M. Rowe, D. L. Forthman, B. Whittington, N. V. Langman, T. J. Roberts, K. Huston, C. Boling, and D. Maloney. 1996. Thermal assessment of zoological exhibits. I. Sea lion enclosure at the Audubon Zoo. *Zoo Biology* 15:403-411.

Lent, P. C. 1974. Mother-infant relationships in ungulates. In *The Behaviour of Ungulates and Its Relation to Management,* vol. 1, ed. V. Geist and F. Walther, 14-55. International Union for the Conservation of Nature (IUCN) publication n.s. no. 24. Morges, Switzerland: IUCN.

Leuthold, W. 1977. *African Ungulates: Zoophysiology and Ecology,* vol. 8. Berlin: Springer.

Lindburg, D. G. 1991. Zoos and the "surplus" problem. *Zoo Biology* 10:1-2.

Lingle, S. 1993. Escape gaits of white-tailed deer, mule deer, and their hybrids: Body configuration, biomechanics, and function. *Canadian Journal of Zoology* 71:708-724.

Markowitz, H. 1982. *Behavioral Enrichment in the Zoo.* New York: Van Nostrand Reinhold.

Mason, G. J. 1991. Stereotypies: A critical review. *Animal Behaviour* 41:1015-1037.

Mellen, J., and S. Ellis-Joseph. 1991. Learning principles as they apply to animal husbandry. In *Proceedings of the Annual Conference of the American Association of Zoological Parks and Aquariums,* 548-552. Wheeling, W.Va.: AAZPA.

Moore, B. R. 1973. The role of directed Pavlovian reactions in simple instrumental learning in the pigeon. In *Constraints on Learning: Limitations and Predispositions,* ed. R. A. Hinde and J. Stevenson-Hinde, 150-188. New York: Academic Press.

Muller-Schwarze, D., and M. M. Mozell, eds. 1977. *Chemical Signals in Vertebrates.* New York: Plenum Press.

Muller-Schwarze, D., and C. Muller-Schwarze. 1973. Behavioral development of hand-reared pronghorn *Antilocapra americana. International Zoo Yearbook* 13:217.

Murphy, D. E. 1976. Enrichment and occupational devices for orangutans and chimpanzees. *International Zoo News* 137:24-26.

Odberg, F. O., and K. Francis-Smith. 1977. Studies on the formation of ungrazed eliminative areas in fields used by horses. *Applied Animal Ethology* 3:27-34.

Owen-Smith, N., and P. Novellie. 1982. What should a clever ungulate eat? *American Naturalist* 119:151-178.

Partridge, J. 1990. Mixed animal exhibits. *International Zoo News* 371:13-18.

Peterson, E. A. 1980. Noise and laboratory animals. *Laboratory Animal Science* 30: 422–439.

Popp, J. W. 1984. Interspecific aggression in mixed ungulate species exhibits. *Zoo Biology* 3:211–219.

Powell, D. M. 1995. Preliminary evaluation of environmental enrichment techniques for African lions *(Panthera leo). Animal Welfare* 4:361–370.

Priest, G. 1991. The methodology for developing animal behavior management programs at the San Diego Zoo and Wild Animal Park. In *Proceedings of the Annual Conference of the American Association of Zoological Parks and Aquariums,* 553–575. Wheeling, W.Va.: AAZPA.

Read, B. 1982. Successful reintroduction of bottle-raised calves to antelope herds at the St. Louis Zoo. *International Zoo Yearbook* 22:269–270.

Robbins, T. W., G. Mittleman, J. O'Brien, and P. Winn. 1989. The neuropsychological significance of stereotypy induced by stimulant drugs. In *Neurobiology of Stereotyped Behaviour,* ed. S. J. Cooper and C. T. Dourish, 25-63. Oxford, U.K.: Clarendon Press.

Rosenzweig, M. R., and E. L. Bennett. 1976. Enriched environments: Facts, factors, and fantasies. In *Knowing, Thinking, and Believing,* ed. L. Petrinovich and Y. J. McGaugh, 179–213. New York: Plenum Press.

Rozin, P., and J. W. Kalat. 1971. Specific hungers and poison avoidance as adaptive specializations of learning. *Psychological Review* 78:459–486.

Rushen, J. 1993. The "coping" hypothesis of stereotypic behaviour. *Animal Behaviour* 45:613–615.

Schenkel, R., and L. Schenkel-Hulliger. 1969. *Ecology and Behavior of the Black Rhinoceros (Diceros bicornis): Mammalia Depicta.* Berlin: Paul Parey.

Schwede, G., H. Hendrichs, and W. McShea. 1993. Social and spatial organization of female white-tailed deer, *Odocoileus virginianus,* during the fawning season. *Animal Behaviour* 45:1007–1017.

Stanley, M. E., and W. P. Aspey. 1984. An ethometric analysis in a zoological garden: Modification of ungulate behavior by the visual presence of a predator. *Zoo Biology* 3:89–109.

Stevenson, M. F. 1983. The captive environment: Its effect on exploratory and related behavioural responses in wild animals. In *Exploration in Animals and Humans,* ed. J. Archer and L. Birke, 176–197. Wokingham, U.K.: Van Nostrand Reinhold.

Stoskopf, M. K., and E. F. Gibbons, Jr. 1993. Quantitative evaluation of the effects of environmental parameters on the physiology, behavior, and health of animals in naturalistic captive environments. In *Naturalistic Environments in Captivity for Animal Behavior Research,* ed. E. F. Gibbons, Jr., E. J. Wyers, E. Waters, and E. W. Menzel, Jr., 140–160. Albany: State University of New York Press.

Thomas, J. A., R. A. Kastelein, and F. T. Awbrey. 1990. Behavior and blood catecholamines of captive belugas during playbacks of noise from an oil drilling platform. *Zoo Biology* 9:393–402.

Thompson, S. D. 1991. Biotelemetric studies of mammalian thermoregulation. In *Biotelemetry Applications for Captive Animal Care and Research,* ed. C. S. Asa, 25–37. Wheeling, W.Va.: American Association of Zoological Parks and Aquariums.

Tong, E. H. 1973. The requirements of ungulates and carnivores in safari parks. In *The Welfare and Management of Wild Animals in Captivity,* 50–55. Potters Bar, U.K.: Universities Federation for Animal Welfare.

Valone, T. J., and L.-A. Giraldeau. 1993. Patch estimation by group foragers: What information is used? *Animal Behaviour* 45:721–728.

Vestal, B. M., and A. Vander Stoep. 1978. Effect of distance between feeders on aggression in captive chamois *(Rupicapra rupicapra). Applied Animal Ethology* 4:253–260.

von Uexkull, J., and G. Kriszat. 1934. *Streifzuge durch die Umwelten von Tieren und Menschen.* Berlin: Springer.

Walther, F. R. 1984. *Communication and Expression in Hoofed Mammals.* Bloomington: Indiana University Press.

Waring, G. H. 1983. *Horse Behavior: The Behavioral Traits and Adaptations of Domestic and Wild Horses, Including Ponies.* Park Ridge, N.J.: Noyes Publications.

Watson, D., and K. Labs. 1983. *Climatic Design.* New York: McGraw-Hill.

Western, D. 1979. Size, life history, and ecology in mammals. *African Journal of Ecology* 17:185–204.

Westoby, M. 1974. An analysis of diet selection by large generalist herbivores. *American Naturalist* 108:290–304.

Wienker, W. R. 1986. Giraffe squeeze cage procedure. *Zoo Biology* 5:371–377.

DONALD G. LINDBURG

ENRICHMENT OF CAPTIVE MAMMALS THROUGH PROVISIONING

The history of animal-keeping is a sobering reminder of the ease with which quality living can be denied to even the most sentient and developmentally sensitive animals (Erwin et al. 1979). Early advocates of enriched living environments such as Robert Yerkes (1925) and Heini Hediger (1950) recognized the need for mental and physical stimulation in captive milieux, but were exceptional figures in an era dominated by considerations for sanitation, convenience to keepers and users, and economies in captive holding (Coe 1987; Segal 1989). In recent years, increased public awareness of conditions often perceived as inhumane has made enrichment fashionable, and ethologists having relevant advice to offer are today enjoying newfound respect (Shepherdson, Chapter 1, this volume). Accordingly, mere intellectual assent to the dynamic interaction between environmental factors and animal behavior and physiology has been replaced by an era in which occupational devices designed to provide stimulation in formerly sterile conditions have greatly proliferated.

Passivity toward the depauperate conditions of captivity thus gave way to a new phase, characterized by the search for relief through instrumentation (Markowitz 1982). This occurred in response to the animal welfare movement and, in the United States, to an act of the U.S. Congress mandating attention to this need (Table 15.1). In fact, most publications on enrichment emanating from U.S. research laboratories today seek scientific buttressing for the effort in this act, seemingly suggesting thereby that, if it were not illegal to do so, many laboratory animals would still be living under conditions of minimal external stimulation. Zoos, on the other hand, although equally slow to respond, have acted in relation to their changing role as conservation entities and to increased

Table 15.1
Phases in the Development of Environmental Enrichment for Captive Animals

1. Passive phase:
   - Intellectual assent to the ideas of Yerkes, Hediger, et al.
   - Emphasis on costs of keeping, on sanitizable environments, and on convenience to users
2. Instrumental phase:
   - Introduction of numerous occupational devices, often without validation of beneficial effects (the era of "show and tell"), in response to concerns about animal welfare
   - Initial remodeling of research labs and exhibits in advance of scientific justification
3. Scientific phase:
   - Attempts at defining "psychological well-being" scientifically
   - Emphasis on strategies and on theoretical issues—i.e., determining why particular approaches work or do not work
   - Emergence of journals and newsletters devoted to reporting on enrichment programs

reliance on captive breeding of wildlife for their exhibition and educational programs (Benirschke 1986). Despite these somewhat different justifications, the occupational devices that have been developed (Shepherdson 1989a,b) often have broad application both to the laboratory cage and to many of the environments found in zoos today.

A third phase in the development of enrichment studies has been signaled by the appearance of papers that go beyond mere descriptions of responses to instrumentation and their presumed benefits to an emphasis on discerning why animals react as they do. As characterized by Shepherdson (1993), immediate responses to an enrichment device or technique may not have generalized effects that are measurable for some time after the act of stimulation itself unless the strategy behind the enrichment taps more effectively into the motivational states of the animals. Marriage of the motivational literature to the useful findings of the instrumental phase offers exciting possibilities for the development of scientific theories of enrichment. Initial examples of the shift in focus from "what" to "why" are papers by Wemelsfelder (1985), Hughes and Duncan (1988), Moodie and Chamove (1990), and Shepherdson et al. (1993).

Although the instrumental phase has been of short duration, it has provided several reminders that simple solutions to complex problems do not always produce the desired effects. For example, it has been repeatedly demonstrated that enrichment that is anthropomorphically defined (Woolverton et al. 1989) is much less effective, if at all, than enrichment that is ecologically relevant (Chamove 1989). Other lessons that we have learned about approaches to enrichment are as follows:

- Beware of a priori assumptions; for example, more is not always better.
- Forget about quick-fix approaches, such as adding autoclavable balls to an otherwise sterile environment.
- To base enrichment strategies on an animal's manifest preferences may be as unhealthy for animals as fast-food availability is for humans.
- Age, sex, and early rearing experience are important in shaping an animal's responses, as seen, for example, in the futility of trying to elicit play from most adult animals.
- Biologically relevant stimuli such as browse are more likely to be effective than toys.
- The unavoidable conditioning (primarily auditory) of captive animals to care and maintenance routines can affect the success of an enrichment effort.
- Novelty is the key to successful enrichment, maintained at adequate levels over the long haul and complementary to stimuli of internal origin. Too little novelty (boredom) or too much novelty (stress) will not produce the desired effect.

RATIONALE FOR ENRICHMENT THROUGH PROVISIONING

Activities concerned with food-getting and food consumption offer excellent opportunities for enriching captive animals. The obvious reason is that obtaining food is a survival activity, requiring effort on the part of each individual at fairly regular intervals. Survival behaviors such as food-getting take precedence, for the most part, over other behaviors that are sex- or age related (e.g., playing, grooming, mating). And, for the many species for which food procurement is a relatively frequent event, the differential investment in this activity between wild and captive environments suggests an obvious route for engaging the interests of those animals that live in captivity.

A second reason for approaching enrichment through food acquisition is that in their natural form food items are inherently interesting to animals, under conditions of good health, and thus invariably elicit a range of manipulative and exploratory acts. These responses are known to occur even when satiation may not be the objective. On the other hand, the rate at which inorganic objects occupy the attention of their intended beneficiaries falls off dramatically unless these items are changed regularly. Achieving an appropriate level of novelty through feeding is of much less concern than when attempting to elicit play or exploratory activity.

THE FEEDING PARADIGM

Predicated on the commonsense notion that in its natural state an animal encounters variation in availability, accessibility, and palatability of food items on a regular basis, enrichment through provisioning most often takes the form of adding variety in the types of food items offered, in their mode of presentation, or in increased frequency of provisioning. As a rule, however, these efforts underestimate the complexity of the feeding endeavor. While distinctions between classes of animals and their foods vary so much that generalizations are not possible, the contrast between strongly herbivorous and strongly carnivorous specialists is illustrative (Table 15.2). A lion *(Panthera leo)*, for example, pursues food that is itself mobile but is concentrated in single, fairly large packages which are somewhat opportunistically available. Although the lion may consume large quantities at each feeding, both the quantity consumed and the time spent in forage-feeding activities are small by comparison with a large-bodied herbivore, such as the gaur *(Bos frontalis)*. The food of the latter, by contrast, comes in small packages, is dispersed, immobile, and more predictably available than prey. The time spent in harvesting and the quantity consumed are, accordingly, greater than for that of the carnivore. This contrast illustrates that, as a starting point, knowledge of the biology of a species is prerequisite to successful enrichment through provisioning.

Despite the variety of food in the natural world, the quest for food has certain fundamental aspects that can be used as a guide in designing meaningful approaches to enrichment. To this end, procurement of food can be divided into four phases that would cover the majority of vertebrates (Table 15.3).

A first requirement is that for feeding to occur there must be contact between the animal and its food source. At one extreme the animal goes in search of food; at the other, the sentinel predator waits for the food to come within its range (Hamilton 1973). For searchers, the activity can be as simple as that of the grazer which seeks the preferred patches of grass, or as demanding as that of the tiger *(Panthera tigris)* which travels many kilometers in a night of hunting. Acquisition of the food, once contacted, is the next step in the food-getting process. Various forms of stalking, ambushing, sprinting, and killing are regular parts of the capture sequences of carnivores, and are in contrast to the harvesting of seeds, fruits, flowers,

Table 15.2
Contrasts in Attributes of the Diets of Herbivores and Carnivores

Food attribute	Herbivore	Carnivore
Packaging	Small, attached	Large, mobile
Distribution	Dispersed	Concentrated
Availability	More predictable	Less predictable
Foraging bouts	Long	Short
Quantity consumed	Large	Small
Protein content	Low	High
Indigestible content	Large percentage	Small percentage
Length of retention time	Long period	Short period

leaves, nectars, gums, blades of grass, twigs, etc., that are dispersed throughout the range of the herbivore. For most vertebrates, locating and acquiring food are aspects that rely primarily on the visual and auditory modalities (although olfaction also may be important in certain taxa). In addition, these are activities that take place in complex social and ecological milieux. For purposes of devising enrichment strategies, these behaviors are grouped under the term foraging.

Less frequently considered in enrichment studies, particularly in an age of commercially produced diets, are the processing and ingesting phases of food acquisition. Processing ranges from being relatively nonexistent for mouthfuls of grass or the swallowing of intact prey (e.g., by snakes) to the vigorous labor of parceling up larger prey items into ingestible portions. Teeth, jaws, tongues, olfactory mechanisms, hands, and paws are employed to one degree or another in such handling and parceling activities. Chemical analysis of the food through contact with the taste buds precedes the consummatory act of swallowing or of expelling, as the case may be. These latter phases of food-getting rely on the more primitive senses of taste and smell and for enrichment purposes may be designated as a "processing phase," in which food is reduced to ingestible portions and evaluated for its palatability. Relatively fewer studies in enrichment have been concerned with this aspect of food consumption.

Before expanding on these points, some practical considerations in designing food enrichment programs may be noted. Approaches that fail to engage the intended beneficiaries, overstimulate activities such as stalking, or result in overfeeding are not useful. In addition, approaches that are simple in design, safe for the animals, practical from the standpoint of caretaking, and have application to a broad range of taxa or situations are preferred.

Table 15.3
Opportunities for Enrichment in the Four Phases of Food Acquisition

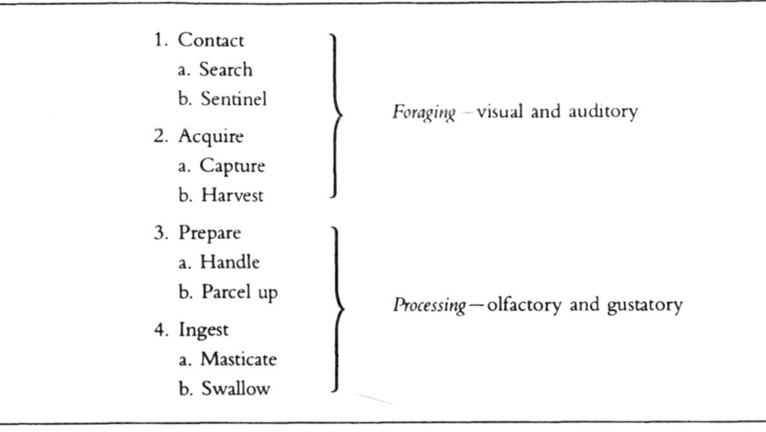

Phase	Mode
1. Contact a. Search b. Sentinel 2. Acquire a. Capture b. Harvest	*Foraging*—visual and auditory
3. Prepare a. Handle b. Parcel up 4. Ingest a. Masticate b. Swallow	*Processing*—olfactory and gustatory

ENRICHMENT THROUGH MODIFICATION OF FORAGING

Captive provisioning usually consists of straightforward presentation of commercially prepared foods that require little more from the animals than approaching the food source and ingesting. While staff scheduling, efficiency in preparation and cleanup, economies in food procurement, and concern for public perceptions place certain constraints on the effort given to provisioning, both humane and health considerations would dictate that even modest investments in enriched feeding routines would be of benefit.

Inducement of foraging activity has been achieved by a variety of paraphernalia too numerous to detail here (for a review, see Shepherdson 1989a). All have as their goal the mental activation that attends the foraging task, an increase in gross levels of motor activity such as locomotion and extraction efforts, and filling more of the hours of the day with species-appropriate behavior. These aims are most readily achieved with foods that can be dispensed through feeders, termite mounds, and puzzle devices or can be hidden or scattered in a complex substrate of hay, straw, or wood chips or in a pile of brush or logs. Among the effects realized from use of these paraphernalia is increased dispersal of individual food items in space, increased difficulty in obtaining (harvesting) items by requiring various acts such as climbing, reaching, "termiting," or locating by touch, increased variation in temporal availability, as through mealworm dispensers, and enhanced search time as in sorting through woodchips or straw or poking under brush piles. From the standpoint of the forager, the result is to add more bits of information to the environment and to allow for more potentially relevant choices (Chamove 1989).

The majority of these applications have been made with primates, whose manual dexterity and climbing abilities are thus exploited in ways crudely similar to procurement processes in the three-dimensional space of the wild (Molzen and French 1989). Opportunities for enriching the lives of grazing animals are somewhat more limited because the procurement process is less profoundly altered by the constraints of captivity. Harvesting by grazers and browsers proceeds a mouthful at a time, whether the food grows in a meadow or is concentrated in bales or in feeders, leaving distances traveled in finding food as the major contrast with the wild state. For hunting animals, in addition to the travel required to locate prey, the final phases of capture entail short-term, high-level expenditures of locomotor energy. A low success rate in capture efforts requires that they be repeated until a kill is made (Lindburg 1988). In stark contrast, the captive hunter is usually provided with dead or processed prey and therefore puts no effort into acquisition. Whether commercial or carcass diets are fed, foraging requirements typically are nil.

Induction of capture-like activities for some of the larger felids has been realized through the use of feeding poles suspended from food boxes attached at the enclosure roof. To procure food, the cats are required to climb the swing-pole and to maintain their hold while reaching for food through an opening in the bottom of the food box (Law et al 1990). Shepherdson et al. (1993) elicited hunting behavior from fishing cats *(Prionailurus viverrinus)* by introduction of live fish into a pool, and found that hiding food items in a brush pile had a similar effect on leopard cats *(Prionailurus bengalensis).* In addition, increased exploratory activity, increased behavioral diversity, and reduced pacing were noted in subsequent periods of observation.

Somewhat more complex in design are the engineered devices used to elicit effort in food procurement, for example, the device developed by Markowitz (1982) for catapulting fish into the pool of a polar bear *(Ursus maritimus).* For several years the Duisberg Zoo in Germany has used a model of a zebra (*Equus* sp.) on a moving line, with a piece of meat attached, to induce running after food in its colony of wild dogs *(Lycaon pictus)* (Gewalt, n.d.). In each of these cases, presumably, the behavior is sustained by being rewarded with the "capture" of food.

At the San Diego Wild Animal Park in California, a lure device is used to induce running in cheetahs *(Acinonyx jubatus),* premised on the need for locomotor activity and mental stimulation in cats deprived of hunting activity. This

Figure 15.1. Cheetahs at the San Diego Wild Animal Park in pursuit of a lure. (Photo courtesy of the Zoological Society of San Diego.)

device is notable in that it simulates the sprint portion of the capture phase, but without food reinforcement. The cheetahs presumably are attracted by the rapid movement of the nonedible lure itself (Figure 15.1). A study in progress is testing the sustainability of this activity over time and its potential for improving body condition.

From the foregoing, it would appear that the best opportunities for enrichment through foraging are realized where subjects normally traverse three-dimensional space in obtaining food, use body structures other than the mouth alone in capture, and pursue movable targets.

FOOD PROCESSING AS AN APPROACH TO ENRICHMENT

Why do callitrichid monkeys prefer unshelled peanuts to shelled ones by 80 percent, when both are equally available (Chamove 1989)? Why do lion-tailed macaques *(Macaca silenus)* orally "peel" foods lacking protective coverings, for example, carrots, when such effort is not essential to feeding (Smith et al. 1989)? Our concern with processing activities arises from these observations and from the fact that many foods in their unaltered form require effort on the part of the animal to reduce them to ingestible portions. The protective coverings of many fruits and vegetables require removal before consumption, for example, and the food carnivores eat is often packaged in quantities substantially larger than bite-size. As noted earlier, mass-produced commercial diets for carnivores have virtually no processing requirements, and pelletized foods for herbivores offer nutrients and mass in concentrated form so that reduced feeding effort is required to reach satiation.

Where garden variety foods are offered, zoos have invested heavily in kitchens to carry out food preparation in the form of peeling, slicing, dicing, chopping, and shelling—activities that are performed by keepers rather than by the animals themselves. The rationale usually given for human preparation is that presentation in small-sized parcels helps to equalize their distribution, especially among animals with social dominance hierarchies. However, this practice, so common to humans, may be an activity that in the case of animals is somewhat anthropomorphically determined (Lindburg and Smith 1988). In any case, the threat to the health of the masticatory apparatus and the greatly reduced manipulative and mental involvement of animals who are not preparing their food for ingestion are aspects that deserve consideration in food enrichment programs.

In carnivores, the consistent feeding of soft-textured foods leads to an array of oral structure problems, including greatly enhanced levels of dental plaque, some of which can be ameliorated by the feeding of hard-substance supplements. Experimental work with timber wolves *(Canis lupus)* (Vosburgh et al. 1982) and Amur tigers *(P. t. altaica)* (Haberstroh et al. 1984) demonstrated that adding hard-fiber foods to the usual soft diet substantially improved gingival health. The occurrence of palatine erosion from dental malocclusion in cheetahs (Fitch and Fagan 1982) has been attributed to deficiencies in jaw and muscle exercise from early life, and the recent documentation of an absence of this condition in wild-living cheetahs (Phillips 1993) offers support to the importance of feeding hard-textured foods. In a comparative study of the skulls of equids and rhinoceroses, Groves (1966, 1982) found that wild and captive specimens could be easily differentiated morphologically, and suggests that muscle disuse in feeding may have contributed to reduced robustness in the captive individuals.

Anecdotes such as the case of an ocelot *(Leopardus pardalis)* that stopped plucking its own fur after addition of whole avian prey to the diet (Hancocks 1980) suggest other situations in which processing activities are important to animal well-being. In a systematic study, Bond and Lindburg (1990) compared responses of cheetahs given commercial and carcass diets, and found that cheetahs receiving carcasses fed longer, spent more time exploring their food, and significantly increased the use of the carnassial dentition in food processing. In addition, greater possessiveness in the form of contests over food was seen when carcasses were fed.

Although ungulates have received very little attention from the standpoint of processing, Dittrich (1976) suggested that the feeding of nutritionally balanced concentrates resulted in long periods of inactivity between feedings and in the development of abnormal levels of licking on fences and other pen structures. It was further suggested by Dittrich that these behaviors might compensate for the absence of foraging and could be alleviated by supplemental feedings of suitable browse.

When presented with intact garden foods, a group of lion-tailed macaques spent more time feeding, increased their food intake, and consumed a greater range of different types of foods than when presented with the same foods in chopped form (Smith et al. 1989). The increased dietary diversity shown by individuals at both ends of the dominance hierarchy was contrary to predictions that hoarding by highest ranking individuals is ameliorated by presentation of foods in small parcels (chopped, peeled, etc.). Although the highest ranking individuals might initially seize as many whole foods as they could hold in hands and feet, they frequently discarded an item after one or two bites, in which case it would be picked up by an individual lower in rank. The increase in feeding time was attributed to the fact that removal of husks, rinds, etc., was performed by the animals themselves. Presenting these same macaques with fresh-cut browse placed in a vertical position and anchored at the base, versus the more traditional style of merely tossing it on the floor, has elicited new kinds of games by the young, such as jumping onto the browse from above, hide-and-seek chases around the feeder, or attempts to break off the branches from their base. In addition, older adults who had ignored browse that was presented flat began to use it as a food source when it was presented upright. Each of these studies in one way or another documents the importance of nonnutritional properties of the food items themselves and their processing requirements, either in terms of potential health benefits to oral structures or in extending opportunities for expression of a normal range of behavior patterns.

The final phase of the food consumption process extends from initial placement of the food item in the mouth to the final act of swallowing or, in some cases, expelling. Activation of olfactory evaluators undoubtedly begins during foraging and initial preparation for consumption, but the final decision about the utility of an item rests on the sense of taste. Avoiding certain toxins and other chemical defenses of plants is important to successful foraging, and reliance on taste in the search for nutrients is, according to Scott (1992, 278), "the most primitive of all motives."

The primacy of nutritional value in captive provisioning is readily acknowledged. However, as stressed by Lang (1970, 263), "unless food is eaten, its nutritional value is of no consequence." An implication of vigorous feeding activity is the animal's experiencing of pleasure, extending from taste bud to gut. While it is obvious that threshold values are affected by hunger or disease, we may safely infer differences in palatability from the eagerness shown in feeding or tenacity in hoarding under conditions of good health and regular provisioning. To the extent that an animal experiences pleasure in feeding ("hedonic appreciation"—Scott 1992), we can say that enhancement of palatability constitutes a form of enrichment. To this end, the taste and odor of food are attributes that, along with form, texture, and nutritional value, merit our attention.

FEEDING SCHEDULES AND BEHAVIORAL CONDITIONING

Two temporal aspects of captive provisioning that appear to be relevant to discussions of food-related enrichment are, first, the collapsing of feeding times into one or two "meals" per day and, second, the conditioned expectation that food will become available at a given time, based on sensory detection of food preparation activity. For browsers and grazers, food-getting is an act of considerable duration because the food items to be harvested are often fairly small and dispersed. Typically, however, the "continuous" feeders will exhibit peak feeding times during the twenty-four-hour day. In a discussion of primate feeding activity, for example, Oates (1986) pointed out that the diurnally active primate begins the day with an intense period of feeding shortly after sunrise, when hunger is most keenly felt, and often shows a secondary peak in the second half of the day. The "banquetters," on the other hand, are those species that consume great quantities in a relatively short period of time, as in the case of the carnivore that makes a kill. While there may be preferred times for hunting, the opportunism that is a part of the hunter's capture of food results in less regular intervals between feedings than in the case of the long-term foragers.

Captive provisioning is often fairly uniform across significantly variant taxa, with staff schedules being a major determinant of feeding styles. In the case of foragers, the presentation of the day's ration in one feeding results in a virtual elimination of all harvesting requirements, such that feeding times become more carnivore like. The obvious result of rapidly consummated feeding is much free time, often leading to boredom. While it is desirable to increase feeding time with foraging opportunities, the first provisioning of the day at a regularly scheduled time is unquestionably the preferred approach, contrary to the contentions of some animal welfare advocates, because withholding food in diurnal foragers would prolong bouts of hunger and lead to increased risk of injury from conflict among group members. A study by Wasserman and Cruikshank (1983) of the behavior of hamadryas baboons *(Papio hamadryas)* in relation to feeding time established, for example, that aggressive behaviors increased while affiliative ones declined during the period immediately preceding provisioning. In these cases, while the temporal pattern of provisioning fits the biology of the animal, the absence of any foraging requirements remains as an area for potentially useful enrichment efforts.

The less regular feeders, on the other hand, become strongly conditioned to regularity after a time in captivity, primarily through learning the routines of staff and associated indications of times when food will be made available. Carnivores are seemingly able to adapt to regularity in feeding, and in fact could be stressed by an irregular schedule once the aforementioned conditioning has taken place. In addition, more frequent feedings of smaller portions could be effective in reducing the amount of stereotyped pacing (Shepherdson et al. 1993). An interesting aspect of carnivore provisioning is withholding of food one day per week from animals such as the felids. This practice appears to have been adopted as an accommodation to weekend staffing, premised on the knowledge that large felids in the wild do not feed daily. Whether fasting at seven-day intervals has any negative psychological or behavioral consequences is unknown. However, in attempting to increase the frequency of fasting in cheetahs, Chi (1992) found that pacing and other behavioral responses to the sounds of food preparation dropped to baseline levels on fast days once the usual time of provisioning had passed.

Temporal factors are of less importance in cases where food is more or less continuously available, as in the case of grazers. They are of utmost importance in all instances where food is available for shorter periods of time during the twenty-four-hour day, in which cases it must be recognized that human-imposed schedules have a conditioning effect that may become a source of stress if ignored.

CONCLUSIONS

The foraging aspects of food-getting provide enriched living for captive animals through inducement of harvesting activities on the part of plant feeders and pursuit activities on the part of predators. Captive provisioning offers further opportunities for enrichment through attention to the nonnutritional properties of food items, such as form, texture, size, and palatability. The success of enrichment through food is in part contingent on recognition of the unavoidable conditioning of captive animals to the activities of caregiving staff.

REFERENCES

Benirschke, K., ed. 1986. *Primates: The Road to Self-Sustaining Populations.* New York: Springer-Verlag.

Bond, J. C., and D. G. Lindburg. 1990. Carcass feeding of captive cheetahs *(Acinonyx jubatus):* The effects of a naturalistic feeding program on oral health and psychological well-being. *Applied Animal Behaviour Science* 26:373–382.

Chamove, A. S. 1989. Environmental enrichment: A review. *Animal Technology* 40:155–178.

Chi, D. 1992. The conditioning of captive cheetahs *(Acinonyx jubatus)* to regularly occurring environmental stimuli. Master's thesis, San Diego State University, San Diego, Calif.

Coe, J. C. 1987. In search of Eden: A brief history of great ape exhibits. In *Proceedings of the American Association of Zoological Parks and Aquariums Annual Conference,* 628–638. Wheeling, W.Va.: AAZPA.

Dittrich, L. 1976. Food presentation in relation to behaviour in ungulates. *International Zoo Yearbook* 16:48–54.

Erwin, J., T. Maple, and G. Mitchell, eds. 1979. *Captivity and Behavior: Primates in Breeding Colonies, Laboratories, and Zoos.* New York: Van Nostrand Reinhold.

Fitch, H. M., and D. A. Fagan. 1982. Focal palatine erosion associated with dental malocclusion in captive cheetahs. *Zoo Biology* 1:295–310.

Gewalt, W. n.d. *Zoo Duisburg Guide Book.* Duisburg, Germany: Zoo Duisburg.

Groves, C. P. 1966. Skull changes due to captivity in certain Equidae. *Zeitschrift für Saeugetierkunde* 31:221–237.

———. 1982. The skulls of Asian rhinoceroses: Wild and captive. *Zoo Biology* 1:251–261.

Haberstroh, L. I., D. E. Ullrey, J. G. Sikarskie, N. A. Richter, B. H. Colmery, and T. D. Myers. 1984. Diet and oral health in captive Amur tigers *(Panthera tigris altaica). Journal of Zoo Animal Medicine* 15:142–146.

Hamilton, W. J. III. 1973. *Life's Color Code.* New York: McGraw-Hill.

Hancocks, D. 1980. Bringing nature into the zoo: Inexpensive solutions for zoo environments. *International Journal for the Study of Animal Problems* 1:170–177.

Hediger, H. 1950. *Wild Animals in Captivity.* London: Butterworths.

Hughes, B. O., and I. J. H. Duncan. 1988. The notion of ethological "need," models of motivation, and animal welfare. *Animal Behaviour* 36:1696–1707.

Lang, C. M. 1970. Organoleptic and other characteristics of diet which influence acceptance by nonhuman primates. In *Feeding and Nutrition of Nonhuman Primates,* ed. R. S. Harris, 263-275. New York: Academic Press.

Law, G., H. Boyle, J. Johnston, and A. MacDonald. 1990. Food presentation. 2. Cats. *Ratel* 17:103-105.

Lindburg, D. G. 1988. Improving the feeding of captive felines through application of field data. *Zoo Biology* 7:211-218.

Lindburg, D. G., and A. Smith. 1988. Organoleptic factors in animal feeding. *Zoonoos* 61 (12): 14-15.

Markowitz, H. 1982. *Behavioral Enrichment in the Zoo.* New York: Van Nostrand Reinhold.

Molzen, E. M., and J. A. French. 1989. The problem of foraging in captive callitrichid primates: Behavioral time budgets and foraging skills. In *Housing, Care, and Psychological Well-Being of Captive and Laboratory Primates,* ed. E. F. Segal, 89-101. Park Ridge, N.J.: Noyes Publications.

Moodie, E. M., and A. S. Chamove. 1990. Brief threatening events beneficial for captive tamarins? *Zoo Biology* 9:275-286.

Oates, J. F. 1986. Food distribution and foraging behavior. In *Primate Societies,* ed. B. S. Smuts, D. L. Cheney, R. M. Seyfarth, R. W. Wrangham, and T. T. Struhsaker, 197-209. Chicago: University of Chicago Press.

Phillips, J. A. 1993. Bone consumption by cheetahs at undisturbed kills: Evidence for a lack of focal-palatine erosion. *Journal of Mammalogy* 74:487-492.

Scott, T. R. 1992. Taste, feeding, and pleasure. *Progress in Psychobiology and Physiological Psychology* 15:231-291.

Segal, E. F., ed. 1989. *Housing, Care, and Psychological Well-Being of Captive and Laboratory Primates.* Park Ridge, N.J.: Noyes Publications.

Shepherdson, D. J. 1989a. Review of environmental enrichment in zoos: 1. *Ratel* 16:35-40.

———. 1989b. Environmental enrichment in zoos: 2. *Ratel* 16:68-72.

———. 1992. Environmental enrichment: An overview. In *Proceedings of the American Association of Zoological Parks and Aquariums Annual Conference,* 100-103. Wheeling, W.Va.: AAZPA.

Shepherdson, D. J., K. Carlstead, J. D. Mellen, and J. Seidensticker. 1993. The influence of food presentation on the behavior of small cats in confined environments. *Zoo Biology* 12:203-216.

Smith, A., D. G. Lindburg, and S. Vehrencamp. 1989. Effect of food preparation on feeding behavior of lion-tailed macaques. *Zoo Biology* 8:57-65.

Vosburgh, K. M., R. B. Barbiers, J. G. Sikarskie, and D. E. Ullrey. 1982. A soft versus hard diet and oral health in captive timber wolves *(Canis lupus). Journal of Zoo Animal Medicine* 13:104-107.

Wasserman, F. E., and W. W. Cruikshank. 1983. The relationship between time of feeding and aggression in a group of captive hamadryas baboons. *Primates* 24:432-435.

Wemelsfelder, F. 1985. Animal boredom: Is a scientific study of the subjective experiences of animals possible? In *Advances in Animal Welfare Science,* ed. M. W. Fox and L. D. Mickley, 115-154. Boston: Martinus Nijhoff.

Woolverton, W. L., N. A. Ator, P. M. Beardsley, and M. E. Carroll. 1989. Effects of environmental conditions on the psychological well-being of primates: A review of the literature. *Life Sciences* 44:901-917.

Yerkes, R. M. 1925. *Almost Human.* New York: Century.

JANET F. BAER

A VETERINARY PERSPECTIVE OF POTENTIAL RISK FACTORS IN ENVIRONMENTAL ENRICHMENT

The role of the veterinary staff in a zoo setting involves the active promotion of animal health and well-being in addition to the treatment of overt disease. The relationship between an animal, its physical environment, and the occurrence of medical problems has been well documented (e.g., Brockman et al. 1988; Munson and Montali 1990). Environmental enrichment influences the physical, mental, and social well-being of captive animals, frequently resulting in beneficial effects on overall animal health. In this way, environmental enrichment should be viewed as an integral component of an active preventive veterinary medicine program.

Captive environments differ significantly from natural environments in a number of ways. In contrast to the dynamic state encountered in the natural environment, the social and physical constraints of captive environments are frequently more static. Physical factors such as temperature, humidity, structural features, and the type, quantity, and availability of food are typically more predictable in a captive environment. As a result of this predictability, captive environments frequently offer less stimulation and opportunity for choice than natural environments. However, not all aspects of the natural environment are directly beneficial to an animal; predation, disease, and malnutrition during periods of food scarcity unquestionably harm animals in the wild. The safety provided by the captive environment is of obvious benefit, but the predictability and monotony of the captive environment derive in large part from attempts to achieve that safety. The challenge of a successful environmental enrichment program is, therefore, to provide stimulation and choice while minimizing potential health risks.

Environmental enrichment may be achieved by modifying any number of aspects of the physical or social environment of the animal. The physical environment encompasses the physical characteristics of the enclosure such as type of substrate, availability of usable horizontal and vertical space, denning or nesting sites and materials, temperature, illumination, humidity, noise exposure, and dietary factors. The social environment consists of potential interactions with conspecifics, other species in multispecies exhibits, and members of the animal care staff. Differences in the degree and type of human-animal interaction may vary with the species, individual animal history, and philosophy of the animal care staff. Modifications of either the physical or the social environment have the potential to directly affect an animal's activity budget, behavioral repertoire, reproductive potential, and overall health and well-being.

ROLE OF STRESS

A discussion of animal well-being from a veterinary perspective would be incomplete without including the role of stress in animal health. Although difficult to define and measure because of its complex nature (Moberg 1987), stress is commonly thought of as the antithesis to well-being. Indeed, it has been suggested that the absence of stress be used as a criterion to assess animal well-being (Moberg 1985a). Alternatively, some authors suggest that a limited amount of stress may be beneficial to an animal's well-being (Breazile 1987; Moodie and Chamove 1990; Kreger et al., Chapter 5, this volume). It is important to point out that well-being, like stress, represents a dynamic state that is extremely complex and therefore difficult to accurately define. Furthermore, it is unrealistic to expect an animal to be in a continuous state of optimal well-being at all times. Certainly, animals living in the wild are not without stressful experiences (Sapolsky 1990). Disease, malnutrition, dehydration, agonistic social encounters, predation, and temperature and humidity extremes represent only a few of the potential external stressors experienced by animals under natural conditions.

In its simplest form, stress involves the interaction of an animal with an external stimulus perceived as a stressor by the animal's central nervous system. Response to a stressor may be through behavioral, autonomic, or neuroendocrine means (Moberg 1985a). Behavioral responses include relatively minor alterations in behavior such as simply moving away from the stressor (e.g., moving from a cold environment to a warmer one or vice versa) or initiating a displacement behavior (i.e., a seemingly irrelevant display behavior given in conflict situations where it has no direct functional significance) (see Wittenberger 1981). Autonomic responses such as elevated heart and respiratory rates, increased blood pressure and peripheral vascular resistance, and secretion of the adrenal hormones epinephrine and norepinephrine represent rapid physiological response to stress (Moberg 1985a). The neuroendocrine response to stress, acting through the hypothalamic-pituitary axis, has been the focus of extensive investigation because of its impact on several major physiological systems. Reproduction, metabolism, immune function, behavior, and growth and development are all regulated or influenced by the neuroendocrine system and thus provide physiological end-points for the study of stress (Moberg 1985a).

External stimuli that may be perceived as stressors can be grouped into three broad categories: (1) those which cause physical stress such as pain or starvation; (2) those that cause psychological stress such as fear, boredom, or separation, and (3) those which cause stress such as physical restraint, noise, and the presence of social conspecifics (National Research Council 1992). Stress that results in pathological changes has been referred to as "distress" (National Research Council 1992). A distressed animal is one that is unable to adaptively cope with external stressors. Interestingly, loneliness in social species and boredom caused by certain husbandry practices are considered by some to be more physiologically and mentally distressful than pain (Wolfle 1987).

The ultimate impact of a stressor on an animal depends on the animal's ability to effectively cope with the stressor. Coping is in turn dependent on the choices available to an animal in conjunction with the stressor. Animals presented with no options during stressful events become passive, disengaged from the environment, and depressed (Seligman 1975). Mediated by the pituitary gland, this type of stress response is characterized by increased adrenocortical activity and vagal tone, decreased reproductive hormonal activity, and chronic blood pressure elevation. Clinical signs associated with this type of stress response include increased susceptibility to disease, gastric ulcers, decreased reproductive performance, and eventually death (Seligman 1975). Alternatively, animals presented with coping strategies in conjunction with an acute stressor may adopt a fight-or-flight response. This type of response, characterized by release of catecholamines, leads to increased heart rate and cardiac output and elevated blood pressure (Moberg 1985a). Repeated exposure to stress resulting in the release of catecholamines may lead to a state of distress characterized by cardiac pathologies, specifically arteriosclerosis (Henry and Stephens-Larson 1985).

From a veterinary perspective, chronic or intermittent stress is undesirable because it has a potentially harmful impact on all aspects of animal health. Chronic or intermittent stress has been associated with impaired reproduction (Moberg 1985b), increased susceptibility to disease (Landi et al. 1982; Kelley 1985), gastric ulcers, cardiovascular pathology, and alterations in basal metabolism

(Klasing 1985). For example, McColl (1981) compared pathology in newly captive and long-term captive platypus *(Ornithorhynchus anatinus),* finding lesions suggestive of chronic stress sequelae in individuals dying after lengthy captivity. Increased susceptibility to disease associated with chronic or intermittent distress is manifested through an increased risk of clinical infections, delayed wound healing, and a diminished vaccination response, all of which stem from impaired immune system function (Kelley 1985).

Appropriately planned and implemented environmental enrichment programs can contribute to improved animal health through creating opportunities for the animal to exert some form of control over its environment. Whether the enrichment allows the animal to avoid a stressful external stimuli altogether or simply provides an opportunity for displacement behavior, the ability of the animal to have some control over its environment appears to be crucial in stress reduction and therefore reduces the risk of associated health problems (Dantzer and Mormède 1985). In addition to reducing stress, a well-designed environmental enrichment program provides further health benefits by creating an opportunity for an animal to exhibit species-typical behaviors and encouraging an increase in physical activity. Conversely, poorly designed or implemented environmental enrichment programs may cause distress to the animal. Examples of inappropriate environmental enrichment include incompatible social partners and intense competition for mates, food, or desired environmental enrichment objects or sites. Frequent monitoring and evaluation of the effects of an environmental enrichment program are essential to ensure that positive elements of the program are maintained and negative elements are eliminated. A team approach contributes to optimum effectiveness of the program.

PHYSICAL ENVIRONMENT

The physical environment of an animal can be modified in a seemingly endless number of ways (Hutchins et al. 1984). Metal, granite, or concrete substrates can be replaced or augmented with dirt, sand, grass, bedding, or browse. The addition of novel objects can increase spatial, visual, and olfactory environmental complexity. The addition of extra perches can increase the usable space within the environment. Opportunities can be engineered to provide the animal a choice in selecting environmental parameters such as light intensity, temperature, noise levels, and food type and availability. However, although each of these modifications may result in increased well-being, they may also pose some, albeit minor, inherent risk to overall animal health. As such, a cost-benefit risk analysis on an individual animal basis should be performed before implementation of any environmental modification that may present health risks.

Usable Space

Modifications in the usable space of an environment can greatly contribute to an animal's well-being. For example, Williams et al. (1988) added perches at several heights for group-housed squirrel monkeys *(Saimiri sciureus).* The additional perches resulted in a direct positive impact on the health of the animals; male squirrel monkeys moved from the floor to the lower perches, thus decreasing the propensity for development of foot and tail ulcers associated with floor contact. Furthermore, overall morbidity was decreased through reduced conspecific aggression associated with incorporation of the additional perches.

Careful consideration of individual and species-specific needs must be taken into account when contemplating modifications in usable space. For example, giving a male-dominated group of *Macaca nemestrina* access to two rooms rather than one allowed part of the group to be out of the dominant male's sight; loss of the male's control over the group resulted in a threefold increase in female aggression (Erwin 1979). This example demonstrates that unless planning is appropriate for species-typical behaviors, changes made in the physical aspects of the environment may negatively impact animal health by leading to an increased risk of injury and social stress secondary to frequent fighting.

Environmental Parameters

Provision of choice in level of illumination, noise, and ambient temperature may be used as a component of an environmental enrichment program. Exposure to temperature extremes results in metabolic changes and, in some species, may cause detrimental reproductive changes (Jainudeen and Hafez 1980). Reptiles exposed to temperatures outside their "preferred optimum" range appear to be more susceptible to infection (Cooper 1986; see also Hayes et al., Chapter 13, this volume). Noise has been shown to be detrimental to animal well-being in a laboratory setting (Pfaff and Stecker 1976); however, appropriately chosen auditory stimulation may serve as a form of environmental enrichment (Shepherdson et al. 1989; Shepherdson, Chapter 1, this volume). Failure to provide appropriate levels of illumination during the dark phase of the light-dark cycle has been shown to lead to a drastic reduction in food intake in *Aotus* (Erkert 1989). Thus, variations in temperature, noise, and illumination may have direct and indirect health consequences. Providing animals with the opportunity to choose from a gradient of these parameters may constitute an important form of environmental enrichment, resulting in direct enhancement of animal health.

Substrate

In addition to being more aesthetically appealing to zoo visitors, natural substrates such as loose dirt, rocks, and grass are frequently beneficial in terms of animal well-being. These substrates provide an opportunity for expression of species-specific behavior such as digging or foraging, and offer a more "giving" surface than concrete, leading to fewer pressure sores on weight-bearing surfaces as well as fewer footpad ulcerations observed in animals exhibiting stereotypic pacing on concrete surfaces (Wallach and Boever 1983, 542).

Although the use of natural substrates is preferred for a variety of reasons, use of these substrates may lead to an increased risk of certain health problems. Current concepts of disease control and eradication promote the use of sanitizable, impervious surfaces in animal housing areas to eliminate potentially pathogenic microorganisms through the use of appropriate disinfectants. Internal and external parasites and microorganisms such as bacteria, viruses, and fungi are known to proliferate under certain environmental conditions. While some parasites and microorganisms are effectively inactivated by sunlight or temperature variation, others have evolved mechanisms that allow them to survive under even the harshest environmental conditions. Inactivation of these organisms must therefore be taken into account when choosing substrates. Surfaces composed of natural materials such as loose rock, dirt, grass, or other bedding types pose problems in effective disinfection and thus may harbor infectious agents for extended periods of time. Although pathogenic microorganisms may not have a readily apparent detrimental impact on healthy adult animals, their presence in the environment may lead to clinical disease in neonatal, aged, or otherwise immunocompromised individuals. In addition, these organisms may result in subclinical disease in apparently healthy animals (Banish et al. 1993a).

Although exceptions exist, under natural conditions excretory wastes are generally confined to one area or distributed throughout an animal's home range. In captivity, animals are confined to a restricted space resulting in heavy contamination of the local environment with animal wastes over time. Moreover, in such limited space conditions, captive animals are often more likely to reencounter their own wastes than would animals in the wild. Additionally, some captive and wild animals are known to manipulate and even ingest their fecal wastes (i.e., coprophagy). Ingestion of oocysts in fecal matter is the most common route of infection for many parasitic diseases; thus, prompt and thorough removal of fecal material from the environment is a key element in breaking the cycle of reinfection. Recognition that many viral and bacterial diseases are also transmitted through fecal-oral contact provides further rationale for attention to proper sanitation of the environment.

Clinical disease transmitted through fecal-oral contact resulted in both morbidity and mortality in a large zoological collection of primates from chronic, frequently subclinical infection with the bacteria *Shigella flexneri* (Banish et al. 1993a). Clinical infection with this bacteria results in diarrhea, leading to heavy contamination of the environment. Eradication of the disease in this case required removal of the animals from their exhibit for extended periods of time to allow thorough disinfection of the environment. Additionally, daily administration of intramuscular antibiotics for ten consecutive days was necessary to control the outbreak. All animals were treated in an attempt to eliminate the organism from clinically healthy carrier animals. Interestingly, with extensive effort the organism was effectively removed from naturalistic exhibits but difficulty was experienced in eradicating it from older, concrete enclosures; surface imperfections such as cracks and holes in the concrete hampered disinfection procedures (Banish et al. 1993a,b).

Pasture rotation, utilized by the livestock industry in an effort to reduce the total number of pathogens present in the environment at any given time, cannot be effectively implemented in most zoos because of space limitations. Frequent fecal parasitology evaluation and treatment with anthelmintics may be necessary to adequately manage parasite control of animals maintained year round on irrigated pasture (Mikolon et al. 1992). Sanitation through the use of chemical disinfectants or the periodic replacement of exhibit substrates such as sand, dirt, or loose rock should be designed into exhibit plans to provide control or elimination of pathogenic organisms.

Animals housed on natural substrates also suffer gastritis and gastrointestinal impaction from consumption of large amounts of the substrate. Dirt, sand, or the chemicals used to sanitize the exhibit area may cause gastrointestinal illness in some species (Schmidt 1986; Honnas et al. 1991). This problem can best be avoided through knowledge of species-typical behavior, careful exhibit planning and maintenance, and sound animal management practices. In many instances the potential health risks associated with natural substrates can be reduced through the use of effective quarantine procedures, routine screening of the environment and animal health, implementation of a pest control program, and the judicious use of anthelmintics and antibiotics. Moreover, naturalistic exhibits should be designed in such a way that thorough sanitation, either through the use of disinfectants or replacement of environmental substrates, can be readily conducted.

Bedding

The use of various types of bedding such as straw, pine shavings, wood chips, pelleted corncobs, and shredded paper has been advocated as a form of environmental enrichment for primates as well as for other taxonomic groups housed in traditional laboratory or zoo environments (Chamove et al. 1982; McKenzie et al. 1986; Duncan 1994). Primates provided with bedding demonstrate positive behavioral patterns including increased activity, decreased social aggression, and decreased stereotypic activity (Chamove et al. 1982; McKenzie et al. 1986). Dispersal of small, discrete food items in bedding promotes foraging behavior in many primates (Chamove et al. 1982). Additional benefits include a reduction in the occurrence of pressure sores and contact surface calluses and, with woodchip litter, inhibition of certain pathogenic bacteria (Chamove et al. 1982).

The use of bedding carries some inherent risks, however. Any item introduced into a captive environment may serve as a vector for disease transmission. Woodchip bedding serves as a good example; wood products coming from the forest and processed at a mill are very likely to have come in contact with wild animals and thus may harbor an infectious disease. Bedding may also be contaminated with *Aspergillus fumigatus,* a fungus associated with clinical disease in many avian species (Chute 1978). To prevent accidental introduction of pathogenic bacteria, protozoa, parasites or fungi into existing laboratory animal colonies, bedding used for rodents is commonly heat treated or autoclaved before use.

Bedding also provides an avenue for accidental exposure to pesticide residues and other chemical contaminants (Foley 1978; Weisbroth 1979). Chemical contaminants may originate from exogenous sources such as exposure to agricultural or industrial chemicals, or from components inherent in the actual product such as volatile hydrocarbons or estrogenic compounds. Coniferous softwoods such as red cedar, white pine, and ponderosa pine contain volatile hydrocarbons that have been implicated in the induction of increased hepatic microsomal enzyme activity in laboratory rodents, contraindicating the use of these materials as direct bedding for these animals (Institute of Laboratory Animal Resources 1996). Laminitis and edema of the lower limbs has been associated with the use of shavings of black walnut *(Fuglans nigra)* shavings as direct bedding for horses (Blood et al. 1983). Furthermore, carcinogenic (i.e., cancer-producing) constituents have been documented in certain types of wood shavings (Schoental 1973).

Physical properties of bedding materials may also adversely affect animal health. The elevated moisture content of certain types of bedding provides a substrate for growth of molds and fungi that may be toxic if ingested. Dust associated with bedding may be irritating to the respiratory mucosa, leading to an increased incidence of respiratory disease. Laboratory rodents directly housed on bedding materials ingest considerable quantities of bedding (Weisbroth 1979), which predisposes them to gastrointestinal impaction. Ingested sharp fragments have been associated with irritation of the oropharyngeal mucosa in guinea pigs *(Cavia procellus),* which predisposes them to streptococcal lymphadenitis (Weisbroth 1979). Apes must be monitored for their aberrant use of bedding to probe their own bodily orifices or those of groupmates or infants; straw bedding placed in the ear canal has been implicated in severe otitis in chimpanzee infants *(Pan troglodytes)* (J. Fritz, personal communication) and in the death from meningoencephalitis of an adult gorilla *(Gorilla gorilla)* (Iverson and Popp 1978). Repeated surgeries (a temporary perineal urethrostomy and subsequent urethral anastomosis) were required to treat urethral perforation of a young adult male chimpanzee who used straw to probe the urethral orifice (Caligiuri et al. 1990). Finally, the absorptive qualities of bedding materials may make assessment of urine production and evaluation of urine characteristics more difficult than in situations where bedding is not provided, hampering early detection of renal disease and collection of urine samples to assess the reproductive condition of breeding females.

Selection of bedding material should take into account the species involved, the individual needs of each animal, and the source, type, and availability of bedding material. Criteria for selecting and evaluating bedding material for rodents and laboratory animals are well established (Kraft 1980; Wirth 1983). Bedding should be free of infectious disease agents, pesticide residues, and toxic contaminants and have physical properties appropriate to the species for which it is intended. The use of coniferous softwoods is contraindicated in small burrowing mammals because of the presence of volatile hydrocarbons. Undoubtedly, the proximity of small burrowing animals to their bedding material and the amount of exposure to potential toxins relative to body size combine to increase the risk of deleterious effects in this group of animals. Instead, corncob, heat-treated hardwood, and other types of artificial bedding substrates are advocated for small burrowing animals (Kraft 1980). In most cases, the behavioral benefits associated with the use of bedding materials far outweigh the potential health risks. Stringent evaluation of each bedding material before its use can effectively reduce these risks.

Browse

The benefits of browse as a form of environmental enrichment have been well documented (Tripp 1985; Gould and Bres 1986; O'Neill 1988). The addition of browse results in increased activity and decreased abnormal or stereotypic behaviors. Gorillas provided with browse spent less time in regurgitation and reingestion behavior (Gould and Bres 1986) and provision of wadge materials (i.e., materials that are not swallowed but are manipulated within the mouth for a period of time before being discarded) such as fruit pits, palm fronds, and corn cobs reduced coprophagic behavior in captive chimpanzees (Fritz et al. 1992).

In addition to providing dietary fiber, provision of specific types of browse may actually provide essential dietary requirements for some species. For instance, lack of tannin-containing browse has been implicated in the occurrence of hemosiderosis (excessive iron deposits in tissue) in lemurs (*Lemur* spp.) (Spelman et al. 1989). Tannins inhibit iron absorption in the gut and thus act to prevent the accumulation of high iron levels, which may lead to hemosiderosis. Provision of lemurs with a dietary source of tannin in the form of *Tamarindus indica* pods or tamarind syrup has been recommended (Spelman et al. 1989).

Browse type must be carefully selected because some plants may be poisonous if ingested or result in contact irritation if touched. Ingestion of several types of relatively common trees and shrubs results in illness ranging from diarrhea to central nervous system disease, renal failure, abortions, and death in livestock species (Blood et al. 1983). For example, ingestion of oak leaves and acorns (*Quercus* spp.) has been associated with gastrointestinal tract and kidney lesions in cattle, sheep, goats, and horses. Ingestion of oleander *(Nerium oleander)* and yew *(Taxus baccata)* causes sudden death in many mammalian species. Ingestion of pine needles (*Pinus* spp.) has been associated with abortion in cattle, and consumption of red maple leaves *(Acer rubrum)* results in acute hemolytic anemia and methemoglobinemia in horses. Ingestion of hairy nightshade *(Solanum sarrachoides)* resulted in the death of two of three black and white ruffed lemurs *(Varecia variegata)* (Drew and Fowler 1991). Some types of browse, although not poisonous, may lead to intestinal obstruction because of their indigestibility. For example, ingestion of *Acacia* spp. by a Douc langur *(Pygathrix nemaeus)* resulted in intestinal obstruction that required surgical correction, and ingestion of *Hedychium flavum* caused the death of a second Douc langur from duodenal perforation and peritonitis (Janssen 1994).

Clearly, selection of browse used as a form of environmental enrichment should include a review of the poisonous potential of all aspects of the plant. Although no single reference guide is available specifically for identification of plants poisonous to exotic species, general reference materials are available (Kingsbury 1964; Hardin and Arena 1974; Lampe and McCann 1985; Ruhr 1986). Additionally, a report on the results of a 1992 toxic plant survey of browse materials used in zoos is available through the North Carolina Zoological Park. The possibility of pesticide and chemical contamination should also be an important consideration in the selection of browse. Adequate assessment of plant materials before use as browse can effectively reduce multiple potential health risks.

Novel Objects

Novel objects have been used widely as a form of environmental enrichment in numerous primate species (Tripp 1985; Bloomstrand et al. 1986; McGrew et al. 1986; O'Neill 1988; Paquette and Prescott 1988; Maki and Bloomsmith 1989; Visalberghi and Vitale 1990) and with increasing frequency in other species (Huls et al. 1991; DeLuca and Kranda 1992). These objects typically stimulate investigative or manipulative behavior that increases when the object is first introduced into the environment but wanes over time as the animal habituates to the presence of the object. Age-related differences in response to novel objects have been reported (O'Neill 1988; Maki and Bloomsmith 1989).

Provision of novel objects has been shown to cause increased activity and decreased frequency of abnormal behaviors exhibited by individual animals (Bloomstrand et al. 1986; Paquette and Prescott 1988). Increased activity may counter obesity and musculoskeletal deterioration secondary to inactivity and provides veterinarians and caregiving personnel with increased opportunities to monitor health status (Schmidt and Markowitz 1977). Clinical signs of disease or injury are more apparent in active animals through either decreased activity or other behavioral changes. In this way, the use of novel objects provides an opportunity for early diagnosis and thus treatment of disease or injury.

From a veterinary perspective, the ideal novel object would have many of the same characteristics required by safety standards in children's toys. Novel objects should be durable, contain no toxic substances, be sanitizable if used repeatedly, have no sharp edges and no parts that might allow an animal to entrap a digit, limb, or its neck, and pose no threat to the animal if ingested. Novel objects with movable parts that may be inadvertently ingested if the item comes apart should be excluded from use. Ingested foreign bodies frequently result in a variety of gastrointestinal disorders such as gastrointestinal obstruction, intussusception, and ulceration or perforation of the gut, most of which require surgical intervention for correction (Jones 1992). Ingestion by birds or small mammals of novel items containing lead frequently causes lead poisoning (Bratton and Kowalczyk 1989). The use of galvanized metal is discouraged because toxicosis in many species may result from ingestion of the associated white, zinc-containing powder (Ogden 1992). Suspended ropes, chains, and nets should be evaluated for the potential for entrapment of limbs or neck, possibly resulting in injury or strangulation (Bielitski 1992). Large-diameter ropes or chains are less likely to form entrapping loops. Small-diameter ropes or chains may be encased within discarded lengths of water hose to prevent looping or kinking. Both ends of ropes should be secured such that there is insufficient slack to allow loop formation. Rope and cloth that fray should be avoided as ingested string may result in a life-threatening gastrointestinal disorder or may become entangled around the tongue or digits, obstructing blood flow and leading to amputation secondary to strangulation of surrounding tissues (personal observation).

In summary, potential health risks associated with provision of novel objects as a form of environmental enrichment may be decreased through a thorough initial evaluation of each object, continual reevaluation for signs of deterioration that may lead to injury of the animal, and routine sanitation of reused objects.

Diet

The nutritional adequacy of many zoo diets has traditionally been based upon historical data documenting the survival, longevity, and reproductive ability of exotic species. Knowledge of the exact nutritional requirements of many exotic species is unavailable; therefore, information is frequently extrapolated from domestic species to infer the adequacy of diets for species with similar digestive physiology (Dierenfeld 1993; Ullrey and Allen 1993). Extensive information is available on response to changes in the level of macro- and micronutrients fed to domestic ruminants and horses (Blood et al. 1983) and companion animals (Lewis et al. 1987). This information suggests that small changes in required nutrients can result in dramatic health consequences. For example, a deficiency in the amino acid taurine in commercial cat foods was determined to be the cause of myocardial failure in thousands of pet (domestic) cats annually (Pion et al. 1987). Similar but more limited data are available for exotic species (Nelson 1981; Ullrey 1993). Environmental enrichment programs involving dietary variation must therefore take into account basic nutritional requirements. Moreover, the dietary needs of individual animals vary depending on their physiological status; for instance, the dietary needs of newborn and geriatric patients differ widely. Sick, pregnant, lactating, or obese animals may also require nutritional support differing from that of other animals.

In addition to nutritional needs, many factors play a role in determining the composition of zoo diets. Economic incentives in terms of reduced labor and cost, ready availability, and ease of use have made preprocessed diets very popular. However, while the nutritive properties of these diets may be adequate, the nonnutritive characteristics may be suboptimal (Lindburg 1988; and see Lindburg,

Chapter 15, this volume). Processed diets differ from natural diets in consistency, texture, size, shape, color, olfactory cues, and the need for manipulation by the animal before ingestion. These diets are traditionally fed to captive animals on a fixed time schedule with limited additional dietary variation. Because these diets tend to contain a high caloric value with little roughage, and because captive animals typically lead a sedentary lifestyle that predisposes them to weight gain and thus obesity, small portions are fed. Provision of more natural diets that meet the animal's nutritional and psychological needs is certainly one of many challenging forms of environmental enrichment.

Dietary modifications involving variation in the timing, frequency, type, quality, and quantity of diet provided have been shown to serve as a form of enrichment for captive animals (Bloomsmith 1989; Forthman et al. 1992; Shepherdson et al. 1993). Although individual variation was observed, in general, animals provided with a dietary enrichment program were more active, engaged in greater behavioral diversity, and exhibited less stereotypic behavior.

In the wild, many animals devote a large portion of their activity budget to foraging, preparing, and ingesting food. The type, quantity, and quality of food ingested at any given time may vary depending on seasonal environmental constraints. Diets under natural conditions tend to vary daily. As pointed out by Lindburg (1988), the application of information acquired through investigation of field data may benefit dietary management of captive animals.

An example of dietary diversity in the wild is available through review of field data on the dietary habits of brown bears *(Ursus arctos)* and black bears *(Ursus americanus)*. These animals reportedly consume a variety of graminoids, forbs, ferns, fruits, nuts, cones, leaves, bark, insects, and mammals (Hamilton and Bunnell 1987; Ohdachi and Aoi 1987; Cicnjak et al. 1987; Eagle and Pelton 1983). Plant species constituted the largest portion of the diet, and animal-origin foods represented less than 20 percent of the diet in all cases. Clearly, based on these field data, dietary variation is standard for many wild ursids. However, many zoo diets for members of this genus are limited in variety, consisting mainly of a pelleted commercial omnivore diet supplemented with vegetables and fish. Although animal material is a readily digestible source of protein for bears, members of this genus appear to be able to extract protein from a wide variety of herbaceous materials (Eagle and Pelton 1983). Because herbaceous material contains a large amount of fiber that the bear cannot digest, a large amount of the material must be consumed to meet the animal's protein requirements. As a result, greater time is spent foraging and feeding when consuming a primarily herbaceous diet. Application of field data such as these to the management of captive species provides many possibilities in terms of environmental enrichment.

Seasonal changes in food consumption were also evident in the aforementioned field studies on bears. However, seasonal variation in food preference was not demonstrated in captive black bears, suggesting that seasonal changes in food consumption of wild bears are related to availability of preferred foods (Bacon and Bughardt 1983). This knowledge may be helpful in designing enrichment programs involving dietary variation. For instance, the data suggest that captive bears will be selective, eating only "treats" when provided with a choice between a processed but nutritionally balanced diet and treat items that may not be nutritionally balanced.

Altering the time of provision between the two food types may be a successful solution to this problem: the processed, formulated diet could be fed in the morning when the animal's appetite is greatest, and the treats fed in the afternoon. This feeding technique is applicable to many species. Its usefulness is emphasized in a report documenting hyperphosphatemia and hypocalcemia in association with severe debilitating clinical disease in a group of captive lemurs simultaneously fed a variety of foods in conjunction with a commercially prepared, nutritionally balanced diet in the form of a biscuit (Tomson and Lotshaw 1978). Although nutritionally balanced biscuits were provided the animals chose instead to eat the other foods. Almost exclusive consumption of the nonbiscuit portion of the diet, which was not nutritionally balanced, resulted in overt clinical disease in several animals.

Alterations in normal oral physiology have been associated with feeding refined pelleted diets. Macropods fed pelleted rations with limited fresh browse develop soft crumbly plaque, which predisposes them to the occurrence of lumpy jaw (Butler 1981). Cheetahs *(Acinonyx jubatus)* fed a refined, formulated diet develop focal palatine erosion, a disease not seen in animals fed a less processed, more natural diet (Fitch and Fagan 1982). Domestic dogs fed soft diets develop a greater incidence of periodontal disease than dogs fed hard diets (Egelberg 1965). In the aforementioned cases, improvement in oral health followed dietary modifications that included addition of fewer or unprocessed foods. If refined diets are the only option available, provision of chewing devices (e.g., nontoxic urethane chew toys) may assist in promoting oral health (Duke 1989).

Although provision of more natural diets has obvious health benefits, certain health risks may be associated with feeding these diets. For example, Dierenfeld (1993) discouraged the use of whole chickens as a dietary staple for cheetahs or other felids because variation in taurine content in chicken carcasses may lead to nutritional deficiencies in the animals consuming them. Data on the incidence of foreign-body obstruction in the domestic dog and cat indicate that ingested bones are the most common foreign body encountered (Jones 1992). These data suggest that an increased risk of foreign-body obstruction may be encountered when feeding carcasses that have been manipulated (e.g., cut up or cooked). Long bones that are splintered through manipulation of a carcass before presentation may pose a greater hazard than if left intact. The risk of foreign-body obstruction may therefore be reduced by providing animals with prey items that are relatively intact and appropriate for the size of the animal being fed. Provision of intact prey items may also provide an avenue for transmission of infectious disease if the food source is not selected carefully (Holzinger and Silberman 1974). If possible, the health status of the prey should be known before it is used as a food item.

In addition to adding dietary variety, preferred food items are often employed to promote the occurrence of desired behaviors such as foraging, dipping in termite mounds, or solving food puzzles (Bloomstrand et al. 1986; Forthman et al. 1992; Shepherdson et al. 1993). Care must be taken to ensure that dominant animals do not limit access of subordinant animals to the preferred foods to avoid malnutrition or obesity of the dominant animal and frustration and increased stress in subordinate animals. Additionally, caution must also be exercised to ensure that foods used as treats or rewards do not spoil from prolonged exposure to warm temperatures or attract rodents or insects that can serve as disease vectors (Calle et al. 1993; Scanga et al. 1993).

In summary, carefully planned and monitored dietary diversity in the frequency, timing, type, quality, and quantity of foods provided to captive species may serve as a unique form of environmental enrichment. This type of enrichment may result in improved oral health, gastrointestinal integrity, and behavioral well-being. However, the potential adverse effects of foods used as environmental enrichment must be carefully evaluated before making dietary changes. Follow-up evaluation should also be performed to ensure that the changes have resulted in the desired improvement in animal health and well-being.

SOCIAL ENVIRONMENT

For social species, the addition of conspecifics to the environment is an obvious as well as cost-effective means of providing a dynamic form of environmental enrichment. The benefits of providing a complex social environment are many. Isolation stress, a well-documented phenomenon in rodent species, is characterized by undesirable behavioral and physiological responses in individual animals held in long-term isolation (Baer 1971). Isolated animals are frequently inactive, exhibit stereotypic behavior, and have impaired immunological function.

In addition to preventing the deleterious aspects of isolation stress, housing social animals in groups imparts a dynamic aspect to what in many cases may be an otherwise static environment. The addition of conspecifics creates opportunities for the expression of species-specific social behavior including courtship, mating, grooming, and playing. Moreover, opportunities for learning parental care, which may be critical for the future reproductive success of the individual, are provided through observation and participation in infant rearing (Swartz and Rosenblum 1981). And finally, overall activity levels may be increased and the frequency of stereotypic and abnormal behaviors reduced.

From a veterinary perspective, however, housing animals in social groups carries some inherent health risks. Certainly, the risk of transmission of an infectious disease increases when animals are in direct contact with one another, sharing the same feeders, watering devices, and resting sites. Crowded conditions exacerbate this possibility through increased contamination of the environment and the potential immunosuppressive consequences of social stress. Mixing groups of animals from different sources also potentially leads to disease transmission; for instance, animals infected with a latent virus may shed the virus when placed under stressful conditions such as introduction to a new social environment (Weigler et al. 1993); the shed virus then provides an avenue of infection for other group members. Difficulty in identifying sick or injured animals is increased in group-housed animals; on occasion significant time must be spent trying to identify which animal is responsible for the presence of blood, vomit, diarrhea, or bloody urine in the enclosure. Individual appetite, water consumption, and fecal and urine production are also difficult to monitor in group-housed animals, further hampering early disease detection.

Treatment of disease in individual animals or a group of animals is made more challenging by the constraints of group housing. Most drug dosages are based on the weight of the animal. Ingestion of a dose calculated for an adult animal may be lethal in an infant; therefore, it is imperative that each animal receive the amount of drug calculated for it. Treatment of an individual animal in a group may necessitate removal of the animal from the group. Removal and isolation of a sick animal may impose further stress (Bobek et al. 1986) on an already debilitated animal, predisposing it to a more lengthy recovery. Additionally, removal of an animal from a group may lead to increased fighting secondary to a change in the social dynamics of the group. In some instances, increased aggression may also occur following reintroduction of the removed animal to its previous group (Thompson 1993).

In some species, knowledge of the reproductive status of a female (i.e., pregnant or nonpregnant) may not be available in group-housed animals because copulation was not observed. If pregnancy goes undetected or detection is delayed, the risk of poor outcome is increased through potential exposure of the mother to procedures, immunizations, anthelmintics, or other medications contraindicated during pregnancy, through failure to provide prenatal care and meet specialized nutritional needs of pregnancy, and through an inability to provide timely treatment of obstetrical complications. To circumvent these problems, animals may be trained to provide urine samples for diagnosis of pregnancy, or facilities may be designed for this purpose.

Aggression culminating in physical combat is not uncommon in some species housed in social groups. Fighting appears to be most prevalent when new animals are introduced to a socially stable group but may also be observed in conjunction with changes in reproductive cycles (e.g., sexual maturity in male rabbits) (Love and Hammond 1991) and a wide variety of changes in the environment (e.g., introduction of a limited amount of a preferred food). Aggressive behavior is associated with the defense of territory, mates, and offspring in wild animals. However, in the wild, socially subordinant animals have the opportunity to avoid potentially aggressive dominant animals and flee if faced with attack. This opportunity is diminished in captivity by the confines of the enclosure. Stress, weight loss, malnutrition, and finally disease may occur in socially subordinant animals if food sites are limited and therefore food competition is created within the group. Furthermore, the presence of an immunologically compromised animal provides an opportunity for entry of disease into a group of animals. One challenge of environmental enrichment programs involving group housing of animals thus is to provide subordinant animals ample opportunities to limit their exposure to dominant animals should they choose to do so. The addition of multiple food and water stations, visual barriers or cover (Erwin 1977), and numerous nest or resting sites may serve to minimize aggressive behavior in group-housed animals.

The presence of other species in multispecies exhibits and the formation of xenospecific pairs (i.e., pairs or groups composed of different species) also provides opportunities for social interaction. Xenospecific pairings or groupings present an obvious risk of disease and injury and may be inappropriate in a zoological setting from an educational standpoint. However, in the absence of available conspecific companions, xenospecific pairs may provide direct benefits, especially during crucial developmental periods (Mason et al. 1968; Thompson et al. 1991).

Although the risk of disease transmission and difficulty in providing veterinary care may arise in group-housed animals, in most cases the benefits outweigh the health risks involved. Many of the health risks can be reduced through thorough evaluation of animal health before formation of social groups, development of a well-planned introduction method, careful monitoring of individual animal health following introductions, and the development of innovative training and treatment methods. Positive reinforcement training has been used to train zoo animals to voluntarily cooperate during veterinary examination, medical treatment, and collection of physiological specimens (Reichard et al. 1993; Desmond and Laule 1994; Dover et al. 1994; Stone et al. 1994; Laule and Desmond, Chapter 17, this volume). These training techniques have also been used to modify aggression between conspecifics (Desmond et al. 1987; Bloomsmith et al. 1992, n.d.). A team approach between the animal care staff, the veterinary staff, and management is essential for optimizing effective medical management of socially housed species.

CONCLUSIONS

In summary, one stated objective of environmental enrichment programs is to promote animal well-being. When this objective is achieved, real benefits in animal welfare and health may be observed. In addition to direct and indirect health benefits, many forms of environmental enrichment result in increased activity and increased behavioral diversity. These behaviors provide a unique opportunity to evaluate the health status of individual animals. However, in some instances environmental enrichment is not without associated risks to animal health. These risks must be carefully and thoroughly evaluated before implementation of an enrichment program. Compromise between an animal's physical and social requirements (as dictated by its natural history), the constraints of the captive environment, and the potential for disease and injury is best achieved through design and implementation of an environmental enrichment program developed through a team approach with input from the animal care staff, behavioral scientists, animal trainers, management, and the veterinary staff. In this way, a balance resulting in maximum benefit to the animal can best be achieved.

REFERENCES

Bacon, E. S., and G. M. Burghardt. 1983. Food preference testing of captive black bears. In *Bears: Their Biology and Management* [proceedings of the 5th International Conference on Bear Research and Management, held in Madison, Wis., February 1980], 102–105. International Association for Bear Research and Management.

Baer, H. 1971. Long-term isolation stress and its effects on drug response in rodents. *Laboratory Animal Science* 21:341–349.

Banish, L. D., R. Sims, M. Bush, D. Sack, and R. J. Montali. 1993a. Clearance of *Shigella flexneri* carriers in a zoological collection of primates. *Journal of the American Veterinary Medical Association* 203:133–136.

Banish, L. D., R. Sims, D. Sack, R. J. Montali, L. Phillips, and M. Bush. 1993b. Prevalence of shigellosis and other enteric pathogens in a zoological collection of primates. *Journal of the American Veterinary Medical Association* 203:126–132.

Bielitski, J. 1992. Letter to the editor: Enrichment hazards. *Lab Primate Newsletter* 31:36.

Blood, D. C., O. M. Radostits, and J. A. Henderson. 1983. *Veterinary Medicine: A Textbook of the Diseases of Cattle, Sheep, Pigs, Goats, and Horses,* 6th ed. East Sussex, U.K.: Bailliere Tindall.

Bloomsmith, M. A. 1989. Feeding enrichment for captive great apes. In *Housing, Care, and Psychological Well-Being of Captive and Laboratory Primates,* ed. E. F. Segal, 336–356. New York: Noyes Publications.

Bloomsmith, M. A., G. E. Laule, P. L. Alford, and R. H. Thurston. n.d. Using training to moderate chimpanzee aggression during feeding. *Zoo Biology* (in press).

Bloomsmith, M. A., G. E. Laule, R. H. Thurston, and P. L. Alford. 1992. Using training to moderate chimpanzee aggression. In *Proceedings of the American Association of Zoological Parks and Aquariums Annual Conference,* 719–722. Wheeling, W.Va.: AAZPA.

Bloomstrand, M., K. Riddle, P. Alford, and T. L. Maple. 1986. Objective evaluation of a behavioral enrichment device for captive chimpanzees *(Pan troglodytes). Zoo Biology* 5:293–300.

Bobeck, S., J. Niezgoda, K. Pierzchala, P. Litynski, and A. Sechman. 1986. Changes in circulating levels of iodothyronines, cortisol, and endogenous thiocyanate in sheep during emotional stress caused by isolation of the animals from the flock. *Journal of Veterinary Medicine* 33:698–705.

Bratton, G. R., and D. F. Kowalczyk. 1989. Lead poisoning. In *Current Veterinary Therapy,* vol. 10, ed. R. W. Kirk, 152–158. Philadelphia: W. B. Saunders.

Breazile, J. E. 1987. Physiologic basis and consequences of distress in animals. *Journal of the American Veterinary Medical Association* 191:1212–1215.

Brockman, D. K., M. S. Willis, and W. B. Karesh. 1988. Management and husbandry of ruffed lemurs, *Varecia variegata,* at the San Diego Zoo. III. Medical considerations and population management. *Zoo Biology* 7:253–262.

Butler, R. 1981. Epidemiology and management of "lumpy jaw" in macropods. In *Wildlife Diseases of the Pacific Basin and Other Countries* [proceedings of the 4th International Conference of the Wildlife Disease Association, held in Sydney, Australia, August 25–29, 1981], ed. M. Fowler, 58–61.

Caligiuri, R., T. Norton, E. Jacobsen, O. J. Hart III, R. Locke, N. Ackerman, and C. Spencer. 1990. Urethral obstruction and abscessation in a chimpanzee *(Pan troglodytes). Journal of Zoo and Wildlife Medicine* 21:206–214.

Calle, P. P., D. L. Bowerman, and W. J. Pape. 1993. Nonhuman primate tularemia *(Francisella tularensis)* epizootic in a zoological park. *Journal of Zoo and Wildlife Medicine* 24:459–468.

Chamove, A. S., J. R. Anderson, S. C. Morgan-James, and S. P. Jones. 1982. Deep woodchip litter: Hygiene, feeding, and behavioral enhancement in eight primate species. *International Journal for the Study of Animal Problems* 3:308–318.

Chute, H. L. 1978. Fungal infections. In *Diseases of Poultry,* ed. M. S. Hofstad, B. W. Calnek, C. F. Helmboldt, W. M. Reid, and H. W. Yoder, 376–381. Ames: Iowa State University Press.

Cicnjak, L., D. Huber, H. U. Roth, R. L. Ruff, and Z. Vinovrski. 1987. Food habits of brown bears in Plitvice Lakes National Park, Yugoslavia. In *Bears: Their Biology and Management* [proceedings of the 7th International Conference on Bear Research and Management], 221–226. Washington, D.C.: Port City Press.

Cooper, J. E. 1986. Reptiles: Physiology. In *Zoo and Wild Animal Medicine,* ed. M. E. Fowler, 883–923. Philadelphia: W. B. Saunders.

Dantzer, R., and P. Mormède. 1985. Stress in domestic animals: A psychoneuroendocrine approach. In *Animal Stress,* ed. G. P. Moberg, 81–95. Bethesda, Md.: American Physiological Society.

DeLuca, A. M., and K. C. Kranda. 1992. Environmental enrichment. *Lab Animal* 21:38–44.

Desmond, T., and G. Laule. 1994. Use of positive reinforcement training in the management of species for reproduction. *Zoo Biology* 13:471–477.

Desmond, T., G. Laule, and J. McNary. 1987. Training to enhance socialization and reproduction in drills. In *Proceedings of the American Association of Zoological Parks and Aquariums Regional Conference,* 352–358. Wheeling, W.Va.: AAZPA.

Dierenfeld, E. S. 1993. Nutrition of captive cheetahs: Food composition and blood parameters. *Zoo Biology* 12:143–150.

Dover, S., L. Fish, T. Turner, and A. Kelley. 1994. Husbandry training as a technique for behavioral enrichment in marine mammals. In *Proceedings of the American Association of Zoo Veterinarians* [conference held in Pittsburgh, Pa., October 22–27, 1994], 284.

Drew, M. L., and M. E. Fowler. 1991. Poisoning of black and white ruffed lemurs *(Varecia variegata variegata)* by hairy nightshade *(Solanum sarrachoides). Journal of Zoo and Wildlife Medicine* 22:494–496.

Duke, A. 1989. How a chewing device affects calculus build-up in dogs. *Veterinary Medicine* 84:1110–1114.

Duncan, A. E. 1994. Lions, tigers, and bears: The road to enrichment. In *Proceedings of the American Association of Zoo Veterinarians* [conference held in Pittsburgh, Pa., October 22–27, 1994], 270–277.

Eagle, T. C., and M. R. Pelton. 1983. Seasonal nutrition of black bears in the Great Smokey Mountains National Park. In *Bears: Their Biology and Management* [proceedings of the 5th International Conference on Bear Research and Management, held in Madison, Wis., February 1980], 94-101. International Association for Bear Research and Management.

Egelberg, J. 1965. Local effect of diet on plaque formation and development of gingivitis in dogs. I. Effect of hard and soft diets. *Odontologisk Revy* 16:31-41.

Erkert, H. G. 1989. Lighting requirements of nocturnal primates in captivity: A chronobiological approach. *Zoo Biology* 8:179-191.

Erwin, J. 1977. Factors influencing aggressive behavior and risk of trauma in the pigtail macaque *(Macaca nemestrina)*. *Laboratory Animal Science* 27:541-547.

———. 1979. Aggression in captive macaques: Interaction of social and spatial factors. In *Captivity and Behavior: Primates in Breeding Colonies, Laboratories, and Zoos,* ed. J. Erwin, T. Maple, and G. Mitchell, 139-171. New York: Van Nostrand Reinhold.

Fitch, H. M., and D. A. Fagan. 1982. Focal palantine erosion associated with dental malocclusion in captive cheetahs. *Zoo Biology* 1:295-310.

Foley, K. 1978. A comparison of the pesticide residues in corncob and wood beddings. Presentation at the 29th Annual Session of the American Association for Laboratory Animal Science, New York, N.Y., September 1978.

Forthman, D. L., S. D. Elder, R. Bakeman, T. W. Kurkowski, C. C. Noble, and S. W. Winslow. 1992. Effects of feeding enrichment on behavior of three species of captive bears. *Zoo Biology* 11:187-195.

Fritz, J., S. Maki, L. T. Nash, T. Martin, and M. Matevia. 1992. The relationship between forage material and levels of coprophagy in captive chimpanzees *(Pan troglodytes)*. *Zoo Biology* 11:313-318.

Gould, E., and M. Bres. 1986. Regurgitation and reingestion in captive gorillas: Description and intervention. *Zoo Biology* 5:241-250.

Hamilton, A. N., and F. L. Bunnell. 1987. Foraging strategies of coastal grizzly bears in the Kimsquit River Valley, British Columbia. In *Bears: Their Biology and Management* [proceedings of the 7th International Conference on Bear Research and Management], 187-197. Washington, D.C.: Port City Press.

Hardin, J. W., and J. M. Arena, eds. 1974. *Human Poisoning from Native and Cultivated Plants,* 2nd ed. Durham, N.C.: Duke University Press.

Henry, J. P., and P. Stephens-Larson. 1985. Specific effects of stress on disease processes. In *Animal Stress,* ed. G. P. Moberg, 161-176. Bethesda, Md.: American Physiological Society.

Holzinger, E. A., and M. S. Silberman. 1974. Salmonellosis in zoo born cheetah cubs. In *American Association of Zoo Veterinarians Annual Proceedings,* 204-205.

Honnas, C. M., J. Jensen, J. L. Cornick, K. Hicks, and B. Kuesis. 1991. Proventriculotomy to relieve foreign body impaction in ostriches. *Journal of the American Veterinary Medical Association* 199:461-465.

Huls, W. L., D. L. Brooks, and D. Bean-Knudsen. 1991. Response of adult New Zealand white rabbits to enrichment objects and paired housing. *Laboratory Animal Science* 41:609-611.

Hutchins, M., D. Hancocks, and C. Crockett. 1984. Naturalistic solutions to the behavioral problems of captive animals. *Zoologische Garten* 54:28–42.
Institute of Laboratory Animal Resources (National Research Council). 1996. *Guide for the Care and Use of Laboratory Animals.* Washington, D.C.: National Academy Press.
Iverson, W. O., and J. A. Popp. 1978. Meningoencephalitis secondary to otitis in a gorilla. *Journal of the American Veterinary Medical Association* 173:1134–1136.
Jainudeen, M. R., and E. S. E. Hafez. 1980. Reproductive failure in males. In *Reproduction in Farm Animals,* ed. E. S. E. Hafez, 471–493. Philadelphia: Lea & Febiger.
Janssen, D. L. 1994. Morbidity and mortality of Douc langurs *(Pygathrix nemaeus)* at the San Diego Zoo. In *Proceedings of the American Association of Zoo Veterinarians,* 221–223.
Jones, B. D. 1992. Management of esophageal foreign bodies. In *Current Veterinary Therapy,* vol. 11, ed. R. W. Kirk and J. D. Bonagura, 577–580. Philadelphia: W. B. Saunders.
Kelley, K. W. 1985. Immunological consequences of changing environmental stimuli. In *Animal Stress,* ed. G. P. Moberg, 193–224. Bethesda, Md.: American Physiological Society.
Kingsbury, J. M. 1964. *Poisonous Plants of the U.S. and Canada.* Englewood Cliffs, N.J.: Prentice-Hall.
Klasing, K. C. 1985. Influence of stress on protein metabolism. In *Animal Stress,* ed. G. P. Moberg, 269–280. Bethesda, Md.: American Physiological Society.
Kraft, L. M. 1980. The manufacture, shipping and receiving, and quality control of rodent bedding materials. *Laboratory Animal Science* 30:366–376.
Lampe, K., and M. A. McCann. 1985. *AMA Handbook of Poisonous and Injurious Plants.* Chicago: American Medical Association.
Landi, M. S., J. W. Kreider, C. M. Lang, and L. P. Bullock. 1982. Effects of shipping on the immune function in mice. *American Journal of Veterinary Research* 43:1654–1657.
Lewis, L. D., M. L. Morris, and M. S. Hand. 1987. *Small Animal Clinical Nutrition,* 3rd ed. Topeka, Kans.: Mark Morris Associates.
Lindburg, D. G. 1988. Improving the feeding of captive felines through application of field data. *Zoo Biology* 7:211–218.
Love, J. A., and K. Hammond. 1991. Group-housing rabbits. *Lab Animal* 9:37–43.
Maki, S., and M. A. Bloomsmith. 1989. Uprooted trees facilitate the psychological well-being of captive chimpanzees. *Zoo Biology* 8:79–87.
Mason, W. A., R. K. Davenport, and E. W. Menzel. 1968. Early experience and the social development of rhesus monkeys and chimpanzees. In *Early Experience and Behavior,* ed. G. Newton and S. Levin, 440–480. Springfield, Ill.: Charles C. Thomas.
McColl, K. A. 1981. Necropsy findings in captive platypus *(Ornithorhynchus anatinus)* in Victoria, Australia. In *Wildlife Diseases of the Pacific Basin and Other Countries* [proceedings of the 4th International Conference of the Wildlife Disease Association, held in Sydney, Australia, August 25–29, 1981], ed. M. Fowler, 238.
McGrew, W. C., J. A. Brennan, and J. Russell. 1986. An artificial "gum-tree" for marmosets *(Callithrix j. jacchus). Zoo Biology* 5:45–50.

McKenzie, S. M., A. S. Chamove, and A. T. C. Feistner. 1986. Floor-coverings and hanging screens alter arboreal monkey behavior. *Zoo Biology* 5:339–348.

Mikolon, A., W. Boyce, J. Allen, I. Gardner, and L. Elliot. 1992. Epidemiology and control of nematode parasites in a collection of captive exotic ungulates. In *Proceedings of a Joint Conference of the American Association of Zoo Veterinarians and the American Association of Wildlife Veterinarians* [held in Oakland, Calif., October 1992], 200.

Moberg, G. P. 1985a. Biological response to stress: Key to assessment of animal well-being? In *Animal Stress,* ed. G. P. Moberg, 27-50. Bethesda, Md.: American Physiological Society.

———. 1985b. Influence of stress on reproduction measure of well-being. In *Animal Stress,* ed. G. P. Moberg, 245-268. Bethesda, Md.: American Physiological Society.

———. 1987. Problems in defining stress and distress in animals. *Journal of the American Veterinary Medical Association* 191:1207–1211.

Moodie, E. M., and A. S. Chamove. 1990. Brief threatening events beneficial for captive tamarins? *Zoo Biology* 9:275–286.

Munson, L., and R. J. Montali. 1990. Pathology and diseases of great apes at the National Zoological Park. *Zoo Biology* 9:99–105.

National Research Council. 1992. *Recognition and Alleviation of Pain and Distress in Laboratory Animals.* Washington, D.C.: National Academy Press.

Nelson, L. S. 1981. Secondary hypocuprosis in an exotic animal park. In *Wildlife Diseases of the Pacific Basin and Other Countries* [proceedings of the 4th International Conference of the Wildlife Disease Association, held in Sydney, Australia, August 25–28, 1981], ed. M. Fowler, 139–145.

Ogden, L. 1992. Zinc toxicosis. In *Current Veterinary Therapy,* vol. 11, ed. R. W. Kirk and J. D. Bonagura, 197–200. Philadelphia: W. B. Saunders.

Ohdachi, S., and T. Aoi. 1987. Food habits of brown bears in Hokkaido, Japan. In *Bears: Their Biology and Management* [proceedings of the 7th International Conference on Bear Research and Management], 215–220. Washington, D.C.: Port City Press.

O'Neill, P. 1988. Developing effective social and environment enrichment strategies for macaques in captive groups. *Lab Animal* 17:23–36.

Paquette, D., and J. Prescott. 1988. Use of novel objects to enhance environments of captive chimpanzees. *Zoo Biology* 7:15–23.

Pfaff, J., and M. Stecker. 1976. Loudness levels and frequency content of noise in the animal house. *Lab Animal* 10:111–117.

Pion, P. D., M. D. Kittleson, Q. R. Rogers, and J. G. Morris. 1987. Myocardial failure in cats associated with low plasma taurine: A reversible cardiomyopathy. *Science* 237:764-768.

Reichard, T., W. Shellabarger, and G. Laule. 1993. Behavioral training of primates and other zoo animals for veterinary procedures. In *Proceedings of the American Association of Zoo Veterinarians,* 65–69. Lawrence, Kans.: American Association of Zoo Veterinarians.

Ruhr, L. P. 1986. Ornamental toxic plants. In *Current Veterinary Therapy,* vol. 9, ed. R. W. Kirk, 216–220. Philadelphia: W. B. Saunders.

Sapolsky, R. M. 1990. Stress in the wild. *Scientific American* 262:116–123.

Scanga, C. A., K. V. Holmes, and R. J. Montali. 1993. Serologic evidence of infection with lymphocytic choriomeningitis virus, the agent of callitrichid hepatitis in primates in zoos, primate research centers, and a natural reserve. *Journal of Zoo and Wildlife Medicine* 24:469–474.

Schmidt, M. J. 1986. Elephants (Proboscidea). In *Zoo and Wild Animal Medicine,* ed. M. E. Fowler, 883–923. Philadelphia: W. B. Saunders.

Schmidt, M. J., and H. Markowitz. 1977. Behavioral engineering as an aid in the maintenance of healthy zoo animals. *Journal of the American Veterinary Medical Association* 171:966–969.

Schoental, R. 1973. Carcinogenicity of woodshavings. *Laboratory Animals* 7:47–49.

Seligman, M. E. P. 1975. *Helplessness: On Depression, Development, and Death.* San Francisco: Freeman.

Shepherdson, D. J., N. Bemment, M. Carman, and S. Reynolds. 1989. Auditory enrichment for Lar gibbons. *International Zoo Yearbook* 28:256–260.

Shepherdson, D. J., K. Carlstead, J. D. Mellen, and J. Seidensticker. 1993. The influence of food presentation on the behavior of small cats in confined environments. *Zoo Biology* 12:203–216.

Spelman, L. H., K. G. Osborn, and M. P. Anderson. 1989. Pathogenesis of hemosiderosis in lemurs: Role of dietary iron, tannin, and ascorbic acid. *Zoo Biology* 8:239–251.

Stone, A. M., M. A. Bloomsmith, G. E. Laule, and P. L. Alford. 1994. Documenting positive reinforcement training for chimpanzee urine collection. *American Journal of Primatology* 33:342.

Swartz, K. B., and L. A. Rosenblum. 1981. The social context of parental behavior: A perspective on primate socialization. In *Parental Care in Mammals,* ed. D. J. Gubernick and P. H. Klopfer, 417–454. New York: Plenum Press.

Thompson, K. V. 1993. Aggressive behavior and dominance hierarchies in female sable antelope, *Hippotragus niger:* Implications for captive management. *Zoo Biology* 12:189–202.

Thompson, M. A., M. A. Bloomsmith, and L. L. Taylor. 1991. A canine companion for a nursery reared infant chimpanzee. *Lab Animal Newsletter* 30:1–4.

Tomson, F. N., and R. R. Lotshaw. 1978. Hyperphosphatemia and hypocalcemia in lemurs. *Journal of the American Veterinary Medical Association* 173:1103–1106.

Tripp, J. K. 1985. Increasing activity in captive orangutans: Provision of manipulable and edible materials. *Zoo Biology* 4:225–234.

Ullrey, D. E. 1993. Nutrition and predisposition to infectious disease. *Journal of Zoo and Wildlife Medicine* 24:304–314.

Ullrey, D. E., and M. E. Allen. 1993. Identification of nutritional problems in captive wild animals. In *Zoo and Wild Animal Medicine: Current Therapy,* ed. M. Fowler, 38–41. Denver, Colo.: W. B. Saunders.

Visalberghi, E., and A. F. Vitale. 1990. Coated nuts as an enrichment device to elicit tool use in tufted capuchins *(Cebus appella). Zoo Biology* 9:65–71.

Wallach, J. D., and W. J. Boever. 1983. *Diseases of Exotic Animals.* Philadelphia: W. B. Saunders.

Weigler, B. J., D. W. Hird, J. K. Hilliard, N. W. Lerche, J. A. Roberts, and L. M. Scott. 1993. Epidemiology of cercopithecine herpesvirus 1 (B virus) infection and shedding in a large breeding cohort of rhesus macaques. *Journal of Infectious Diseases* 167:257–263.

Weisbroth, S. H. 1979. Chemical contamination of lab animal beddings: Problems and recommendations. *Lab Animal* 8:24–34.

Williams, L. E., C. R. Abee, S. R. Barnes, and R. B. Ricker. 1988. Cage design and configuration for an arboreal species of primate. *Laboratory Animal Science* 38:289–291.

Wirth, H. 1983. Criteria for the evaluation of laboratory animal bedding. *Laboratory Animals* 17:81–84.

Wittenberger, J. F. 1981. *Animal Social Behavior.* Boston: Duxbury Press.

Wolfle, T. L. 1987. Control of stress using non-drug approaches. *Journal of the American Veterinary Medical Association* 191:1219–1221.

17

GAIL LAULE AND TIM DESMOND

POSITIVE REINFORCEMENT TRAINING AS AN ENRICHMENT STRATEGY

There is a growing trend in the zoological, aquarium, and biomedical communities to recognize the use of operant conditioning techniques as a valuable tool for animal care and management (Kirkwood et al. 1989; Laule and Desmond 1990; Priest 1991; Reichard and Shellabarger 1992; Laule 1993a; Kuczaj et al., Chapter 18, this volume). Operant conditioning offers three basic alternatives for influencing behavior: positive reinforcement, escape or avoidance (i.e., negative reinforcement), and punishment (Reynolds 1975; Pryor 1984). In assessing the benefits of training to animals, particularly in regard to psychological well-being, it is important to distinguish the type of training being used and the specific techniques employed.

The training that is referred to throughout this chapter and recommended as the approach of choice is positive reinforcement training. Animals are reinforced with pleasurable rewards for the desired behavioral response. Operationally, this means that the positive alternatives are exhausted before any kind of negative reinforcement is used. We further suggest that on the rare occasions when an escape-avoidance technique (i.e., negative reinforcement) is necessary, it is kept to a minimum and balanced by positive reinforcement at all other times. We consider that punishment, which by definition is used to eliminate a behavior, is only appropriate in a situation that is life threatening for person or animal. To dispel a common misconception, positive reinforcement training does not require any food deprivation. Animals are fed their daily allotment of food. Rewards for training either use a part of that diet or consist of extra treats. Finally, this training regime relies on voluntary cooperation by the animal to be successful.

Using these techniques, an impressive array of benefits has been demonstrated. Training animals to voluntarily cooperate in husbandry and veterinary procedures contributes to a decreased use of immobilizing drugs. In addition, trained animals maintain a high degree of reliability in participating in these procedures and appear less stressed while doing so (Turkkan et al. 1989; Reinhardt and Cowley 1990; Reinhardt et al. 1990; Priest 1991; Laule et al. 1992; Luttrell et al. 1994).

Positive reinforcement training has also proven to be effective in addressing socialization issues in a variety of species (Laule and Desmond 1991). One study documented a reduction in excessive aggressive behavior in a male chimpanzee *(Pan troglodytes)* toward other group members during feeding time through the use of training techniques (Bloomsmith et al. 1994). Primates trained with positive reinforcement techniques (Heath 1988) and elephants *(Elephas maximus* and *Loxodonta africana)* trained through protected contact techniques (Desmond and Laule 1991; Maddox 1992) showed a significant reduction in aggressive behavior toward trainers. Finally, training can be a useful tool for addressing novel situations, for example, in improving maternal behavior in animals that lack proper skills (Joines 1977; Desmond 1985) and obtaining voluntary cooperation in physiological studies (Rogers et al. 1992).

The preceding examples are only a small sampling of the benefits training can offer animals, staff, and the zoological or research institution. The value is apparent and measurable, although there have been few formal evaluations to date. We focus on another benefit to animals that is not so apparent: the use of positive reinforcement training as an enrichment strategy to enhance the psychological well-being of captive animals.

WORKING DEFINITION OF PSYCHOLOGICAL WELL-BEING

In the continuing quest to define psychological well-being and to evaluate the effectiveness of various enrichment strategies, many approaches have been discussed. It has been suggested that psychological well-being be generally defined as the ability to adapt, that is, to respond and adjust to changing situations (Petto et al. 1990). Of the many observable measurable features that relate to this concept, such as behavior, health, reproduction, and longevity, these authors suggested that a combination of two or more criteria should be used in assessing psychological well-being. Using this approach, one can argue that specific training techniques can enhance psychological well-being.

DESENSITIZATION TRAINING

Through a process termed desensitization, animals can learn to tolerate presumably frightening or uncomfortable stimuli. In basic terms, desensitization is a process designed to "train out," or overcome, fear. By pairing positive rewards with any action or object that elicits fear, that fearsome entity slowly becomes less negative, less frightening, and presumably less likely to produce a stressful response. Using this technique, animals have been desensitized to husbandry and veterinary procedures, new enclosures, unfamiliar people, negatively perceived people such as veterinarians, novel objects, strange noises, and so on. In fact, we have previously reported that animals being desensitized to specific stimuli can, over time, become generally desensitized to anything novel or unexpected (Laule 1983; Laule and Desmond 1991).

COOPERATION DURING FEEDING

One of the most desirable forms of enrichment is the housing of naturally social animals in pairs or groups (Reinhardt et al. 1987; de Waal 1991). However, because of the dynamic nature of social interactions and also the constraints captivity imposes upon animals and their ability to avoid or escape aggressive behavior, social housing can be a double-edged proposition. In fact, if not carefully implemented and monitored, social housing can become a stressful, negative experience for subordinate animals (Coe 1991; also see Crockett, Chapter 9, this volume).

Using a training technique we call "cooperative feeding," it is possible to enhance introductions, mitigate dominance-related problems, and reduce aggression. Operationally, this entails reinforcing two events within the group simultaneously: Dominant animals are reinforced for allowing subdominant animals to receive food or attention, while the subdominant animals are reinforced for being brave enough to accept food or attention in the presence of these more aggressive animals.

This technique was one successful strategy used with a group of five drills *(Mandrillus leucophaeus)* at the Los Angeles Zoo in California (Desmond et al. 1987). The primary goal of the project was to increase positive social interactions and reproduction among the group members; both these features have been identified as indicators of psychological well-being (Petto et al. 1990). Animals cooperatively fed in different dyads and triads were reinforced when they ate and appeared relaxed in close proximity to one another. To encourage reproductive behavior the dominant male was reinforced for touching the dominant female, and she was simultaneously reinforced for allowing him to touch her. Results of the seven-month project showed significant increases in all forms of affiliative behavior, including grooming, inspection, and mounting, during and following the project (Cox 1987).

WORKING FOR A LIVING

Another avenue for defining psychological well-being is to look at findings of studies on human well-being as a model for nonhuman primates. In discussing several factors associated with human well-being, Sackett (1991) identified monetary income as a major correlate and suggested, "Perhaps the opportunity to engage in, and succeed at, problem solving could affect a primate's well-being much as income affects well-being in humans" (p. 39). Hediger (1950) pointed out that captivity deprives wild animals of the need and opportunity to engage in the tasks of survival, that is, finding food and avoiding enemies. He suggested, "The captive animal must be given a new interest in life, an adequate substitute for the chief occupations of freedom . . . this substitute can take the form of biologically suitable training and assumes the importance of occupational therapy" (p. 158).

Positive reinforcement training offers animals a chance to "work" for their food, to perform certain tasks or behaviors for a food reward. Studies have shown that given a choice animals most often voluntarily work for their food even if the same food is freely available (Neuringer 1969; Anderson and Chamove 1984). In 365 protected contact training sessions between two Asian and two African elephants, the animals chose to work for the extra treats 99 percent of the time (Laule and Desmond 1992). Mineka et al. (1986) found that rhesus monkey *(Macaca mulatta)* infants that were given the opportunity to work for their food showed less fearfulness when exposed to threatening stimuli and demonstrated better coping responses when separated from cage mates than did monkeys that received food ad libitum. The authors discussed their results in terms of the importance of providing animals opportunities to exercise control over their environment.

GREATER CHOICE AND CONTROL

Hanson et al. (1976) measured cortisol levels in groups of rhesus macaques who were either given or denied control over loud, continuous white noise. Striking results showed that levels of cortisol and aggression did not differ in the group having control over noise compared to the control group (i.e., a third group receiving no noise). In contrast, the group having no control over noise had significantly higher levels of blood cortisol and showed more frequent aggression. Furthermore, when control over the noise was taken away from the first experimental group, their cortisol and aggression levels were the highest of any condition.

The conditions and restrictions of captivity typically offer animals little choice or control over their lives (see Markowitz and Aday, Chapter 4, this volume). Furthermore, the traditional dependence on the use of escape or avoidance techniques (negative reinforcement) contributes to an animal's lack of control. In our experience, positive reinforcement training provides one of the best opportunities for animals to gain greater, albeit not total, control over events through their actions. In a positive reinforcement environment, animals are free to experiment with a broader range of behavioral responses because there are no negative consequences to that experimentation. In fact, skilled trainers consistently reward animals not just for overt correct responses but for more subtle and subjective actions like problem-solving, offering creative solutions, and trying hard.

Most positive reinforcement training is based on voluntary cooperation by the animals. However, even in those instances when compliance is mandatory, positive reinforcement still increases their choice and control to a great degree. For example, consider the animal that must receive an injection for its physical well-being. Without training, the animal has no choice in how that event occurs. If escape-avoidance training is used, offering a choice (i.e., present an arm for the injection) requires the threat of another more negative stimulus. This exposes the animal to stress from both stimuli. In contrast, by using a positive reinforcement approach, the animal is trained through shaping and rewards to voluntarily present an arm for an injection and is concurrently desensitized to the procedure. When the injection is needed, it would seem logical to argue that having a clearer choice in how that event happens, and being less fearful of it, contributes to that animal's psychological well-being.

ABNORMAL BEHAVIOR AND WELL-BEING

In Sackett's (1991) discussion of the relationship between personality factors and human well-being, neurotic behavior correlates negatively with well-being. Studies of captive animals to date have not demonstrated an unequivocal relationship between the presence or absence of abnormal behavior and animal well-being.

However, in her discussion of stereotypic behavior, Carlstead (Chapter 11, this volume) points out that despite the absence of scientific evidence, stereotypic behavior is generally viewed as abnormal and further is considered a possible indicator of poor well-being. In several cases, well-chosen environmental enrichment has been successful in reducing stereotypies.

Training, too, has proven useful in reducing abnormal behavior (Laule 1993b). For example, a bottlenose dolphin *(Tursiops truncatus)* exhibited two behaviors—swallowing foreign objects and regurgitation—at a rate and frequency that indicated an abnormal behavior pattern. Several strategies were implemented to eliminate these behaviors: one was training the animal to retrieve objects for a reward as a behavior incompatible with swallowing them, and another was specific reinforcement of the absence of regurgitation during times observations had shown regurgitation was most likely to occur. Overall activity and stimulation were also increased through multiple daily training sessions. As a result, the dolphin completely ceased swallowing objects and dramatically decreased its rate of regurgitation (Laule 1984). Similar results were obtained in the study of a group of drills described earlier. Although the focus of the training was to enhance positive socialization, results showed that neurotic and self-directed behaviors were significantly reduced as a result of the training program (Cox 1987).

Sackett (1991) reported that personality factors like extroversion and approach-oriented behavior correlate positively with human well-being. As previously described, skilled trainers opportunistically reinforce animals for extroverted, exploratory behavior.

INCREASING MENTAL STIMULATION AND PHYSICAL ACTIVITY

In the literature on psychological well-being and enrichment strategies, commonly cited goals include raising the level of mental stimulation and physical activity for captive animals (e.g., Markowitz 1982; Dittrich 1984; Shepherdson 1989; Carlstead et al. 1991; Markowitz and Aday, Chapter 4, this volume; Poole, Chapter 6, this volume). Training can be used to address both. First, training is teaching; being trained is learning. It is a problem-solving process that can easily be as challenging and rewarding as the most complex enrichment device. To a great extent, this is because training provides a stimulating human-animal interface (Heath 1988; Reinhardt 1992). One recent study documented the impact of human-animal interaction, with positive outcomes such as reduction of abnormal behavior resulting from as little as six minutes of interaction per week (Bayne et al. 1993).

Second, work with marine mammals over the years has demonstrated the value of training in increasing overall activity levels in captive animals. Animals engaged in training programs spend their day in a variety of activities: performance in training sessions and shows, moving between holding areas and show pools, going outside enclosures, as in the case of pinnipeds, and engaging in play and social interactions with trainers and other animals.

Training can contribute to increased activity simply by expanding an animal's behavioral repertoire. Trainers anecdotally cite examples of animals utilizing newly trained behaviors outside of training sessions, spontaneously, and in novel ways. For example, bottlenose dolphins trained to slide out on a platform as both a show and a husbandry behavior were observed exhibiting the behavior during free time, and adding their own variations—twisting around and returning into the water head first, lying with head or tail in the water, spinning around on the deck, or sliding all the way across the platform and back into the water. Animals were only observed using this behavior after its basic form had been trained (Laule and Desmond 1992). One training exercise called "innovation training" reinforces animals for inventing new and creative behaviors (Pryor 1969; Kreiger 1989).

Training a simple behavior like retrieval (i.e., bringing an item to the trainer for a reward) can create multiple benefits. In the case of a female drill, training her to retrieve provided the opportunity to give her many novel objects, because trainers were confident she would not eat them but would return them to the trainer when asked (Laule and Desmond 1992).

TRAINING TO ENHANCE ENVIRONMENTAL ENRICHMENT

Positive reinforcement techniques are useful in training animals to perform peripheral behaviors that allow greater enrichment opportunities. These behaviors include separation from other group members for individual veterinary attention or enrichment, entering transport cages to access remote play or exercise areas, and shifting between enclosures and on and off exhibit to allow environmental manipulation of unoccupied areas. Bringing animals off exhibit several times each day provides maximum opportunity to rotate enrichment devices and toys, seed exhibits, make minor changes to the environment, add new exhibit furniture, and access animals for daily training sessions. By utilizing flexible and frequent shifting for enrichment purposes, the physical activity of animals is increased, and shifting them for routine husbandry purposes becomes easier and more reliable (Laule and Desmond 1992).

Training can also be used in conjunction with enrichment activities to enhance their effectiveness. Enrichment devices have sometimes been discarded because the animals failed to use them, which may have indicated that the animals simply did not know how to use them. In one case, an adult male chimpanzee never used his pipe feeder (PVC pipes attached to the outside of the cage, filled with apple sauce, Jell-O, or some other treat that required the animal to use a stick to access the treat). Eventually the keeper stopped giving him the feeders. However, when this animal entered a training program, he was taught how to use the feeder. Caregivers now report it is a preferred enrichment device (M. Bloomsmith, personal communication).

TRAINING AS ENRICHMENT

Finally, positive reinforcement training, in itself, has enrichment value for animals. A study conducted at the M. D. Anderson Science Park chimpanzee breeding facility was intended to assess the enrichment value of positive reinforcement training (Bloomsmith 1992). Four group-housed adult male chimpanzees were observed during a baseline period before training, during training sessions (where animals were trained to present body parts and accept injections), and during nontraining times. Preliminary results show that during training sessions approximately 40 percent of each animal's time was spent in positive interactions with the trainer; less than 1 percent of their time was spent ignoring or displaying aggressive behavior toward the trainer. In fact, animals remained attentive in the sessions, even when they were not directly being trained. Three positive changes also occurred during training: reduced self-directed behavior, increased activity, and increased social play between group members. Each of these behavioral changes is typically considered to be a positive outcome of an enrichment procedure.

LIMITATIONS OF TRAINING

Despite the many benefits training offers, it is not a panacea for solving every behavioral problem. It is simply a useful tool, with some limitations. First, even the most basic training skills take time and practice to develop. Poorly planned and implemented training can create more problems than it solves. It can also result in confused and frustrated animals, certainly not a state indicative of enhanced well-being. Second, training is time- and labor intensive, which can limit its practicality when used exclusively as an enrichment strategy. However, if training is integrated into a comprehensive animal management program, the long-term benefits can outweigh these costs.

CONCLUSIONS

Positive reinforcement training is gaining stature in the zoological, aquarium, and biomedical community as a useful tool for enhancing animal welfare. Through desensitization and by promoting cooperation during feeding, animal caregivers have a proactive means of improving the well-being of animals under stress from environmental and social factors. Training sessions focus on the problem-solving process, presenting animals with mental and physical challenges that allow them to control events in their lives—"occupational therapy," as Hediger (1950) described it. Training also provides a means of addressing sensory deprivation and abnormal behavior. By utilizing control behaviors, such as shifting, in a frequent and flexible manner, training maximizes the effectiveness of environmental enrichment activities by increasing opportunities to provide more diverse enrichment on a more random basis. It can also be used to teach animals how to use enrichment apparatuses.

Finally, while care must be taken to ensure good planning and proper implementation by skilled personnel, it is clear that training can be an important, if not critical, element in a comprehensive approach to enhancing the psychological well-being of captive animals.

ACKNOWLEDGMENTS

We gratefully acknowledge Michael Keeling, Mollie Bloomsmith, and the staff at M. D. Anderson Science Park. Work at the M. D. Anderson chimpanzee breeding facility was supported by NIH/DRR grants R01-RR03578 and U42-RR03489.

REFERENCES

Anderson, J., and A. Chamove. 1984. Allowing captive primates to forage. In *Standards in Laboratory Animal Science,* vol. 2, 253–256. Potters Bar, U.K.: Universities Federation for Animal Welfare.

Bayne, K., S. Dexter, and G. Strange. 1993. The effects of food provisioning and human interaction on the behavioral well-being of rhesus monkeys *(Macaca mulatta)*. *Contemporary Topics* (AALAS) 32 (2): 6-9.

Bloomsmith, M. 1992. Chimpanzee training and behavioral research: A symbiotic relationship. In *Proceedings of the American Association of Zoological Parks and Aquariums Annual Conference,* 403-410. Wheeling, W.Va.: AAZPA.

Bloomsmith, M., G. Laule, R. Thurston, and P. Alford. 1994. Using training to modify chimpanzee aggression during feeding. *Zoo Biology* 13:557-566.

Carlstead, K., J. Seidensticker, and R. Baldwin. 1991. Environmental enrichment for zoo bears. *Zoo Biology* 10:3-16.

Coe, C. 1991. Is social housing of primates always the optimal choice? In *Through the Looking Glass: Issues of Psychological Well-Being in Captive Non-human Primates,* ed. M. Novak and A. Petto, 78-92. Washington, D.C.: American Psychological Association.

Cox, C. 1987. Increase in the frequency of social interactions and the likelihood of reproduction among drills. In *Proceedings of the American Association of Zoological Parks and Aquariums Annual Conference,* 321-328. Wheeling, W.Va.: AAZPA.

Desmond, T. 1985. Surrogate training with a pregnant *Orcinus orca.* In *Proceedings of the International Marine Animal Trainers Association Annual Conference* [held in Orlando, Fla., October 1985], 1-6.

Desmond, T., and G. Laule. 1991. Protected contact elephant training. In *Proceedings of the American Association of Zoological Parks and Aquariums Annual Conference,* 606-613. Wheeling, W.Va.: AAZPA.

Desmond, T., G. Laule, and J. McNary. 1987. Training for socialization and reproduction with drills. In *Proceedings of the American Association of Zoological Parks and Aquariums Annual Conference,* 435-441. Wheeling, W.Va.: AAZPA.

de Waal, F. 1991. The social nature of primates. In *Through the Looking Glass: Issues of Psychological Well-Being in Captive Non-human Primates,* ed. M. Novak and A. Petto, 69-77. Washington, D.C.: American Psychological Association.

Dittrich, L. 1984. On the necessity to promote activity of zoo-kept wild animals by artificial stimuli. In *Proceedings of the International Congress on Applied Ethology in Farm Animals* [meeting held in Kiel, Germany].

Hanson, J., M. Larson, and C. Snowdon. 1976. The effects of control over high intensity noise on plasma cortisol levels in rhesus monkeys. *Behavioral Biology* 16: 333-340.

Heath, M. 1988. The training of cynomolgus monkeys and how the human/animal relationship improves with environmental and mental enrichment. *Animal Technology* 40 (1): 11-21.

Hediger, H. 1950. *Wild Animals in Captivity.* New York: Dover.

Joines, S. A. 1977. Training programme designed to induce maternal behaviour in a multiparous female lowland gorilla. *International Zoo Yearbook* 17:185-188.

Kirkwood, J., C. Kichenside, and W. James. 1989. Training zoo animals. In *Proceedings of Animal Training Symposium: A Review and Commentary on Current Practices,* 93-99. Cambridge, U.K.: Universities Federation for Animal Welfare.

Kreiger, K. 1989. The lighter side of training. In *Proceedings of the International Marine Animal Trainers Association Annual Conference* [held in Amsterdam, October 1989], 138–142.

Laule, G. 1983. Training pinnipeds to work without walls. In *Proceedings of the International Marine Animal Trainers Association Annual Conference* [held in Minneapolis, Minn., October 1983], 6–10.

———. 1984. Behavioral intervention in the case of a hybrid *Tursiops* sp. In *Proceedings of the International Marine Animal Trainers Association Annual Conference* [held in Los Angeles, Calif.], 23–29.

———. 1993a. Using training to enhance animal care and welfare. *Animal Welfare Information Center Newsletter* 4 (1): 1–9.

———. 1993b. The use of behavioral management techniques to reduce or eliminate abnormal behavior. *Animal Welfare Information Center Newsletter* 4 (4): 1–11.

Laule, G., and T. Desmond. 1990. Use of positive behavioral techniques in primates for husbandry and handling. In *Proceedings of the American Association of Zoo Veterinarians Annual Conference* [held on South Padre Island, Tex.], 269–273.

———. 1991. Meeting behavioral objectives while maintaining healthy social behavior and dominance: A delicate balance. In *Proceedings of the International Marine Animal Trainers Association Annual Conference* [held in San Francisco, Calif.], 19–25.

———. 1992. Addressing psychological well-being: Training as enrichment. In *Proceedings of the American Association of Zoological Parks and Aquariums Annual Conference,* 415–422. Wheeling, W.Va.: AAZPA.

Laule, G., M. Keeling, P. Alford, R. Thurston, and T. Beck. 1992. Positive reinforcement techniques and chimpanzees: An innovative training program. In *Proceedings of the American Association of Zoological Parks and Aquariums Annual Conference,* 713–718. Wheeling, W.Va.: AAZPA.

Luttrell, L., L. Acker, M. Urben, and V. Reinhardt. 1994. Training a large troop of rhesus macaques to co-operate during catching: Analysis of the time investment. *Animal Welfare* 3:135–140.

Maddox, S. 1992. Bull elephant management: A safe alternative. In *Proceedings of the American Association of Zoological Parks and Aquariums Regional Conference,* 376–384. Wheeling, W.Va.: AAZPA.

Markowitz, H. 1982. *Behavioral Enrichment in the Zoo.* New York: Van Nostrand Reinhold.

Mineka, S., M. Gunnar, and M. Champoux. 1986. The effects of control in the early social and emotional development of rhesus monkeys. *Child Development* 57:1241–1256.

Neuringer, A. 1969. Animals respond for food in the presence of free food. *Science* 166:339–341.

Petto, A., M. Novak, S. Fingold, and A. Walsh. 1990. The search for psychological well-being in captive nonhuman primates: Information sources. *Science and Technology Libraries* 10 (2): 101–127.

Priest, G. 1991. Training a diabetic drill *(Mandrillus leucophaeus)* to accept insulin injections and venipuncture. *Laboratory Primate Newsletter* 30 (1): 1–4.

Pryor, K. 1969. Behavior modification: The porpoise caper. *Psychology Today* 3 (7): 47–49.

———. 1984. *Don't Shoot the Dog.* New York: Simon and Schuster.

Reichard, T., and W. Shellabarger. 1992. Training for husbandry and medical purposes. In *Proceedings of the American Association of Zoological Parks and Aquariums Annual Conference,* 396–402. Wheeling, W.Va.: AAZPA.

Reinhardt, V. 1992. Improved handling of experimental rhesus monkeys. In *The Inevitable Bond: Examining Scientist-Animal Interactions,* ed. H. Davis and A. Balfour, 171–177. Cambridge: Cambridge University Press.

Reinhardt, V., and D. Cowley. 1990. Training stumptailed monkeys *(Macaca arctoides)* to cooperate during in homecage treatment. *Laboratory Primate Newsletter* 29 (4): 9–10.

Reinhardt, V., D. Cowley, J. Scheffler, R. Vertein, and F. Wegner. 1990. Cortisol response of female rhesus monkeys to venipuncture in homecage versus venipuncture in restraint apparatus. *Journal of Medical Primatology* 19:601–606.

Reinhardt, V., W. Houser, S. Eisele, and M. Champoux. 1987. Social enrichment of the environment with infants for singly caged adult rhesus monkeys. *Zoo Biology* 6:365–371.

Reynolds, G. 1975. *A Primer of Operant Conditioning.* Chicago: Scott, Foresman.

Rogers, W., A. Coelho, Jr., K. Carey, J. Ivy, R. Shade, and S. Easley. 1992. Conditioned exercise method for use with nonhuman primates. *American Journal of Primatology* 27:215–224.

Sackett, G. 1991. The human model of psychological well-being in primates. In *Through the Looking Glass: Issues of Psychological Well-Being in Captive Non-human Primates,* ed. M. Novak and A. Petto, 35–42. Washington, D.C.: American Psychological Association.

Shepherdson, D. 1989. Environmental enrichment. *Ratel* 16 (1): 4–9.

Turkkan, J., N. Ator, J. Brady, and K. Craven. 1989. Beyond chronic catheterization in laboratory primates. In *Housing, Care, and Psychological Well-Being of Captive and Laboratory Primates,* ed. E. Segal, 305–322. New York: Noyes Publications.

18

STAN A. KUCZAJ II, C. THAD LACINAK, AND TED N. TURNER

ENVIRONMENTAL ENRICHMENT FOR MARINE MAMMALS AT SEA WORLD

The goals of environmental enrichment include providing a more varied environment and allowing captive animals opportunities to exhibit species-typical behavior to control aspects of their environment. Therefore, it is important to understand the animal in question. For the sake of illustration, we consider killer whales *(Orcinus orca)* here. Studying this species in the wild is difficult, but longitudinal fieldwork has determined that killer whales, or orcas, are usually social creatures that live and travel in stable groups (Bigg et al. 1990). Despite these general findings, some individuals do appear to spend most of their lives alone (Baird et al. 1992). Given that most killer whales (at least those that have been studied) seem to prefer the company of conspecifics, Sea World Marine Parks have opted to build facilities that can accommodate groups of whales. Such facilities provide animals with opportunities for social interaction, which is an enriching aspect of their environment. The programs developed at Sea World are described in this chapter.

LEARNING AND TRAINING

The opportunity for social interaction is an essential part of any environmental enrichment program for social animals, but it is far from adequate. Even when housed in groups, captive orcas quickly habituate to known situations and also learn to accurately anticipate predictable situations and outcomes. *Habituation* refers to the waning of a behavior because of constant or repeated stimulation, a waning that is not caused by fatigue of the response process or inactivation of the sensory mechanism. Habituation, then, is a form of learning in that it represents a relatively permanent change in behavior that is the result of experience (Flaherty 1985). *Anticipation* (or *expectancy*) is similar to habituation in that aspects of the environment have become predictable (Capaldi et al. 1995). If such expectations are not fulfilled, behavioral problems can occur.

For orcas and other species that can become habituated to redundant environments and learn to anticipate repeated outcomes, environmental enrichment should be planned to reduce unnecessary habituation and unwanted expectations. This aim can be accomplished by making their world less predictable and presumably more interesting. As one example, consider feeding behavior. If an animal is fed the same amount at the same time (or times) in the same location every day, the feeding schedule becomes predictable. By being predictable, the schedule may make feeding uninteresting, and behavioral problems may ensue (see Lindburg, Chapter 15, this volume). For example, the animal may eat less than it should, which could lead to health problems. In a wild environment, feeding is not a predictable event, but instead varies from day to day as the result of many factors. For example, weather and season may affect availability of prey, the presence or absence of other predators may influence the success of a hunt, and the animal's own skills may vary from day to day, as might those of the prey in avoiding capture.

The success rate for feeding in the wild corresponds to a variable-ratio schedule of reinforcement (Ferster and Skinner 1957). Sometimes an animal succeeds on its first attempt to obtain food; other times, it may fail on its first, second, and third attempts but succeed on its fourth try. Whenever reinforcement (in this case, obtaining food) occurs after a variable number of behaviors (in this case, attempts to obtain food), a variable-ratio schedule exists. If *every* fourth attempt to obtain food succeeded, a fixed-ratio schedule of reinforcement would exist. In terms of foraging for food in the wild, fixed-ratio schedules of reinforcement are rare, as are fixed-interval schedules of reinforcement (for example obtaining food every twelve hours). In a continuing effort to eliminate habituation and anticipation from their feeding, orcas at Sea World are fed on a variable-ratio schedule. This schedule reduces the possibility that particular behaviors become associated with feeding, thus lowering the likelihood of anticipatory responses. The whales are fed at different times on different days, with no predictable interval occurring between feedings. Feeding also takes place at different locations, which lessens the likelihood that the killer whales link particular sites with feeding.

Although the foregoing discussion has focused on feeding, the same principles apply to environmental enrichment in general. For orcas, a totally predictable environment is not desirable. However, a totally unpredictable environment is also undesirable. Events that are too novel can cause aversive reactions in animals (Pfister 1979). An animal's reaction to a novel aspect of its environment depends partly on the nature of the novel stimulus or situation and partly on the animal's internal state (McFarland 1987). Animals that have the capacity to habituate and anticipate are most likely to thrive in a moderately discrepant environment. It is just this sort of environment that Sea World provides for its animals. Again, we use orcas here to illustrate the type of environment to which we refer.

In general, the environment provided to the orcas is varied along a number of dimensions. We have already mentioned the variability that is built into the feeding situation. The whales are also given ample opportunities to engage in a wide range of activities. These can be grouped into six broad categories: exercise, learning, show, husbandry and veterinary care, research, and play. These categories are not mutually exclusive.

EXERCISE

For killer whales, exercise is an important part of the environmental enrichment program. Exercise sessions consist of high-energy behaviors performed for varying lengths of time. These sessions are conducted at irregular intervals to avoid predictability. The behaviors used in each session also vary from session to session. Reinforcement is given on a variable-ratio schedule. In addition, the type of reinforcement given is varied throughout each session. Reinforcers include food, various types of tactile stimulation, vocal approval, toys, and opportunities for interaction with humans. The use of reinforcement-type variability helps to ensure that the reinforcers remain effective, because the animals cannot reliably predict the type of reinforcer that will occur.

Similarly, the variable-ratio schedule of reinforcement makes it difficult for the animals to predict when reinforcement will occur, which also helps to maintain the effectiveness of the reinforcers. We should also note that the whales are fed, stimulated tactually, provided with toys, and allowed to interact with humans regardless of whether they choose to engage in any of our structured activities. In fact, the whales are never forced to participate in exercise sessions, learning sessions, research sessions, or any other scheduled activities. The fact that the orcas almost always choose to participate in these types of sessions suggests that the sessions are enriching experiences for them.

LEARNING

During our learning sessions, whales practice known behaviors and we also attempt to teach the whales new behaviors. Both the teaching and the practice incorporate the principles of variable-ratio reinforcement and reinforcement-type variability. In addition, when a new behavior is being taught, the principle of *shaping* (Skinner 1951), which is based on the notion of *successive approximations,* is used. For example, if one wishes to teach a new behavior such as presenting the pectoral fin, one would initially reinforce a behavior such as having the left or right side of the whale face the trainer. Once the orca has learned that this behavior receives reinforcement, the trainer then requires a behavior somewhat closer to the objective, such as having the whale roll slightly on its side. Once this has been learned, the trainer next requires a behavior even closer to the objective, such as lifting the pectoral fin slightly out of the water. This process of shaping the desired behavior by reinforcing successive approximations (each closer to the desired behavior than the last) is an effective technique for teaching an animal new behaviors. When this technique is combined with the variable-ratio schedule of reinforcement and reinforcement-type variability, the acquisition of new behaviors in learning sessions becomes an important part of our environmental enrichment program. Just as shaping helps to ensure that the whales do not find the learning sessions overly demanding, the use of learning sessions to practice known behaviors ensures that they do not find the sessions too novel. In essence, we have made the learning sessions moderately discrepant events. Such events are most likely to result in learning because they are designed to motivate individuals to learn. Consequently, we are able to teach new behaviors for husbandry, show, and research purposes. Before moving on to these and other topics, we wish to mention two other aspects of our learning sessions that are relevant to the notion of environmental enrichment.

The first aspect involves social interaction, which is a valuable component of environmental enrichment. If two or more orcas are involved in a learning session, it is possible that one whale has already learned the behavior that is being taught to another whale. If so, the whale that has acquired the target behavior can help to teach the behavior to the "student" whale. This can be accomplished by asking the "teacher" whale to perform the behavior, which then may be spontaneously imitated by the "student" whale. Imitation is a powerful learning mechanism (Bandura 1977; Kuczaj 1987), and need not immediately follow the target behavior for successful mimicry to occur (Piaget 1952; Kuczaj 1987). We have observed orcas imitating other whales in both learning and play sessions, and know that these whales will imitate both known and novel behaviors. Attempts at mimicry initially occur in the first year of life of orca calves (Turner et al. 1992). Therefore, we design learning sessions to increase the opportunity for calves to learn novel behaviors through imitation.

Another important aspect of our learning sessions focuses on *desensitization*. This process incorporates aspects of habituation and aspects of shaping in that it conditions animals to react in specific ways to specific types of novel situations by gradually familiarizing the animals with the situations. At Sea World, orcas are desensitized to humans swimming with the whales. Initially, the animal is taught to ignore any human activity that occurs in the water, such as splashing or placing a body part (for example, a hand or foot) in the water. The whale is taught to ignore such behavior by being asked to perform a specific known behavior (for example, presenting a pectoral fin) and is reinforced for performing this behavior rather than investigating the human activity in the water. Once the whale has learned this, it is then taught to ignore humans swimming or floating in the water. The next step is to teach the whale to interact with a swimming or floating human, but only if the human initiates the interaction.

Orcas are also taught an underwater recall tone. When this tone is sounded, whales are reinforced for interrupting any interaction with a human in the water and proceeding to the side of the pool. The use of the recall tone is incorporated into the variable nature of the learning sessions, which lessens the predictability of the tone. In turn, the whales learn to consistently and reliably respond to the recall tone, a fact that significantly increases the safety of all water-based whale-human interactions. The desensitization process allows us to successfully interact with killer whales in their aquatic environment, which has proven to be invaluable for a variety of husbandry procedures in public and educational presentations. We also believe that the opportunity to interact with humans in the water is environmentally enriching to the whales. Desensitization has also proven to be a valuable technique for reducing fear of specific objects or situations (Wolpe 1973). In our discussion of husbandry, we describe examples of the use of desensitization to reduce fear responses in marine mammals.

SHOWS

One of our responsibilities as a zoological institution is to educate the public about the behavior, physiology, and ecology of marine mammals. Shows are one method by which we fulfill this responsibility. Although each show follows a basic format, they differ in terms of the specific animals being used, the behaviors being displayed, and the humans who orchestrate the overall presentation. Each show also uses the variable-ratio schedule of reinforcement and reinforcement-type variability. Shows, then, are designed to be moderately discrepant events. As a deliberate consequence, shows are not routines but are instead part of our overall environmental enrichment program.

HUSBANDRY AND VETERINARY CARE

An important aspect of our environmental enrichment program concerns behaviors that augment Sea World's health maintenance program. Husbandry and veterinary behaviors are taught with the same procedures that are used in exercise, learning, and show sessions.

The first step in teaching an animal to allow husbandry procedures to be performed consists of desensitizing the animal to human touch. Initially, a human touches the animal, who is reinforced for not flinching or turning away. Through successive approximations and reinforcement of a relaxed response to human touch, animals learn to allow tactual contact to all parts of the body. Using the same procedures, animals can then be taught to present specific body parts to a human when that area of the body is touched. The acceptance of human touch is the foundation on which all husbandry behaviors with marine mammals are based.

Another important aspect of our husbandry procedures involves using desensitization techniques to reduce an animal's fear responses to unfamiliar objects and uncomfortable procedures. By gradually familiarizing animals to strange objects that might produce fear responses, the animals become habituated to the objects, which in turn allows techniques such as X-rays and sonograms to be used in husbandry and veterinary procedures.

As a result of using the same procedures to teach husbandry and veterinary care behaviors that we use in our environmental enrichment program, we have made many procedures rewarding to our animals and reduced the aversive nature of other procedures. Consequently, Sea World personnel routinely perform physical examinations, draw blood, obtain urine samples, gather milk samples from lactating females, and weigh and measure marine mammals. In addition, more specialized procedures are possible: X-rays can be obtained to investigate possible skeletal and dental problems. For animals such as orcas that are known to suffer dental problems both in the wild (Carl 1945) and in zoological settings (Graham and Dow 1990), the use of X-rays can pinpoint dental problems before they become too severe. Desensitization allows problem teeth to be cleaned, brushed, and even drilled if necessary (an aversive situation for all species, we imagine). Sonograms can be used to monitor prenatal development in pregnant females and to diagnose certain health problems in animals of both sexes.

To sum up, our husbandry procedures enhance the health of our animals because we are better able to monitor their condition and to diagnose and treat their health problems should they arise. Because husbandry is strongly associated with our environmental enrichment program, the animals do not exhibit fearful responses to these procedures, which significantly reduces the chance of injury to animal or human during a husbandry or veterinary procedure (McHugh et al. 1989).

COGNITIVE RESEARCH

Although the research program at Sea World encompasses a wide array of projects (ranging from genetic analyses of familial groups to biochemical analyses of the milk of lactating animals to observations of social interactions and analyses of activity levels), we focus here on a project that is intended to stimulate the animals as well as increase our understanding of their cognitive and sensory capabilities.

As part of our research sessions with orcas, we expose the whales to underwater tones and teach them to associate each tone with a specific behavior. As always, this involves shaping, variable-ratio reinforcement schedules, and reinforcement-type variability. One research project with this tone system has compared the learning rates of individual whales. To date, the results demonstrate that calves learn new tones much more quickly than do adult whales (McHugh et al. 1991). The calves also seem better able than adult orcas to retain what they have learned about the tones after the learning environment has undergone a significant change (Turner et al. 1991).

Research projects such as this can be integral parts of an environmental enrichment program. If the projects are conducted with the species-specific capabilities of the animal in mind, then the animal should be stimulated by the experience. A repetitive predictable research design presumably will not maintain the animal's interest, nor will an approach that asks the animal to do something it cannot understand. On the other hand, a research project that incorporates the principles of moderately discrepant events, variable-ratio reinforcement schedules, and reinforcement-type variability can undoubtedly contribute to an environmental enrichment program.

Equally important from a scientific point of view, the use of moderately discrepant events can be used to better understand the mechanisms that underlie an animal's cognitive abilities. For example, the principles of moderate discrepancy were employed to teach a bottlenose dolphin *(Tursiops truncatus)* to respond to sequences of visual gestures (Herman et al. 1984). The meaning of these sequences depended on the order in which the gestures occurred. For example, the gestural sequence *surfboard-basket-fetch* instructed the dolphin to take a basket to a surfboard, while *basket-surfboard-fetch* instructed the dolphin to take a surfboard to a basket. Although the results demonstrated that the dolphin had learned to correctly interpret gestural sequences, the manner in which the dolphin did this was not clear. In an attempt to ascertain the cognitive processes used by the dolphin, a set of anomalous gestural sequences was constructed.

The anomalous sequences violated semantic rules and syntactic constraints that had been built into the system of gestural sequences taught to the dolphin. For example, the dolphin was presented with *Phoenix-surfboard-basket-fetch,* which had too many nouns before the verb. (The use of terms such as noun and verb is a matter of convenience, and is not intended to imply that the dolphin understood the concepts of "noun" and "verb.") When presented with an anomalous sequence, the dolphin could fail to respond or could interpret the sequence in terms of what it knew about "normal" sequences (Herman et al. 1993). Because the anomalous sequences were moderately discrepant from the "normal" sequences, the dolphin usually responded. Analyses of these responses revealed that the dolphin had developed several strategies for interpreting gestural sequences (Kuczaj et al. 1989). For example, the noun closest to the verb was almost always incorporated into the dolphin's response, suggesting that the dolphin had learned to pay particular attention to verbs and to the nouns that immediately preceded them. The use of moderately discrepant gestural sequences provided a better window for looking at the mechanisms that underlie dolphin behavior.

PLAY

Although we have saved this topic for last, it is one of the most important aspects of an environmental enrichment program. Play has long been thought to be an essential part of development (Piaget 1962; Lancy 1980; Ford 1983; McFarland 1987; Fagan 1993; Brown 1994), and is now known to be an important aspect of social and cognitive development (Kuczaj 1982, 1983; Bretherton 1984). Play is also now well recognized as a critical component of creativity that is displayed by both children and adults (Amabile 1983; Finke et al. 1992).

Despite the growing consensus about the significance of play in the life of humans and nonhumans, it is notoriously difficult to provide an adequate operational definition of play. Generally, play is thought to be spontaneous activity in which an animal engages for the sake of the activity itself. In other words, play is not goal oriented, at least not in the sense of striving to achieve an immediate goal. Thus, McFarland (1987) described play as leaping, jumping, bucking, or running when there is no obstacle to overcome, no enemy to flee, not no object to attain. Most investigators of play would approve this definition, although each investigator would likely add specific behaviors to the list of playful non-goal-oriented activities.

We admit that this definition of play is far from ideal but offer it as a starting point for our discussion of the role of play in an environmental enrichment program. Even though play is a concept that is difficult to define in unequivocal terms, it is an important component of life for many species. And even though play is usually defined as behavior for its own sake, it is clear that play facilitates development in the young and fosters well-being in adults (McFarland 1987; Fagan 1993; Brown 1994).

Again, we use orcas to illustrate the role of play in our environmental enrichment program. Orcas appear to play in the wild (Ford and Ford 1981; Jacobsen 1986; Osborne 1986). Play often involves object manipulation. For example, kelp may be held in the mouth and dragged alongside the whale while swimming or "spyhopping." Food is also an object of play. Orcas have been observed to capture a fish, mouth it for a short while, let the fish escape and swim a short distance, recapture the fish, and repeat the entire process again and again (Ford and Ford 1981). Like other cetaceans, orcas have been observed to ride the wave produced by the bow of large ships (Dalheim 1980).

The orcas at Sea World are provided with ample opportunities to play. The fact that the whales are kept in groups provides them with ready access to their favorite playthings—each other. Such interactive play includes chases, races, imitation, playful sexual behavior, and object manipulation. Interactive play seems to be an important part of their lives, and likely helps the adults to socialize the calves and the calves to learn from adults and from each other. Interactive behaviors are considered to be play only if the whales appear to be engaging in the behaviors for their own sake rather than for some specific goal. For example, sexual behavior would not be considered playful if coitus occurred. A chase would not be considered playful if one whale caught and raked another whale with its teeth (in this case we are assuming that raking is associated with aggression, although it is possible that "gentle" raking could be part of a play scenario).

The orcas are also provided with objects to manipulate (for example, large ropes, air-filled or water-filled barrels). These toys are sometimes shared with other whales and at other times are hoarded by a single whale. Using the principles outlined previously, the whales are taught to return the toys when requested to do so. This allows us to reduce the amount of clutter in the orca's environment, and also helps to ensure that the toys remain novel and a source of reinforcement to the whales. By varying the toys given to the whales, we are better able to maintain a moderately discrepant environment, which is the goal of our environmental enrichment program.

CONCLUSION

Although we have used orcas to illustrate Sea World's environmental enrichment program, we must emphasize that the principles and procedures that we have discussed are used with many of the marine mammals in our care. Our experiences with these species suggest the following procedures for establishing a successful environmental enrichment program for marine mammals.

1. *Learn as much as possible about the species in question.* Become familiar with published reports on the behavior of the species in the wild and in captivity. Systematically observe the animals in their existing zoological setting before making changes. Plan to evaluate the initial changes before further changes are made.
2. *Always remember that just as species differ, so do individual members of a species.* What is enriching for one animal may not be enriching for another. Moreover, an experience that is enriching at one time for a given animal may not be enriching at another time for the same animal. Thus, it is important to know as much as possible about each individual animal when designing, implementing, and evaluating an environmental enrichment program. Related to this concern for individual variability is the ability to recognize the appropriate time to introduce a change in the environment. For example, one might wish to decrease the frequency with which an animal engages in a particular behavior (e.g., spitting water at visitors). Let us assume that it has been decided to enrich the animal's environment with a variety of toys in an attempt to decrease the frequency of the behavior. If the animal is given a toy whenever it is observed spitting water at visitors, the frequency of the behavior may not decrease. Providing the animal with an interesting toy *after* the animal has begun to

spit water at visitors will most likely cause the animal to associate spitting with reinforcement (interesting toys), which will lead to an increased frequency of spitting. If enrichment is used to distract an animal once an undesirable behavior has commenced, the behavior may cease when the enriching distracter is noticed by the animal but will likely occur with greater and greater frequency because the animal has learned that the undesirable behavior has positive consequences. Rather than enhancing an animal's environment while it is engaging in an undesirable behavior, an environmental enrichment program should strive to reinforce desirable behaviors and provide ample stimulation to all animals.

3. *The key element of any environmental enrichment program is moderately discrepant events.* An enriched environment should provide novelty on a consistent but unpredictable basis. It should also provide a sufficient amount of familiarity so that the animal has a base from which to deal with the requisite novelty. This combination of familiarity and novelty constitutes a moderately discrepant event. There is no more enriching environment than one that is replete with moderately discrepant events.

The components of a moderately discrepant environment will vary from species to species, from individual to individual, and from time to time for a given individual. For social animals, the presence of other members of the species should be enriching. The extent to which opportunities for social interaction are enriching varies from animal to animal and from day to day for individual animals. Other aspects of environmental enrichment programs will also vary in their effectiveness Therefore, it is important to have a multifaceted enrichment program. Multiple enrichment opportunities that are presented in an unpredictable fashion increase the probability that the events will actually be enriching for the animals. This is the reason that we strive to ensure that all interactions between animals and humans at Sea World vary in all possible dimensions: type of session (for example, exercise, research, show), time of day, length of session, other animals present, specific humans who interact with the animals, frequency with which reinforcement is given, type of reinforcement, location at which reinforcement is provided, and time at which animals receive reinforcement.

Although we have emphasized the importance of variability for environmental enrichment programs, this does not mean that all change is good. Changes must occur in the context of the animal's experience so that the animal has a basis from which to interpret the change. Change that is moderately discrepant is most likely to be enriching. Another way to think of moderately discrepant events is in terms of desensitization. In desensitization, aversive events become more acceptable through gradual familiarization, the process consisting of a series of moderately discrepant events.

EVALUATION OF ENVIRONMENTAL ENRICHMENT PROGRAMS

In addition to some general suggestions for implementing an environmental enrichment program, we also offer some strategies for evaluating the effectiveness of such programs.

1. Obtain preenrichment baselines of the behavioral patterns and health of the animals. Compare these baselines to their behavioral patterns and health as the environmental enrichment program progresses. Also ask the individuals who have the greatest opportunities to observe and interact with the animals to record their impressions of behavioral patterns, moods, and well-being. Although this will lead to some subjective opinions and conclusions, these comments may provide clues about the strengths and weaknesses of the environmental enrichment program, clues that can then be evaluated in a more objective and systematic manner.
2. If an environmental enrichment program is effective, the animals should be healthier than animals in a sterile predictable environment. Breeding, live births, and successful rearing of offspring should also improve if the enrichment program is effective.
3. Socially unacceptable behaviors should decrease in frequency in a successful enrichment program. For example, unwarranted physical attacks on other animals should decrease if the animal's environment is providing ample stimulation.
4. Socially acceptable behaviors should increase in frequency in a successful enrichment program. For example, orcas seem to exhibit more interactive play and mimicry when new toys are introduced.

In closing, we wish to add that attempts to enrich an animal's experience invariably have a wonderful by-product: We learn more about the animal as we learn how to enrich its life. For those of us entrusted with the care of animals, the possibility of learning more about their needs and capabilities and the possibility of learning how to best fulfill those needs and maintain those capabilities is a most enriching challenge.

ACKNOWLEDGMENTS

We thank Dave Force, Mark McHugh, Mike Scarpuzzi, and Chuck Tompkins for sharing their valuable insights and experiences about the nuts and bolts of environmental enrichment programs. We also thank Daniel Odell for his written comments on an earlier version of this paper. This is Sea World of Florida Technical Contribution No. 9501-F.

REFERENCES

Amabile, T. M. 1983. *The Social Psychology of Creativity.* New York: Springer-Verlag.

Baird, R. W., P. A. Abrams, and L. M. Dill. 1992. Possible indirect interactions between resident and transient killer whales: Implications for the evolution of foraging specializations in the genus *Orcinus. Oecologia* 89:125–132.

Bandura, A. 1977. *Social Learning Theory.* Englewood Cliffs, N.J.: Prentice-Hall.

Bigg, M. A., P. F. Olesiuk, G. M. Ellis, J. K. B. Ford, and K. C. Balcomb. 1990. Social organization and genealogy of resident killer whales *(Orcinus orca)* in the coastal waters of British Columbia and Washington State. In *Individual Recognition of Cetaceans: Use of Photo-Identification and Other Techniques to Estimate Population Parameters,* ed. P. S. Hammond, S. A. Mizroch, and G. P. Donovan, 383–405. Report of the International Whaling Commission (Special Issue 12). Cambridge, U.K.: International Whaling Commission.

Bretherton, I., ed. 1984. *Symbolic Play: The Development of Social Understanding.* New York: Academic Press.

Brown, S. L. 1994. Animals at play. *National Geographic* 186:2–35.

Capaldi, E. J., K. M. Birmingham, and S. Alptekins. 1995. Memories of reward events and expectancies of reward events may work in tandem. *Animal Learning and Behavior* 23 (1): 40–48.

Carl, G. C. 1945. *A School of Killer Whales Stranded at Estevan Point, Vancouver Island.* Report of the Provincial Museum of Natural History and Anthropology, B21-28. Victoria, B.C.: Provincial Museum.

Dahlheim, M. E. 1980. Killer whales observed bowriding. *Murrelet* 62:78.

Fagan, R. 1993. Primate juveniles and primate play. In *Juvenile Primates,* ed. M. E. Pereira and L. A. Fairbanks, 182–196. New York: Oxford University Press.

Ferster, C. B., and B. F. Skinner. 1957. *Schedules of Reinforcement.* New York: Appleton-Century-Crofts.

Finke, R. A., T. B. Ward, and S. M. Smith. 1992. *Creative Cognition.* Cambridge: M.I.T. Press.

Flaherty, C. F. 1985. *Animal Learning and Cognition.* New York: Alfred A. Knopf.

Ford, B. 1983. Learning to play, playing to learn. *National Wildlife* 21:12–15.

Ford, J. K. B., and D. Ford. 1981. The killer whales of B.C. *Waters (Journal of the Vancouver Aquarium)* 5:3–32.

Graham, M. S., and P. R. Dow. 1990. Dental care for a captive killer whale, *Orcinus orca*. *Zoo Biology* 9:325–330.

Herman, L. M., S. A. Kuczaj, and M. D. Holder. 1993. Responses to anomalous gestural sequences by a language-trained dolphin: Evidence for processing of semantic relations and syntactic information. *Journal of Experimental Psychology (General)* 122:184–194.

Herman, L. M., D. G. Richards, and J. P. Wolz. 1984. Comprehension of sentences by bottlenosed dolphins. *Cognition* 16:129–219.

Jacobsen, J. K. 1986. The behavior of *Orcinus orca* in the Johnstone Strait, British Columbia. In *Behavioral Biology of Killer Whales,* ed. B. C. Kirkevold and J. S. Lockard, 135–185. New York: Alan R. Liss.

Kuczaj, S. A. 1982. Language play and language acquisition. In *Advances in Child Development and Behavior,* ed. H. Reese, 197–233. New York: Academic Press.

———. 1983. *Crib Speech and Language Play.* New York: Springer-Verlag.

———. 1987. Deferred imitation and the acquisition of novel lexical items. *First Language* 7:177–182.

Kuczaj, S. A., L. M. Herman, and M. D. Holder. 1989. A dolphin's processing of sequential information in an artificial language. Paper presented at the 8th Biennial Conference on the Biology of Marine Mammals, Pacific Grove, Calif., December 3–7, 1989.

Lancy, D. F. 1980. Play in species adaptation. In *Annual Review of Anthropology,* ed. B. J. Siegel, A. R. Beals, and S. A. Tyler, 471–495. Palo Alto, Calif.: Annual Reviews, Inc.

McFarland, D., ed., 1987. *The Oxford Companion to Animal Behavior.* New York: Oxford University Press.

McHugh, M. B., S. A. Kuczaj, C. T. Lacinak, and D. L. Force. 1991. Evidence of a critical period in the acquisition of symbolic auditory cueing system by killer whales *(Orcinus orca).* Paper presented at the 9th Biennial Conference on the Biology of Marine Mammals, Chicago, December 5–9, 1991.

McHugh, M. B., C. T. Lacinak, and D. L. Force. 1989. Husbandry training as a tool for marine mammal management. In *Proceedings of the American Association of Zoological Parks and Aquariums Regional Conference,* 799–802. Wheeling, W.Va: AAZPA.

Osborne, R. W. 1986. A behavioral budget of Puget Sound killer whales. In *Behavioral Biology of Killer Whales,* ed. B. C. Kirkevold and J. S. Lockard, 211–249. New York: Alan R. Liss.

Pfister, H. P. 1979. The glucocorticosterone response to novelty as a psychological stressor. *Physiology and Behavior* 23:649–652.

Piaget, J. 1952. *The Origins of Intelligence in Children.* New York: W. W. Norton.

———. 1962. *Play, Dreams, and Imitation in Childhood.* New York: W. W. Norton.

Skinner, B. F. 1951. How to teach animals. *Scientific American* 185:26–29.

Turner, T. N., C. T. Lacinak, S. A. Kuczaj, M. B. McHugh, D. L. Force, M. R. Scarpuzzi, and C. D. Tompkins. 1992. Behavioral development of captive born killer whale *(Orcinus orca)* calves: The first year. Paper presented at the 20th Annual Meeting of the International Marine Animal Trainers Association, Freeport, Grand Bahamas, November 1-6, 1992.

Turner, T. N., S. G. Stafford, M. B. McHugh, L. Surovik, D. Delgross, and O. Fad. 1991. The effects of context shift in whales *(Orcinus orca)*. Paper presented at the 19th Annual Conference of the International Marine Animal Trainers Association, Concord, Calif., November 3-8, 1991.

Wolpe, J. 1973. *The Practice of Behavior Therapy.* New York: Pergamon Press.

19

JILL D. MELLEN, DAVID J. SHEPHERDSON, AND MICHAEL HUTCHINS

Epilogue

THE FUTURE OF ENVIRONMENTAL ENRICHMENT

Many of Heini Hediger's insights into the psychological and physical needs of captive animals are as valid now as when he wrote them (e.g., 1950, 1955, 1969). In one sense, we can admire his foresight and his vision and applaud the many advances that have been made in the exhibition, care, and propagation of a wide variety of captive animals (Gibbons et al. 1995; Norton et al. 1995; Kleiman et al. 1996). However, the lack of progress made in addressing the psychological needs of captive animals remains frustrating. Almost fifty years have passed since Hediger's early writings, yet zoo biologists are still finding it necessary to encourage professionally managed zoos and aquaria to implement environmental enrichment programs and to use scientific knowledge of behavior to formulate effective animal management strategies (Kleiman 1994a).

There are several factors that might account for the slow movement in this direction. Traditionally, animal management has been more of an art than a science (Thompson 1993; Mellen 1994a; Read 1995). While the magical connection that many keepers have with animals is critical to good animal care, it is imperative that we continue to establish a strong empirical foundation for the science of animal management as well (Thompson 1993; Kleiman et al. 1996). We do not see science as replacing the art of animal management, but rather we suggest a blending of the two approaches, combining the best of both.

With respect specifically to psychological well-being, the financial cost of implementing environmental enrichment programs is probably also a factor that has slowed implementation. This may however just be another way of saying that psychological well-being has not been given sufficiently high priority by zoo managers. In his foreword to this book, Terry Maple states, "We are limited only by our imaginations and our budgets[.]" While our imaginations seem to be taking us forward rapidly, zoo and aquarium budgets are not necessarily increasing at a similar pace, and the costs of enrichment need to be balanced against other legitimate goals such as education and conservation (Kreger et al., Chapter 5, this volume). Maple goes on to say, "I can only hope that committed administrators will identify the financial resources to keep pace with our opportunities." Some zoos have developed successful fund-raising efforts around their environmental enrichment programs (Maas 1993); others have made resources available by reevaluating their fiscal and husbandry priorities in light of the need for enrichment.

There is considerable hope for the future, however. In the past decade environmental enrichment has made substantial progress toward becoming an integral part of captive animal management. This movement is clearly reflected in many relevant zoo publications. The American Zoo and Aquarium Association (AZA) requires that all Species Survival Plan (SSP) management groups create husbandry manuals; included in the guidelines for husbandry manual development is a section on environmental enrichment. A review of *Animal Keepers' Forum,* a publication of the American Association of Zoo Keepers (AAZK), shows that almost every issue since 1990 has contained an article on enrichment; since 1993, each issue has featured a column on enrichment ideas (e.g., Grams and Ziegler 1996). The same trend is apparent in laboratory-oriented publications such as *Laboratory Animal Science* and *Lab Animal.* The newsletter *Shape of Enrichment* also includes many practical examples of relevant techniques. The major science journal related to zoo research, *Zoo Biology,* also continues to publish increasing numbers of papers relating to environmental enrichment, as does *Animal Welfare.*

DIVERSIFYING THE TAXONOMIC APPLICATION OF ENVIRONMENTAL ENRICHMENT

To date, nonhuman primates have been the primary focus of enrichment and its systematic evaluation. One of the first books that addressed the psychological needs of captive animals (Yerkes 1925) focused on nonhuman primates, and many recent symposia and the resulting books have focused on the systematic evaluation of enrichment for various primate taxa (e.g., Segal 1989; Novak and Petto 1991). In 1992, the U.S. Department of Agriculture's Animal and Plant Health Inspection Service (APHIS) mandated a plan to promote the psychological well-being of nonhuman primates in U.S. facilities (Holden 1988). While primates certainly benefit from enhanced environments, they are by no means unique in this regard.

In this volume, we have provided examples and ideas for the enrichment of a broader range of vertebrate species. For example, King (1993) and Mench (Chapter 3, this volume) provide new insights into the "how" and "why" of avian enrichment. Enrichment concepts for reptiles and amphibians have also begun to appear (e.g., Chiszar et al. 1995; Burghardt et al. 1996; and see Hayes et al., Chapter 13, this volume), and we suspect that the concept will be applicable to fish and invertebrates as well. We would suggest that an effective way to increase the taxonomic scope of enrichment in zoos, aquaria, and laboratories is to urge managers to reevaluate husbandry protocols and to work toward applying enrichment techniques to daily care protocols and to the design of new exhibits and holding areas for all taxa under their care.

LEARNING ABOUT THE BEHAVIORS OF WILD ANIMALS

Through technologies such as satellite tracking and radiotelemetry, our ability to learn more about animals in their natural habitats has increased dramatically. These technological advances unfortunately have also been coupled with global habitat loss and a shortage of funds to pursue field studies. Nevertheless, knowledge of how animals behave in the wild remains as the foundation upon which the theory and practice of environmental enrichment stands. Existing and future field studies will always be a primary source of enrichment ideas, and researchers, curators, keepers, and zoo exhibit designers should study this literature carefully when formulating new approaches to animal exhibition, management, and enrichment.

RECOGNIZING THE IMPORTANCE OF LABORATORY STUDIES

Using the analogy of a telephoto lens on a camera to describe levels of analysis in the study of animal behavior, Lehner (1979) speaks of "zooming in" and "zooming out." If zooming out means studying animals in their natural habitats, zooming in applies to laboratory studies in which variables can be controlled and carefully measured. Although often criticized by animal rights advocates, Harlow's laboratory work on attachment in infant rhesus monkeys (Harlow and Harlow 1962) provided the foundation for a dramatic reduction in hand-rearing of primate infants in zoos and laboratories. Their results as well as those of colleagues and students (Mason 1960) provided innovative alternatives to rearing zoo "orphans" in nurseries (Mellen 1992). Carefully controlled studies have also begun to elucidate the causes of stereotypy in captive animals (see Mason 1991 for a review) and its relationship to well-being and motivation.

REFINING METHODS OF MEASURING WELL-BEING

We need to improve and refine the methods by which we measure psychological well-being in captive animals. Scientists have spent at least two decades in attempts to define the term operationally (Segal 1989; Novak and Petto 1991). Many approaches involve the measurement of stress defined in terms of behavioral and physiological parameters. Not only are some of these parameters difficult to measure in captive animals, but their interpretation is often problematic. Another potential indicator of well-being is the similarity of the captive animal's behavior to that of its wild counterparts. As Shepherdson (Chapter 1, this volume) and Veasey et al. (1996) have pointed out, this measure has both practical and philosophical limitations. Other measures cited are whether an animal reproduces successfully or if it can be successfully reintroduced back into the wild. Conceptual problems not withstanding, these two measurements are often difficult to apply because zoo and aquarium breeding programs (e.g., Species Survival Plans) often limit reproduction, and there are few situations in which habitat, finances, or political climates are suitable to attempt reintroductions (however, see Castro et al., Chapter 8, and Miller et al., Chapter 7, in this volume).

Measurements that combine various behavioral and physiological parameters represent the most promising attempts to develop an objective method of assessing psychological well-being. Future research is likely to focus on refinement of these techniques and the development of additional objective measures that are taxon- and situation specific (see Forthman, Chapter 14; Mellen et al., Chapter 12; and Seidensticker and Forthman, Chapter 2, in this volume). Again, a combination of field studies and laboratory experiments will be necessary to refine and act upon these assessments (see Carlstead, Chapter 11; Crockett, Chapter 9; and Morgan et al., Chapter 10, in this volume for examples).

Future studies should also address the effects of human caretakers on the behavior and well-being of animals in captivity. Completed studies have suggested that the type of relationship that a caretaker has with the animals under his or her care strongly affects their well-being (Terdal 1996), age at sexual maturation (Signoret 1970), and reproductive success (Mellen 1991).

INCREASING SAMPLE SIZES THROUGH MULTIZOO STUDIES

Systematic studies of enrichment and well-being of zoo and aquarium animals are often confounded by small sample sizes. Increasingly, zoo researchers are utilizing multiinstitutional studies to obtain a larger and more representative sample of the captive population. A current example of such an interzoo study, called Methods of Behavioral Assessment (MBA), is coordinated by Kathy Carlstead of the National Zoological Park. The study involves more than fifty zoos and focuses on six species that are difficult to breed. The goals of the MBA project are to develop standardized methods of assessing and solving behavioral problems through the use of simplified and generalizable data collection and analysis techniques. Individual behavioral profiles are generated and then correlated with reproductive success and other measurements of well-being.

Results may allow predictions about the reproductive potential of individuals with different temperaments and developmental histories. Reproductive success will also be correlated with aspects of husbandry and the physical and social environment to generate the best recipe for managing particular species and individuals (Kleiman 1994b) (see Mellen 1994b for a review of other multizoo studies). Future studies on environmental enrichment would benefit from this interinstitutional approach.

CONSIDERING THE ANIMAL'S SENSORY MODALITIES IN EXHIBIT DESIGN

Exhibit design continues to center around anthropocentric sensory modalities and two-dimensional space (Kleiman 1994a). Consider the primary sensory modalities of elephants: They communicate primarily with sounds humans cannot hear (Payne et al. 1986) and with chemical signals humans cannot smell (Rasmussen et al. 1982). Similarly, the range of colors detectable by animals is highly variable. We need to evaluate the environments we have created to determine if they facilitate or interfere with these sensory modes. For example, many building fans and heating or ventilation units produce low-frequency sounds that could interfere with the ability of an animal to communicate with conspecifics. A component of the MBA study includes measurements of some of these variables (Fraser and Carlstead, personal communication). Similarly, a recent study on the properties of gunite (a common construction material used in zoos) has raised concerns about the thermal qualities of exhibits and their possible effects on animal physiology and behavior (Langman et al. 1996). This type of research is still in its infancy.

CONCLUSIONS

The goal of this book has been to provide an overview of the evolving art and science of environmental enrichment and to define and expand upon its theoretical framework as well as its practical applications. Animal curators and keepers have long been committed to providing enhanced environments for the animals under their care. However, many of the ideas about enrichment have been based on personal insight and trial and error, rather than empirical evaluation. We believe that continuing efforts to develop and expand on the theoretical framework and systematic evaluation of environmental enrichment are essential to its continued integration into the daily practices of animal husbandry. We hope that the information and ideas presented in this volume will facilitate this process.

By encouraging and supporting the pervasive use of environmental enrichment for captive animals, zoos, aquaria, and laboratories are providing opportunities for animals to have some control over their environments and to occupy their time with species-appropriate activities. By most measures developed and by most definitions used, environmental enrichment increases an animal's psychological well-being. We believe that animals whose lives are enriched represent true ambassadors for their species; in turn, their enhancement facilitates the role of zoos and aquaria as conservation educators.

Finally, as these practices are implemented, we hope that Hediger's work will come to be perceived less as timely and more as historical.

REFERENCES

Animal and Plant Health Inspection Service (APHIS). 1992. *Animal Welfare Regulations.* 311-364/50538. Washington, D.C.: U.S. Government Printing Office.

Burghardt, G. M., B. Ward, and R. Rosscoe. 1996. Problems of reptile play: Environmental enrichment and play behavior in a captive Nile soft-shelled turtle, *Trionyx triunguis. Zoo Biology* 15:223–238.

Chiszar, D., W. T. Tomlinson, H. B. Smith, J. B. Murphy, and C. W. Radcliffe. 1995. Behavioural consequences of husbandry manipulations: Indicators of arousal, quiescence, and environmental awareness. In *Health and Welfare of Captive Reptiles,* ed. F. L. Warwick and J. B. Frye, 186–204. London: Chapman & Hall.

Gibbons, E. F., B. S. Durrant, and J. Demarest, eds. 1995. *Conservation of Endangered Species in Captivity.* Albany: State University of New York Press.

Grams, K., and G. Ziegler. 1996. Enrichment options. *Animal Keepers' Forum* 23:406–407.

Harlow, H. F., and M. K. Harlow. 1962. Social deprivation in monkeys. *Scientific American* 207:136–146.

Hediger, H. 1950. *Wild Animals in Captivity.* London: Butterworths.

———. 1955. *The Psychology and Behaviour of Animals in Zoos and Circuses.* London: Butterworths.

———. 1969. *Man and Animal in the Zoo.* London: Routledge and Kegon Paul.

Holden, C. 1988. Experts ponder simian well-being. *Science* 241:1753–1755.

King, C. E. 1993. Environmental enrichment: Is it for the birds? *Zoo Biology* 12:509–512.

Kleiman, D. G. 1994a. Mammalian sociobiology and zoo breeding programs. *Zoo Biology* 13:423–432.

———. 1994b. Foreword: Animal behavior studies and zoo propagation programs. *Zoo Biology* 13:411–412.

Kleiman, D. G., M. E. Allen, K. V. Thompson, and S. Lumpkin, eds. 1996. *Wild Mammals in Captivity: Principles and Techniques.* Chicago: University of Chicago Press.

Langman, V. A., M. Rowe, D. Forthman, B. Whitton, N. Langman, T. Roberts, K. Huston, C. Boling, and D. Maloney. 1996. Thermal assessment of zoological exhibits. I. Sea lion enclosures at the Audubon Zoo. *Zoo Biology* 15:403–412.

Lehner, P. 1979. *Handbook of Ethological Methods.* New York: Garland STPM Press.

Maas, T. 1993. Phone books and boxes and balls, oh my! *A to Z* (Philadelphia Zoo) 2:12–15.

Mason, G. J. 1991. Stereotypies: A critical review. *Animal Behaviour* 41:1015–1037.

Mason, W. A. 1960. The effects of social restriction on the behavior of rhesus monkeys. I. Free social behavior. *Journal of Comparative and Physiological Psychology* 53:582–589.

Mellen, J. D. 1991. Factors influencing reproductive success in small captive exotic felids (*Felis* spp.): A multiple regression analysis. *Zoo Biology* 10:95–110.

———. 1992. Effects of early rearing experience on subsequent adult sexual behavior using domestic cats *(Felis catus)* as a model for exotic cats. *Zoo Biology* 11:17–32.

———. 1994a. Husbandry, Section 103. In *AZA Conservation Academy 1994 SSP Coordinator Training Course Workbook.* St. Louis, Mo.: AZA Conservation Academy.

———. 1994b. Survey and interzoo studies used to address husbandry problems in some zoo vertebrates. *Zoo Biology* 13:459–470.

Norton, B. G., M. Hutchins, E. F. Stevens, and T. L. Maple, eds. 1995. *Ethics on the Ark: Zoos, Animal Welfare, and Wildlife Conservation.* Washington, D.C.: Smithsonian Institution Press.

Novak, M. A., and A. J. Petto, eds. 1991. *Through the Looking Glass: Issues of Psychological Well-Being in Captive Non-human Primates.* Washington, D.C.: American Psychological Association.

Payne, K., W. R. Langbauer, and E. M. Thomas. 1986. Infrasonic calls of the Asian elephant *(Elephas maximus). Behavioral Ecology and Sociobiology* 18:297–301.

Rasmussen, L. E., M. J. Schmidt, R. Henneous, D. Groves, and G. D. Daves. 1982. Asian bull elephants: Flehmen-like responses to extractable components in female elephant estrous cycle urine. *Science* 217:159.

Read, B. 1995. Training zoo professionals for studbook and species survival plan programs. *Zoo Biology* 14:149–158.

Signoret, J. P. 1970. Sexual behaviour patterns in female domestic pigs (*Sus scrofa* L.) reared in isolation from males. *Animal Behaviour* 18:165–168.

Terdal, E. 1996. Captive environmental influences on behavior in zoo drills and mandrills *(Mandrillus),* a threatened genus of primate. Ph.D. dissertation, Portland State University, Portland, Ore.

Thompson, S. 1993. Zoo research and conservation: Beyond sperm and eggs toward a science of animal management. *Zoo Biology* 12:155–159.

Veasey, J. S., N. K. Waran, and R. J. Young. 1996. On comparing the behaviour of zoo housed animals with wild conspecifics as a welfare indicator. *Animal Welfare* 5:13–24.

Yerkes, R. 1925. *Almost Human.* New York: Century.

INDEX

Page numbers followed by **t** refer to tables, and those followed by *f* refer to figures.